T0211966

Boston Studies in the Philosophy and History of Science

Volume 313

More information about this series at http://www.springer.com/series/5710

Ilie Pârvu • Gabriel Sandu • Iulian D. Toader
Editors

Romanian Studies in Philosophy of Science

 Springer

Editors
Ilie Pârvu
Department of Theoretical Philosophy
University of Bucharest
Bucharest, Romania

Gabriel Sandu
Department of Philosophy,
    History, Culture and Art Studies
University of Helsinki
Helsinki, Finland

Iulian D. Toader
Department of Theoretical Philosophy
University of Bucharest
Bucharest, Romania

ISSN 0068-0346   ISSN 2214-7942   (electronic)
Boston Studies in the Philosophy and History of Science
ISBN 978-3-319-37801-5   ISBN 978-3-319-16655-1   (eBook)
DOI 10.1007/978-3-319-16655-1

Springer Cham Heidelberg New York Dordrecht London
© Springer International Publishing Switzerland 2015
Softcover reprint of the hardcover 1st edition 2015

Printed on acid-free paper

Springer International Publishing AG Switzerland is part of Springer Science+Business Media (www.springer.com)

Contents

Contributors

Sorin Bangu Department of Philosophy, University of Bergen, Bergen, Norway

Tudor M. Băetu Programa de Filosofia, Universidade do Vale do Rio dos Sinos, São Leopoldo, Brazil

Radu J. Bogdan Department of Philosophy, Tulane University, New Orleans, LA, USA

Mircea Dumitru Department of Theoretical Philosophy, University of Bucharest, Bucharest, Romania

Andreea Eșanu Department of Theoretical Philosophy, University of Bucharest, Bucharest, Romania

Mircea Flonta Department of Theoretical Philosophy, University of Bucharest, Bucharest, Romania

Mihai Ganea Department of Philosophy, University of Toronto, Toronto, Canada

Radu Ionicioiu Department of Theoretical Physics, National Institute of Physics and Nuclear Engineering, Bucharest-Măgurele, Romania

Alexandru Manafu Institute for History and Philosophy of Science and Technology, University of Paris-1 Panthéon-Sorbonne, Paris, France

Adrian Miroiu National School of Political and Administrative Studies, SNSPA, Bucharest, Romania

Ioan Muntean The Reilly Center for Science, Technology, and Values, University of Notre, Dame, IN, USA

Gheorghe S. Paraoanu Low Temperature Laboratory, Department of Applied Physics, Aalto University, Espoo, Finland

Ilie Pârvu Department of Theoretical Philosophy, University of Bucharest, Bucharest, Romania

Viorel Pâslaru Department of Philosophy, University of Dayton, Dayton, OH, USA

Gabriel Sandu Department of Philosophy, History, Culture and Art Studies, University of Helsinki, Helsinki, Finland

Gheorghe Ştefanov Department of Theoretical Philosophy, University of Bucharest, Bucharest, Romania

Iulian D. Toader Department of Theoretical Philosophy, University of Bucharest, Bucharest, Romania

Manuela L. Ungureanu Department of Philosophy, University of British Columbia, Okanagan, Kelowna, BC, Canada

Part I
Scientific Practices and Philosophical Traditions

Chapter 1
The Tradition of Scientific Philosophy in Romania

Ilie Pârvu

1.1 Introduction

In a rather schematic way, one can distinguish three main directions in the development of Romanian philosophy:

(a) a systematic-theoretic one, having as its model of excellence the great European tradition in metaphysics and logic, and aiming to build general ontological world-views. This direction was represented by Titu Maiorescu, Vasile Conta, Constantin Rădulescu-Motru, P. P. Negulescu, Ion Petrovici and Mircea Florian.
(b) an essayistic philosophy, inspired mainly by the works of Romanian historian Vasile Pârvan and by the ideas of the controversial philosopher Nae Ionescu, centered on the particularities of national culture, on the mode of representing the world as expressed in Romanian language. The main representatives have been Emil Cioran, Mircea Eliade, Mircea Vulcănescu and Constantin Noica.
(c) a scientific philosophy, correlated with the new theoretical and methodological developments of contemporary science. This kind of philosophy was promoted, with few exceptions, by very important philosophically minded scientists, actively involved in the practices of the real science.

There is, however, a singular figure among Romanian philosophers who seems to defy the above distinction: Lucian Blaga was the author of an original philosophical system based on a philosophical cosmology, a brilliant essayist and a major poet. Also, his posthumously published book, "*The Experiment and the Mathematical*

I. Pârvu (✉)
Department of Theoretical Philosophy, University of Bucharest, Bucharest, Romania
e-mail: ilieparvu1@yahoo.com

© Springer International Publishing Switzerland 2015
I. Pârvu et al. (eds.), *Romanian Studies in Philosophy of Science*, Boston Studies
in the Philosophy and History of Science 313, DOI 10.1007/978-3-319-16655-1_1

Spirit" (1969), shows him a prescient advocate of historically based epistemology, anticipating the historical turn in philosophy of science.

Despite the fact that the "scientific philosophy" of the philosophers-scientists produced many of the most original, profound and enduring works, some of them internationally recognized as genuine paradigms, inspiring new ways of philosophical thinking, in the common histories of Romanian philosophy and of Romanian culture in general, this is typically understated, if not completely ignored. But this is not a "local fact", peculiar only to Romanian intellectual history. Rather, this seems to be a quasi-general perspective of the historians of philosophy: it is enough to observe the attitude in the standard philosophical histories towards such great philosophers as Whitehead, Husserl, Peirce, Einstein, Hilbert, Bohr, Heisenberg, Brouwer, and Gödel, as compared with the attention conferred to more "popular" philosophers, such as Nietzsche, Wittgenstein, Heidegger and Derrida.

The prevalence of essayistic philosophy in Romania, not only in the general cultural milieu but also in the professional philosophical discourse, can be explained by the dominance of the cultural model of belles-lettres and the role of literary critics as the main exponents of critical and reflexive thinking. It was an empirical fact that "in Romanian intellectual life after 1848 literary critics played an exceptional role. In every generation, one or two literary critics emerged who acquired an enormous prestige and whose directives on literary orientations were widely seen as judgments on society's value choices and even authoritative commentaries on the rhythms and models of its development" (Nemoianu 1990: 591). The best known examples of such "public intellectuals" are Titu Maiorescu, Garabet Ibrăileanu, Bogdan Petriceicu Hasdeu, Eugen Lovinescu, George Călinescu et al. The reason for this fact is, according to Virgil Nemoianu, that "inside Romanian culture, aesthetic values preserved a rank and prestige superior to political or ethical values" (Ibidem: 592).

The present study can be considered as a first attempt towards a more balanced perspective on Romanian philosophical history and contemporaneity. In my survey of the main components of the "theoretical corpus" of Romanian scientific philosophy, I intend to present two very important complementary "moments", or two kinds of philosophical thinking, conceived of as intrinsic parts of scientific construction. One is represented by various theoretical programs and methodological models in the new domains of "concrete" sciences, the sciences of complex organized and historically evolving domains, the emergent sciences of history, sociology, geography, economics and linguistics. These new domains of organized complexity cannot be theoretically represented, in the view of some Romanian working scientists, without a clear epistemological insight and a new metaphysical world view, both of them conceived of as genuine components of theoretical structure and interpretation of the new sciences. In their endeavors to build new hypotheses, explanatory theories and fundamental programs in such "soft" sciences, the Romanian scientists were essentially involved in epistemological and methodological reflection, which resulted in philosophical treatises of great depth, originality and significance, transcending the boundaries of the "new sciences", some of them becoming standard works of theoretical and philosophical thinking in science.

The second main kind of philosophical contributions of Romanian scientists was centered on the "categorical change" (Moisil 1942: 145) that occurred in the first half of the last century in mathematics and in "classical", mature science, as a result and at the same time as a precondition of a series of theoretical and methodological revolutions, the most important being the structural construction and reconstruction of mathematics and of the exact sciences of nature. The structural (re)construction of mathematics and physics represented a deep transformation of the relation between science and philosophy, which led up to a possibility of a new "scientific philosophy", a philosophy not only inspired by fundamental scientific achievements, but also internally constituted with the help of scientific (mathematical) "technique of reason". This philosophy of science is not an armchair philosophizing about science, but a new sort of theoretical practice, a fundamental and foundational research developed in the frameworks of the new abstract-structural theories. Before the second world-war, a group of Romanian mathematicians and physicists from the University of Bucharest, known as the "Onicescu seminar in the philosophy of sciences", was a very productive center of philosophical thinking in science (which they also called "axiomatic philosophy" or "structural philosophy of science"). This can be regarded as an important center of "scientific philosophy" with contributions comparable with those of the Vienna Circle, the Berlin Group, the Lvov-Warsaw School, Uppsala or Prague centers. In the second part of my chapter, I will try to justify this bold assertion.

1.2   Epistemological Programs and Methodological Models for Emerging Sciences

In this chapter I will present, without any claim to completeness, some important philosophical theories and research programs formulated by Romanian scientists in the new fields of the emerging sciences. The starting point for the entire development of the philosophical thinking of Romanian working scientists was the big debate on the epistemological status of human and social disciplines, on the possibilities and limits of scientific explanation in hermeneutic disciplines, known as "die Methodenstreit". The first renowned Romanian historian, Alexandru D. Xenopol (1847–1920), a member of the French Academy of the Social and Political Sciences, the author of the first great treatise on Romanian history, was also an important participant in this debate, alongside with Dilthey, Rickert, Windelband, H. Berr, B. Croce, P. Lacombe, G. Gentile et al. Regarded by his contemporaries as "one of the renowned epistemologists of history" (B. Croce) and one of the "founders of the new critical historical theory" (E. Breisach), Xenopol took part in this debate at a very fundamental level: he introduced a new metaphysical horizon, subjacent to the methodological one, a realist, non-Kantian ontology of space and time (conceived of as "formes réelles et existantes; … sans cette conception fondamentale, l'histoire ne serait qu'une immense fantasmagorie" – Xenopol 1908:

90), and a new topology and architectonics of science. In his much discussed work, *Théorie de l'histoire* (reviewed by H. Rickert, H. Berr, G. Gentile, B. Croce, R. Aron, E. Troeltsch et al.), Xenopol launched a new conception of scientific explanation in historical sciences based not on causal deterministic laws, but on a special kind of historical dependencies termed "historical sequences", which was applied as an analytic and interpretive instrument in the reconstruction of some important periods of Romanian history. This conception was based on the distinction between two kinds of objects of science or "phenomena", classified from the spatial-temporal perspective: "La science universelle se partagera, donc, en deux branches: la première comprendra la science des phénomènes sur lesquels le temps n'exerce aucune influence = *les phénomènes de répétition*; La seconde, les sciences qui auront pour objet des forces agissant dans le temps = *les phénomènes successifs*" (Ibidem: 20).

Alongside of many partisans of the irreducible nature of historical science, Xenopol argued for the autonomy of this science, based not on particular method-ological reasons, but on a very general epistemological "fact": history is a unique and irreducible science, a "new mode of knowledge" in the first place, which studies the world of succession, ontologically different from the world of repetition. The former possesses a new kind of causality (an intrinsic one, peculiar to historical development) and consequently necessitates a new form of scientific explanation, based on the "sequence of historical facts", this "forme sérielle de causalité ... la seule forme que le temps lui permette d'embarasser" (71). The idea of an historical sequence, which represents a chain of the individual facts based on an original phenomenon (a *kern*) and on a particular kind of integration of individual phenomena, gives Xenopol the possibility to reject the most influential philosophical construal of the originality of *Geisteswissenschaften*, based exclusively on subjec-tive, epistemological and axiological factors (i.e., on values and understanding, as formulated by Windelband and Rickert), and allows him to introduce a more general conception of the human and natural history: "L'élément de la série de development occupe tout la domaine de la succession. Du point de vue logique, cette circonstance rend la série très apte a constituer l'élément distinct de la succession ... Et si pour la science de la répétition on trouve un élément universel qui les caractérise, celui de la loi, pour science de la succession il en faut aussi un qui soit applicable à tous, condition que ne serrait remplir la notion morale de la valeur, qui ne peut convenir qu'au développement humain, pendant que nous avons vu que la série se trouve dans tout le courant de l'évolution" (125–126). Historical sequences are not isolated fragments of empirical reality; they are successively and hierarchically integrated in even more extended and complex units. Only the progressive integration confers signification and necessity to the evolution of human world, and in this sense the criterion for establishing the significance of particular phenomena can be offered in the last instance only by the integrated history of humankind.

Whereas the science of historical phenomena requires, according to Xenopol, a new theoretical perspective in order to reconstruct the peculiarities of the history conceived essentially as a new and irreducible modality of knowledge, for his contemporary Spiru Haret (1851–1912), a brilliant mathematician and theoretical physicist, the modern ideal of a science constituted by mathematical methods represented the leading methodological instrument and theoretical framework for the new sciences of social phenomena. This was the central methodological point of his sociological treatise "*Mécanique sociale*" (Paris, 1910), conceived as an essay "d'exposer une méthode qui permettra ... d'introduire peu à peu, dans l'étude des questions sociales cette rigueur du raisonnement qui donne de brillants résulté dans ce que l'on appelle 'les sciences exactes'" (Haret 1910: 1–2). Spiru Haret, a first rank mathematician, with important contributions to the study of the n-body problem in celestial mechanics, particularly of the problem of stability of the great semiaxes of orbital path of planets (considered by H. Poincaré as "a great surprise"), was one of the founders of mathematical modeling of social phenomena. In his now classical treatise *Méchanique sociale*, Haret constructed a mathematical model of social statics and dynamics in analogy with the basic categories and fundamental laws of rational mechanics. This analogy represented a starting point for the formulation of fundamental principles of social equilibrium and social dynamics, the law of minimal action, etc. As a mathematical theory of social phenomena, *Méchanique sociale* should be considered not as a mathematical rephrasing or a reconstruction of some pre-existent social theory, but rather as a rational instrument that can make possible a social science as a theoretical description of some domains of existence. This interpretation relies heavily on the role attributed by Haret to the social laws that are analogous to the principles of rational mechanics. These laws are, in Haret's conception, not merely mathematical expressions of some empirical regularities, but rather the main constraints, determinative for the inner structure and development of the social world, being essentially involved in the theoretical construction of the social phenomena as a unified genuine world.

The program of "social mechanics" (with its fundamental matrix: economic position/intellectual power/moral value) was important because the "mechanical analogy" was used by the author essentially as a leading thread for finding the right conditions for a conceptual construction in social theorizing. In this respect, Haret's idea of building an ideal theoretical model as the necessary precondition for the mathematical construction of theories in social disciplines remains a permanent achievement of the epistemology of these sciences, with great general philosophical significance. Haret's social mechanics can be considered an eminent methodological approach to the complexity of idealization in the emerging sciences. Furthermore, as has been remarked, Haret's analyses take into account actual sociological practices and manifest a deep understanding of the nature and requirements of idealization's procedures and of the role of the theoretical models in science. In the same direction, we can observe the following requirements (formulated explicitly by Haret) for the adequate mathematical-theoretical construction of social science: a set of preconditions for the mathematical treatment of social phenomena (including, for example a clear meaning of basic concepts), a hierarchically ordered set of idealizing steps, the possibility for new generalizations (Haret himself for example, introduced the concept of hyperspace in order to better represent the social complex systems and social evolution), etc.

In a very sophisticated manner, Haret extends to social phenomena the methodological "style" of his own researches in celestial mechanics. From this perspective, the main methodological peculiarity of Haret's mathematical modeling of social phenomena consists, in my opinion, in the architectonic manner in which he conceived of the function of mathematical-mechanical analogy in edifying a social theory. Every successful theoretical construction in this complex domain must also observe, on Haret's view, the hierarchical order of laws and the determinative role of the fundamental law(s) for the entire conceptual system. These contributions justify the appreciation of the first reviewer of Haret's treatise, the American sociologist Lesly Ward: "*Mécanique sociale*, the book we are to consider, may be characterized as the result of a keenly felt need for exact methods in the solution of social problems. The mathematical genius of the writer has made it possible for him to conceive of certain quantitative elements in social life as standing in exact relationships to each other, which may be traced to social processes determined by well-defined and mathematically measurable forces … The conclusions reached by the author are as interesting as they are significant" (Ward 1913: 815). As an example of these significant results, the American sociologist noted Haret's endeavor to "demonstrate mathematically that democracy is an essential condition of civilization", the task being "well met by Haret" (Ibidem: 816).

Despite the fact that classical rational mechanics was substituted by relativity theory and quantum mechanics as fundamental and foundational theories of physical sciences, with new basic laws and ontological models, the principles of classical mechanics retain their "transcendental function and significance" (as the first expression and original model) for the process of physical theory-construction, and as a first formulation of the structural conditions, indefinitely generalizable, for the very idea of a natural world. The Newtonian laws (with their fundamental "moments" expressed by the principles of conservation, causality and interdependence) represented the first structural pattern of lawfulness, determinative for an *ontos* as a complex system of entities, endowed with stability, ontological dependence and integrality. We can understand and evaluate the modernity of Haret's approach along these lines. The analogy with rational mechanics, the theory-core for the entire classical mathematical science of nature, represented, for Haret, not simply a transposition of the physical laws into a new domain, but an attempt to imitate the classical mechanics at two levels: (i) the first level concerns the methodological strategy of mathematical modeling or of scientific idealization; in this sense, Haret extends to social phenomena his own type of scientific practice, as represented by this work in celestial mechanics; (ii) the second level concerns the epistemological and metaphysical dimensions of this procedure: the Newtonian laws of general mechanics constitute, for Haret, a model of the theoretical construction of a world, the first effective realization of the structural constraints necessarily involved in every "law-determined world", or "law-constituted world". In this epistemological sense, the general Newtonian laws retain their perennial significance not as a "foundational" basis for every theoretical development in physics or other natural and social sciences, but as a methodological paradigm (as structural principles of a world construction) for every law-like model of the universe.

The social mechanics constructed mathematically by Haret didn't solve all the problems of social theorizing (the problem of free choice, for example, or the nature of individuals and individuality of social events, etc.), but it produced the theoretical premises for solving them. As a general abstract theory, the social mechanics didn't produce by itself predictions of special phenomena. Haret was aware of this fact, and his predictions were presented explicitly as internal "theoretical possibilities": the necessity of a new kind of the representation space for social phenomena ("hyperspace"), the possibility to use the diffusion laws for explaining some social events, the idea of a future "integral civilization" etc. These, and other such ideas, give us some ground for understanding the modernity and profundity of Haret's pioneering work in the complex field of social disciplines.

While the works of Alexandru Xenopol and Spiru Haret were known and commented upon in the scientific and philosophical community during their time or later (for example in the works of H. Rickert, B. Croce, E. Troeltsch, P. A. Hiemstra, E. Breisach, A. Portado, L. Ward, T. Lalescu et al.), and they became standard works in the philosophy and methodology of social sciences, the program launched for the theoretical construction of "concrete sciences" by another Romanian scientist, Simion Mehedinți (1868–1962), was practically ignored by his contemporaries. He was a Romanian geographer and the author of a monumental theoretical and epistemological treatise on geography, conceived of as an original and genuine science, "*Terra. Introduction into Geography as a Science*" (two volumes), published in Romanian in 1931, and of some other books in which he developed and applied to various domains (anthropology, ethnography) the "architectonic model of science" introduced in his masterwork.

The great originality of Mehedinți consists in the fact that his work represents "a model of philosophical construction of a science" (Geană 2002–2003: 14). In this sense, Mehedinti explicitly assumed as the leading thread for his entire project of a new science the Kantian conception of the architectonic construction of science. The "Idea of a science", in one of its Kantian senses as the "reason's scientific concept", signifies in its reflective hypostasis the "reason's concept of the form of the whole insofar as this concept determines a priori both the range of the manifold and the relative position that the parts have among one another" (Immanuel Kant, *Critique of pure reason* A832/B860). This meaning corresponds to the philosophical well foundation of a pre-existing science. In another context, it can be endowed with a "positive" or active meaning. In this determination, it refers to the "schema, as the original germ" for a new system of knowledge. This meaning, presented by Mehedinți in the form in which it occurs in Kant's Preface to the *Physical Geography* ("The Idea is architectonic. It creates science"), was instrumental in the construction of the new theoretical model of the science of geography which represented the main objective of Mehedinți's *Terra*. It was assumed as a fundamental prerequisite of a geography constituted, in Kant's terms, in accordance with "the genuine Idea of science".

The architectonic ideal represented for Mehedinți not only a general constraint for a "pre-paradigmatic" discipline, but mainly a fundamental condition for an adequate answer to the challenge of complexity in the new domains of "concrete" sciences. The modular-architectonic construction of science is a necessary condition for the theoretical construal of the "organicity" of the domain of research, of the interaction of its "levels of reality", each of them endowed with distinct types of causality and forms of lawfulness, and in the last instance for underlying the role of the so-called "centers" of scientific representation.

As a first example of a philosophically constructed science in this new domain, *Terra* effectively presented the science of geography as a theoretical program in the form of an architectonic system. This program itself is a theoretical complex, irreducible to the classical (axiomatic-deductive) model of science. It contains a general formal principle, or a new theoretical model of "organicity" (the "descendent subordination of planetary shelves on the basis of their progressive complexity" Mehedinți 1931, vol. 1: 24), a new "categorical scheme", some mediating models and hypotheses and, finally, specific laws or theories for concrete empirically determined domains. In the methodological perspective, Mehedinți stressed the integrality of the entire architectonics of science and the irreducibility of its different theoretical and empirical levels. In other words, he argued for the necessity of a new mode of theoretical articulation of this "concrete" science, with an immanent epistemology and a new, characteristic concept of method.

This concept of method, as it emerges from the real development of science, contains essentially the sum total of "directive ideas", which are involved in gathering the empirical data, in explaining their relatedness, and in the search for causal relations between facts, and which support, in the last instance, the entire architectonic of the system (Ibidem: 24n). Mehedinți's concept of method represents a rational generalization, at a new level of theoretical sophistication, of the Cartesian rule-centered idea of method. In this sense, the "concrete science" of geography demands a reconstruction of the classical model of the scientific method (centered on operations and rules) and the introduction of a new (system-oriented) meaning of method. Beside the new theoretical paradigm, Mehedinți introduced a new "elementary unit of scientific knowledge", required by its architectonic form, and a new internal logic.

The architectonic model of theory-construction is instantiated in *Terra* by the integrative function of the set of general constraints (or formal-methodological, interdependent principles) such as the progressive complexity of spheres, the causal subordination of planetary covers and causal subordination of geographical zones, of the matrix of static categories (form, dimension and position) and of dynamic categories (composition, density and color). This theory-core of the entire program also contains a general methodological requirement: the order of scientific description should follow the order of causality in natural phenomena.

Mehedinți was not only a scientist who discovered the philosophical significance of the science of earth, as an integrated, hierarchically ordered study of all the "spheres" which define the earth. For *Terra* contains not only a new and revolutionary "vision", or a new epistemological outlook, but develops a new analytical aparatus as well. In this sense, Mehedinti's paradigmatic work contains an exemplary monographic treatment of the most important research instrument of geography (explanatory, predictive and practical), namely the map. The cartographic theory formulated by Mehedinți is conceived of in the same integrative, "organicist" manner as are the thematic structures of the general theory of earth.

An eminent, Romanian-American, mathematician was Nicholas Georgescu-Roegen (1906–1994). A statistician (he authored an important book in this field, *The Statistical Method*, published in Romanian in 1933) and economist (called by Paul Samuelson "a scholar's scholar and economist's economist"), Georgescu-Roegen was a very original and profound thinker, with seminal contributions in several fields of economic theory, and "one of the founders of the field of multidisciplinary studies known today as ecological economics, which he himself, however, defined as bioeconomics" (Bonaiuti 2011: x). He created a new conceptual framework for economic studies by "opening economics to natural science, especially to thermodynamics and biology, [which] led him towards the elaboration of a new economic approach, which was the first to point out, on a sound scientific basis, the biophysical limits to growth" (Ibidem: xi). Georgescu-Roegen not only questioned the theoretical and methodological fundamentals of the neoclassic, "standard" paradigm of contemporary economics, but he also formulated a new alternative research program, based on a new epistemological model with profound scientific and philosophical implications.

On this basis, he was also widely recognized as a revolutionary epistemologist. In his opus magnum, *The Entropy Law and the Economic Process* (1971), he formulated an alternative program to the contemporary mainstream economy, the neoclassical paradigm. In this book, Georgescu-Roegen introduced a new conceptual framework for understanding the economic processes in contemporary society and, for this purpose, he formulated a new epistemology of economics based on a very general conception on science. I attempt to describe, below, only some of his ideas that are significant from the perspective of a general philosophy of science.

The most important metascientific concept introduced by Georgescu-Roegen is the idea of a "determined system of knowledge production", which is able to provide an integrative conceptual model of cognitive activity and of social dimensions of science, including a new approach of the historiography of science. This thematic metatheoretical concept allows us to represent the multidimensional reality of science. As an example of this general idea, Georgescu-Roegen developed the concept of the so-called "arithmomorphic system of knowledge production", based on a particular type of scientific concepts (like the concepts of classical mathematics, well defined and with sharp conditions of applicability), involving specific methodological procedures and a determinate structure and evolution type. In this sense, the very idea of "arithmomorphic system of scientific growth" characterizes not only a complex type of scientific knowledge production or epistemic activity, defined by a specific internal logic and a particular dynamics and evaluation of knowledge, but it also determines a specific type of social integration of scientific results and some peculiar "ideological commitments".

This kind of concepts are among those "thematic ideas" or "*durchlaufender Kategorien*", as the anthropologist Arnold Gehlen called them, which can structure the problematic of vast and multi-level fields of study, by welding whole disciplines into one coherent system, or by defining the guidelines for developing a certain discipline over a long period of time. On the metatheoretical level, these concepts can mediate between the "internalist" and "externalist" approaches of science, providing a framework or a basic interpretive scheme for examining an entire "system of scientific activity" that exists at some point in the historical development of science. By using the idea of an "arithmomorphic system of science", Georgescu-Roegen identified certain types of cognitive structures, defined up to concepts of internal logic (as "discretely distinct" concepts): a generalized ontological framework, a methodological ideal-type (dominated by formalism), a model that causes subtle ironies like the following: no doubt, formal considerations are often inspirational spring, but past that, we tend to forget that they are not actually grounded on anything, and this is their real danger. At the same time, the concept of "arithmomorphic system" includes an analytical scheme, a peculiar kind of evolution required by "theory-based-science", as well as a determinate kind of the "institutionalization" of the arithmetic ideal-type as a paradigm of professionalized science, containing a definite type of the "authority structure" of science, a certain hierarchy based on arithmetic "ideal types".

The major purpose of Georgescu-Roegen is not limited to criticizing the arithmomorphic methodologies, but rather it aims at exploring new modalities to overcome this socio-cognitive system based on "arithmomorphic cognitive structures" and to open new ways of scientific growth beyond those permitted by the arithmomorphic structures, starting from a new kind of concepts, called "dialectical concepts" (taking inspiration by A.N. Whitehead, who was also a philosophical source for the ecological reconstruction of economics). This kind of epistemology introduces a conceptual vocabulary that should allow a wider perspective on the development of knowledge. The main target of Georgescu-Roegen critique is the "mechanistic epistemology" (see Georgescu-Roegen 1974) and the inappropriate use of classical logic in scientific reasoning. In this sense, Georgescu-Roegen declared: "I believe that what social sciences, nay, all sciences need is not so much a new Galileo or a new Newton as a new Aristotle who would prescribe new rules for handling those notions that logic cannot deal with" (Georgescu-Roegen 1971: 41). In his new approach, theoretical science must be conceived as "a living organism precisely because it emerged from an amorphous structure – the taxonomic science – just as life emerged from inert matter" (Ibidem: 36).

The fruitfulness of the categorical system introduced by Georgescu-Roegen was revealed by a case study of the significance of J. von Neumann's famous theorem regarding the interdiction of "hidden variables" in quantum mechanics, which was instrumental in blocking alternative constructions and interpretations in quantum theory (Pinch 1977). In the same direction, we can explain some of the moments in the history of science in which the theoretical-methodological aspects were strongly "connected" to socio-political ones. Thus, a theorem like von Neumann's, which blocked for about two decades and a half any competent attempt to criticize the Copenhagen interpretation of quantum mechanics (the so-called "standard" interpretation), can be understood within the arithmomorphic pattern as being dominated by the ideal of a super-formalization of theories. This could be the only way to explain why, during this entire period, the physicists didn't criticize its logical structure, or its semantic and methodological significance. Only in the context of increasing doubt among mathematicians about the high virtues of formalism and axiomatization were the subtle "dialectic" arguments successfully brought up among the "great priests of science" by David Bohm. By overcoming the "arithmomorphic rigidity", Bohm reopened the "case" of physical and ontological interpretation of quantum mechanics. This "historiographical experiment" proves the epistemological potential and effectiveness of the new general interpretive system proposed by Georgescu-Roegen for understanding and improving the actual development of knowledge, as an important model for a philosophy of real science.

In a certain sense, the works of Haret and Georgescu-Rogen in social sciences instantiate the methodological duality between "mechanistic" and "organicist" paradigms of theoretical construction in this field of science. It is important to underline the fact that both scientists attempted to build a non-exclusive, complementary approach of these two methodological strategies. This attitude can be exemplified by the Haret's laws-centered approach towards the mathematical modeling of social phenomena and, at the same time, by Georgescu-Roegen's acceptance of the organicist mode of thinking in economics, not as a local fact of a special domain of reality, but rather as one with general significance, founded on strict analogy between physical laws, such as the entropy law or the second law of thermodynamics, and the economic processes.

The last Romanian scientist-philosopher to be discussed here is Eugenio Coseriu (1921–2002), a leading theoretical linguist of the last century and a philosopher of language trained in phenomenology. My reasons for presenting his ideas in linguistic theory are related not only to their innovative character and their potential for further generalization, but also to their relevance for the general philosophical significance of contemporary linguistics, which has become, after the successive structuralist and cognitive paradigmatic changes, a "pilot discipline" (R. Thom) for many emerging sciences.

Eugenio Coseriu was the proponent of a new type of theory-construction in linguistics, called "integral linguistics", and the founder of a new school in the contemporary science of language. The theory-core or the theoretical framework of his scientific program, known as "Coseriu's matrix", is determined by the "intersection of three levels of language and three 'points of view' on linguistic (and cognitive) reality, giving (at least) nine different ways in which the phenomenon of language can appear for us, and in which it can be systematically studied" (Zlatev, 2011: 132). In this sense, Coseriu's matrix is a scientific model which has at the same time an intrinsic philosophical significance, being "a helpful epistemological frame of reference for the interpretation not only of the various linguistic problems ranging from the linguistic change to that of translation and of linguistic correctness, but also of the structure of the linguistic disciplines themselves and of recent developments in linguistics" (Coseriu 1985: xxv).

This theoretical framework, which represents for Coseriu his "permanent frame of reference", is based on an implicit fundamental principle "underlying [his] treatment of the different, general, or particular, linguistic problems" (Idem), and which concerns the levels of language (universal, historical and individual), and what has lately been called "linguistic competence" or, as Coseriu has called it, "linguistic knowledge" (*saber linguistico*). The distinction between different levels of language is instrumental in determining the theoretical topos of different linguistic problems and of the different theoretical perspectives in the study of language. The second "dimension" of the theory matrix is represented by the three "points of view" or perspectives in approaching the phenomenon of language, respectively, as activity (*energeia*), knowledge (*dynamis*) or product (*ergon*).

In his "Presidential Address" to the *Modern Humanities Research Association* at University College, London (on January 11, 1985), Coseriu presented what he considers his "main contribution to the study of language and consequently to the foundation of linguistics or, to put it in another words, what constitutes my permanent frame of reference" (Ibidem: xxv), namely the conceptual matrix generated by crossing the points of view with the levels of language (representing a sui generis "categorical scheme" for the study of language with a duality of functions: theoretical/ metatheoretical or fundamental/foundational):

Coseriu's matrix is represented in the following way:

		Points of view	
Levels	*energeia*	*dynamis*	*ergon*
	Activity	Knowledge	Product
Universal	Speaking in general	Elocutional knowledge	Totality of utterances
Historical	Concrete particular language	Idiomatic knowledge	(Abstracted particular language)
Individual	Discourse	Expressive knowledge	Text

The most important distinction in this complex matrix, indispensable, in Coseriu's view, for the understanding of the very structure of language, is that between activity, knowledge and product. This tripartite distinction, operative at all levels of language, and the essential emphasis put on *energeia* (activity), are the basis for a new understanding of the nature and function of the much discussed linguistic competence. For Coseriu, the primary distinction between activity and cognition, which is missing in the great majority of contemporary theories of language, enables us to better understand the multifarious problems of linguistic competence, as well as equally important problems of linguistic change, translation and language learning. For Coseriu, linguistic competence, which is the basis or the medium for the creativity of human language, must be placed at another level of the language structure, and consequently, must be conceived of in a different theoretical manner than, for example, in Chomsky's conception of linguistic competence. In contradistinction to the latter, which locates this essential capacity of language nature and functioning at the level of the "knowledge of language", Coseriu considers that linguistic competence is thereby "under-theorized" and its main characteristic – creativity – is "mistaken for productivity, for the production of infinitely many 'correct' sentences by means of the application of a fixed and finite system of rules" (Ibidem: xxix).

Coseriu's new theory of linguistic competence conceives of its proper dimension – creativity – as irreducible to something else, to something not creative, but regards it rather as a "primary *Faktum*" of human language, which corresponds to the special constitution of human beings (Coseriu 1988: 202). On Coseriu's view, Chomsky and his followers "are not prepared to establish and to accept creativity as a *Faktum*, but they try to reduce it to another thing which is not more creativity, namely the facts of the same kind" (Ibidem: 201). In the best case, they try to reduce creativity to some aspects of experience. This reduction is contradicted by the impossibility to derive from a particular combination of the items of experience the abstract (a priori, for Kant) structure of linguistic competence. In fact, in the case of the constitution of concepts, we find a permanent overcoming of experience and a creation of universals. In the case of a "positivistic thinking", the linguistic competence in general "is not founded on a special possibility of humans to create linguistic knowledge, but it is reduced again to knowledge (*Wissen*)" (Ibidem: 202). As a new kind of competence, the linguistic competence is conceived by Coseriu as a "capacity to create projects of the possible", as is the case, for example, with the construction of concepts. This particular situation is informative for the general idea of linguistic competence, because it proves the impossibility to represent this type of human capacity according to the Aristotelian kind of abstraction.

When applied to explaining language meanings, Coseriu's conception is completely opposed to a current view, shared among philosophers, which considers that "the meanings are deduced from contexts"; this perspective has nothing to do, in Coseriu's conception, with the nature of language meanings, but only with the modality in which they are learned (It is possible that this opinion has been one of the reasons for a non-sympathetic attitude of the neo-wittgensteinians towards Coseriu's philosophy of language). At the same time, the learning of language represents itself "a continuous creation of meanings, i.e., a creation of projects and systems", having the same internal structure with the language competence in general. In the most positive perspective, the creativity of language represents for Coseriu "a capacity to dispose on a system of possibilities", which is not reducible to a system of predetermined rules, but represents a second order capacity, which designates the determinative matrix for the generation or production of different systems of rules: "Es geht nicht darum," Coseriu said in an interview, "dass man unendliche Fakten realisiert aufgrund eines System von Regeln, sondern darum dass man neue Regeln enstehen lässt und neue Regeln schafft" (Kabatek and Murguia 1997: 162). Creativity of language, as a construction of new rules or systems of rules, and not as a repeated application of already given rules, can define a new paradigm for many humanities and social sciences. The theoretical and philosophical potential of Coseriu's matrix and of his conception of an "integral linguistics" can justify the following claim: "it is possible to surmise that linguistics would not have been in its present fragmented state if, sometime half a century ago, it had followed the lead of thinkers such as Coseriu rather than Chomsky" (Zlatev 2011: 132).

I have dedicated a little more space to the exposition of Coseriu's fundamental ideas in linguistics and philosophy of language not only because I think that it has an important epistemological potential for generalization, but also because his conception can contribute essentially towards fulfilling the desideratum of a "theoretical contemporaneity" of the human disciplines with the most developed classical sciences, dominated by the so-called abstract structural mode of theory-construction. This kind of scientific theorizing (as "epistemologically projected" by some Romanian scientists) is the subject matter of the next section of my chapter.

1.3 The "Categorical Change" in the Exact Sciences and the Project of an "Axiomatic Philosophy of Science"

In this section, I will concentrate on a very important project of philosophical construction developed in the 1930s and 1940s by a group of philosophically minded scientists from the University of Bucharest, led by the mathematician Octav Onicescu. This kind of contribution of Romanian scientists to the philosophy of science centered on the "categorical change" (Gr. C. Moisil) in mathematics and physical science in the first part of the last century, a change determined by the structural (re)construction of mathematical sciences, which involved a deep transformation of the relation between science and philosophy and opened the possibility of a new scientific philosophy, a philosophy not only inspired by the new fundamental scientific achievements, but also internally consolidated with the help of scientific concepts and methodological procedures.

In what follows, I will try to present briefly some of the philosophical ideas and metatheoretical concepts formulated by members of the Onicescu seminar. This seminar was attended by some well-established scientists, as well as by a series of young scholars in mathematics, physics, linguistics and philosophy: Grigore C. Moisil, Dan Barbilian, Şerban Ţiteica, Nicolae Teodorescu, Gheorghe Vrănceanu, Alexandru Ghika, Petre Sergescu, Mihai Neculcea, Nicolae Georgescu-Roegen, Alice Botez, et al. Their works were published in a great variety of forms: as monographs or thematic collective volumes (e.g., O. Onicescu, ed., *The Problem of Determinism*, 1940; O. Onicescu, *Principles of Scientific Knowledge*), but also as research papers in scientific and philosophical journals from Romania, France, Germany, Italy, USA, and as academic treatises, university or popular lectures, etc. Many of these works were reviewed by A. Church, E. Nagel, G. Birkhoff, R. Feys, C.H. Langford, P. Henle, A. Turquette, etc.

The Onicescu seminar was organized each year around a fundamental theme, such as the ideas of determinism, space, time, infinity, the object of scientific theory, the problem of language in contemporary science, etc. The "scientific philosophy" of the Onicescu group (alternatively termed, "axiomatic philosophy", "mathematical epistemology" or "the philosophy of structural science") can be characterized by the following important traits:

(i) The philosophical reflection of the Romanian scientists represented in an essential way a product of the critical examination of *abstract-structural theories* in mathematics and physics. The conceptual framework for philosophical analysis was provided by the probabilistic and structuralist revolutions that have occurred in the first half of the last century and produced a sort of "categorical change" in the exact disciplines (Moisil 1942). This "new kind of science", stylistically dominated by the structural research and the construction of abstract theories, offered the methodological instruments, fundamental principles, and patterns of explanation for the philosophical programs or for the new interpretive frameworks.

In this direction, very significant were Onicescu's efforts to build a general philosophy of structural science, in which the most fundamental, thematic ideas of science like object, causality and determinism, time and space, etc., can be creatively re-interpreted. At a time in which the philosophical field was dominated by logical empiricist conception on science, Onicescu rejected the "nominalist" or formalist reduction of science (and especially of mathematics) to language, the instrumentalist interpretation of the relation between mathematics and reality, and the general anti-metaphysical outlook characteristic for many new schools in the philosophical interpretation of science. He also formulated a realist conception about the nature and function of theoretical concepts and laws of science. In his sense, we must note his remarkable analysis of the most complex concept, theoretical *par excellence*, which was operative in science and philosophy: the concept of infinite. Onicescu extended Hilbert's program, launched in the latter's famous 1926 lecture "On the Infinite", to a solution of the problem of the infinite at the logical level, and explained the essential role of the infinite in the constitution of abstract concepts and of new levels of theoretical architectonics of science.

Throughout his research, Onicescu was inspired by the change of perspective produced in science by the two fundamental theories of the last century, Relativity Theory and Quantum Mechanics. For Onicescu (as well as for Dan Barbilian), the main transformation produced by these two theories consists in the introduction of the invariant-theoretical approach, which must be continued at the epistemological and ontological levels of the philosophical reflection on science. The new "science of structures" opened, on Onicescu's view, not only new theoretical perspectives, but it produced also important changes in the "technique of reason", improving at the same time the analytical apparatus of the scientific methodology.

Starting from the same structural construction of mathematical sciences, Grigore C. Moisil launched the program of a rigorous formal "mathematical epistemology", having as central themes the ideas of truth, objectivity, and stability of knowledge (In connection with this last concept, we must note the fact that Moisil was one of the first scientists and philosophers of science who emphasized the epistemological significance of theoretical stability of mature science, especially in understanding the role and function of theoretical laws). Moisil proposed to avoid the traditional empiricist concerns with epistemic certainty of beliefs and the refutation of skepticism, and to redirect the epistemological research to the more fundamental problem of the objectivity of knowledge. In his project for a new "mathematical epistemology", Moisil indicated the possibility and necessity to use the new formal instruments offered by the concepts and theoretical procedures of structural mathematics in order to rend intelligible the problem of knowledge. The new mathematical analysis of the problem of knowledge can reveal "the very formal acts of thought" and the rational connections between thought and reality. Moisil considered his project of a mathematical epistemology as an extension of Poincaré's reflections concerning the role of the idea of group as a "form of the understanding" to all structural concepts and abstract theories of contemporary mathematics.

(ii) Most Romanian scientists benefited from their mastery of mathematics in developing many of the scientific and meta-theoretical instruments for the analysis and theoretical reconstruction of science and philosophy. Thus, for example, Gr. C. Moisil, in a very influential, paradigmatic study, "Determinism and enchainment", applied and further developed the new ideas from the probability theory formulated by O. Onicescu and Gh. Mihoc, in order to reconstruct the internal forms of the principle of determinism as theoretical constraints on the physical laws, characteristic for every type of deterministic theories. The great significance of the new concepts and methodological approaches formulated by Onicescu and Mihoc in probability theory is testified by the fact that Onicescu was invited in 1937 to present the new concept of a statistical chain with complete connections to the international congress on probability at the University of Geneva (where among the invited guests were W. Heisenberg, A. Kolmogorov, B. de Finetti, R. von Misess, J. Nyman etc.).

In the same direction, the extension of the "Erlangen program" in geometry and physics proposed by D. Barbilian (a forerunner of contemporary invariantist approaches to articulating a fundamental theory) resulted in a new axiomatization of classical mechanics (Barbilian 1937). He explicitly formulated the fundamental role in this theoretical construction of science of the fundamental group of a theory, as the key instrument for attaining the objectivity of knowledge at the abstract-structural level of scientific theorizing (Barbilian 1940). The new axiomatization of classical mechanics formulated by Barbilian, as an extension in physics of the group-theoretical reconstruction of mathematical theories (proposed initially by Felix Klein for geometric theories), can also be considered as a step in the realization of the second Hilbert's program – the mathematical axiomatization of physical theories (Hilbert's 6th problem), and in this sense, as a partial fulfillment of Hilbert's scientific-epistemological ideal expressed in his famous lecture "Axiomatical Thinking" (1917): "Alles, was Gegenstand des wissenschaftlichen Denken überhaupt sein kann, verfält, sobald es zur Bildung einer Theorie reif ist, der axiomatischen Methode und damit mittelbar der Mathematik". Along the same lines, Barbilian emphasized the epistemological significance of his axiomatization of classical mechanics: "Jetzt, wo die klassische Mechanik von den Physikern den Geometern überlassen wurde, war es vielleicht nicht unnötig zu zeigen, dass man mit ihr noch etwas machen kann, sie nämlich mit einer durchsichtigen Struktur ausstatten, die an ihr Urbild, die bewegliche Himmelsphäre, erinnert" (Barbilian 1937, in *Opera matematica*, vol. 1: 210).

We should also note that, at about the same time, Alexandru Froda, who was considered in the 1950s one of the most eminent philosophers of physics (according to Suppes 1994), anticipated the new forms of physical axiomatics, developed from an algebraic point of view, as well as their meta-theoretical requirements.

(iii) Conceptual analyses and theoretical reconstructions were performed by Onicescu, Moisil, Barbilian, Ţiteica et al. in the field of real science as an internal aspect of the *"theoretical practice"* of articulating and developing a complex research program, and not as a series of "nominal definitions" or "elucidations" of the meta-scientific concepts or of the principles and methodological procedures considered *in abstracto*, as elements of "rarefied", simplified "models" of scientific activity. The main intention of the Onicescu group was the re-investment of the analytical results of the metatheoretical studies into the fundamental research in order to open new horizons for the constructive extension of science and philosophy. We can illustrate this approach to the analytical research of these Romanian scientists with Barbilian's idea of the role of the fundamental group of a theory, which was remarkably suited, by its capacity to bring into prominence the general features of the structural type of theory-construction, to represent the main instrument for the mathematical generalization of an abstract theory and for generating new alternative research directions by the different modes of the "paradigm's articulation".

The metascientific results of the Onicescu group were not so much logical elucidations of the concepts, arguments and interpretive principles of a general character, but rather new theorems and theories with a genuine creative potential for the extension of exact science and at the same time for "deepening the foundations" and enhancing philosophical understanding, the essential task for axiomatic thinking as defined by David Hilbert.

The constructive nature of the conceptual analyses characteristic of the Bucharest school in scientific philosophy, was, in part, determined by the very essence of the science that constituted the field of metascientific research: the structural-abstract mathematics and natural science, that kind of science for which the axiomatic analysis represented not so much a matter of logical systematization of pre-existing empirical science, but rather a *sui generis* mathematical construction by which a general structure with multiple possible realizations was determined for the first time. In this sense, on the one hand, Moisil understood philosophical analysis as a part of fundamental research in science. The philosophical principles of determinism, for example, were projected as internal constraints, mathematically defined, on the theoretical articulation of science. On the other hand, it is this kind of science, best represented by structural mathematics, which makes it possible to employ the technical instruments, concepts, and constructive methods of science for philosophical reflection, in order to obtain new ways of presenting and solving epistemological problems. In Moisil's own words, "if mathematics of quantity was a wonderful instrument for the knowledge of physical world", structural mathematics "opens the possibility to organize a mathematical epistemology" (Moisil 1937, in Ath. Joja et al (eds) 1971: 144).

(iv) The same invariant-theoretical approach in fundamental research can explain the "mathematical way" (as distinct from the proof-theoretical one) of conceiving the main *metatheoretical constraints* of axiomatic theories. The Romanian scientists intended to bring forth, by means of mathematical concepts, a deeper level of the "logical structure" of a scientific theory, one revealed not by the deductive organization of its set of statements (as in the case of the standard formalization of theories), but rather by the formative and determinative mathematical substructures of the theory. This logical structure or internal logic can be revealed by the axiomatic conceived by Barbilian, for example, as a general and formal theory of "scientific doctrines" (very similar to Carnap's "*allgemeine Axiomatik*"). Consequently, they were interested in exploring other kinds of metatheoretical conditions, all of them strongly related to the internal constructive function of the fundamental group of a structural theory (Thus we can explain their alleged insensitivity to the famous metatheoretical results concerning the deductive completeness of first order axiomatization).

On the group-theoretic or invariantive approach, the operations presented by the abstract theory are, in the axiomatic philosophy of Dan Barbilian, of a cardinal significance from the metatheoretical point of view. The fundamental group, as a formal matrix determinative for an entire research program, is at the same time responsible for the new kind of the metatheoretical conditions of "axiomatic doctrines" (completeness, categoricity, axiomatizability), which constrain essentially the relation between theory and reality, not only the deductive closure of a set of sentences that represents only possible formulations of the theory. In this sense, in the case of axiomatic doctrines, as interpreted by Dan Barbilian, we can speak of an "immanent approach" in the metatheory of exact sciences: constraints of this kind are imposed not by some external, logical conditions, but by the very formal structural core of the theory – the fundamental group. At the same time, given the fact that the fundamental group of a theory is essentially involved in all possible extensions of the general theory, the explicit formulation of the subsidiary mathematical structure of a scientific theory constitutes not only an analytic but also a constructive, "creative" procedure (Barbilian 1940). This methodological procedure can be exemplified by the case of classical mechanics: after being included by Einstein and Poincaré in the "Erlangen Program", the "absolute" character of this doctrine (univalent theory) was abandoned in favor of a general spectrum of theoretical alternatives.

(v) This "*mathematical way*" of the philosophical analysis of science represented, for the Bucharest group, the most suitable possibility for building a new "*scientific philosophy*". It constituted the common core of all three projects of the theoretical reconstruction of philosophy, respectively, "mathematical epistemology" (Moisil), "structural philosophy of science" (Onicescu) and the mathematical meta-mathematics with its "inseparable" structural ontology (Barbilian). The technical results and the philosophical significance of the invariantive perspective in foundational research, as a genuine form of the "logical analysis of science", can be better understood in the light of the recent reevaluations of the instruments of metatheoretical investigation that appeal essentially to the most advanced theories of the structural mathematics, especially to category theory, sometimes considered as the contemporary form of the "Erlangen Program". From this perspective, we have now a better reference frame for evaluating the insistent critique by the Romanian scientists of the formalist approach in the philosophical analysis of science and their project to re-think the whole task of such an analysis as a constitutive part of, and a fundamental/foundational investigation in, the real science, and as an original form of the scientific philosophy.

The Onicescu group was enlarged in the years 1940–1945, when other scientists and philosophers (Simion Stoilow, Anton Dumitriu, Alexandru Mironescu, Constantin Noica, et al.) joined in, forming the "Science and Knowledge Group" and becoming thus the most active center of research in philosophy of science in Romania. It was continued after the Second World War, at the University of Bucharest, mainly through the studies and university lectures by O. Onicescu, Gr. C. Moisil, Al. Froda and M. Neculcea. At other Romanian universities, this kind of foundational, philosophically informed research in science was undertaken by Remus Răduleț, at the Polytechnic University of Bucharest and by Emil Tocaci, at the University of Ploiești.

If we consider the whole development of the ideas described above, we can distinguish in the works of members of the Onicescu seminar and of their followers some common, integrative traits: a set of thematic concepts and methodological principles, some paradigmatic studies that served as models of scientific philosophy, a new technique of philosophical analysis, and some new metatheoretical requirements. All these features represent the most important traits of the constitution of a genuine school in philosophy of science, which further develops at a higher level the individual efforts of such great forerunners. Due to all its achievements and projects, this philosophical school has been appreciated as "the most important philosophical fact in Romanian culture" (Dumitriu 1942: 113).

Bibliography

Albrecht J (ed) (1988) Energeia und Ergon, Sprachliche Variationen– Sprachgeschichte– Sprachtypology: Studia in Honorem Eugenio Coserio, 3 Bde. Gunter Narr, Tubingen

Albrecht J (2003) El paradigm Incompleto de Eugenio Coseriu: Tares Pendiente para la Tercera Generation. Odissea 3:41–54

Barbilian D (1937) Eine Axiomatisierung der klassischen Mechanik. Comptes Rendus de l'Inst Des Sc De Roum II(2):1–7

Barbilian D (1940) Determinism and order. In: Onicescu O (ed) (1940)

Barbilian D (1942) The place of axiomatics (in Romanian). In: Barbilian D (ed) Opera didactică. Editura tehnică, Bucuresti

Beard TR, Gabriel LA (1999) Economics, entropy and the environment: the extraordinary economics of Nicholas Georgescu-Roegen. Edward Elgar, Cheltenham

Bernardo Paniagua JM (1995) La construccion de la linguistica. Un debate epistemologico. Ed. Universitat de Valencia

Berr H Sur Bernheim et Xenopol. Revue Synth Hist XVIII:354–358

Bonaiuti M (ed) (2011) From bioeconomics to degrowth: Georgescu-Roegen's "New Economics" in eight studies. Taylor & Francis, New York

Boutroux E (1908) La théorie de l'histoire par Alexandre D. Xenopol. Académie des sciences morale et politiques, Paris

Colvin P (1977) Ontological and epistemological commitment and social relations in science. The case of the Arithmomorphic System of Science Production. In: Mendelsohn E et al (eds) (1977)

Coseriu E (1955–1956) Determinacion y entorno. Dos problemas de una linguitica del hablar. Rom Jahrb 7:29–54

Coseriu E (1957) Sincronia, diacronia e historia. El problema del cambio linguistico. Rev Fac Humanidades y Cienc (Montevideo) 15:201–255

Coseriu E (1978) Les universaux linguistiques (et les autres). In: Proceedings of the eleventh international congress of Linguists, I, Bologna, pp 47–73 (trans: Linguistic (and other) Universals. In: Makkai A, Heilmann L (eds) (1977) Linguistics at the crossroads. Padua, pp 317–346)

Coseriu E (1985) Linguistic competence: what is it really? The presidential address of the Modern Humanities Research Association, read at University College, London, 11 Jan 1985

Coseriu E (1988) Sprachkompetenz. Grundzuge einer Theorie des Sprechens. Bearbeitet von Heinrich Weber, Franke Verlag, Tübingen

Coseriu E (2000a) The principles of linguistics as a cultural science. Transylvanian Rev (Cluj) IX(1):108–115

Coseriu E (2000b) Structural semantics and 'Cognitive' semantics. Logos Lang (Tübingen) I(1):19–42

Coseriu E (2003) Geschichte der Sprachphilosophie. Von dem Anfangen bis Rousseau. neu bearbeitet und erweitert von Jörn Albrecht, Tübingen

Croce B (1939) Conversazioni critiche. Bari (serie quinta)

Dumitriu A (ed) (1941–1945) Caiete de filosofie. Atelierul de Arte Grafice Independenta, Bucuresti

Dumitriu A (1942) Foreword. Caiete de filosofie 4:111–113

Froda A (1959) La finitude en mécanique classique, ses axiomes et leur implications. In: Henkin L, Suppes P, Tarski A (eds) The axiomatic method with special reference to geometry and physics. North-Holland, Amsterdam

Froda A (1960) Analyse du principe de continuité en physique. In: Suppes P et al (eds) Logic, methodology and philosophy of science. Stanford, North-Holland/Amsterdam

Froda A (1971) Error and paradox in mathematics (in Romanian). Editura enciclopedică română, Bucuresti

Geană G (2003–2004) The actuality of Simion Mehedinți's thought (in Romanian). Studii si cercetări de geografie xlix:13–18

Georgescu-Roegen N (1933) Statistical method (in Romanian). Institutul de statistică generală a statului, Bucuresti

Georgescu-Roegen N (1966) Analytical economics: issues and problems. Harvard University Press, Cambridge, MA

Georgescu-Roegen N (1967) An epistemological analysis of statistics as the science of rational guessing. Analele Universității din București, Seria Acta Logica X:61–91

Georgescu-Roegen N (1971) The entropy law and the economic process. Harvard University Press, Cambridge, MA

Georgescu-Roegen N (1974) Mechanistic dogma and economics. Methodol Sci vii(3):178–184

Georgescu-Roegen N (1986) The entropy law and the economic process in retrospect. East Econ J 12:3–25

Haret S (1910) Mécanique sociale. Gauthier-Villars, Paris

Hiemstra P (1987) Alexandru D. Xenopol and the development of Romanian historiography. Taylor & Francis, New York

Itkonnen E (2011) On Coseriu's legacy. Energeia III:1–29

Joja A et al (eds) (1971) Recherches sur la philosophie des sciences. Editions de l'Académie Roumaine, Bucharest

Kabatek J, Murguia A (1997) "Die Sachen sage wie sie sind . . ." Eugenio Coseriu im Gespräch. Gunter Narr Verlag, Tübingen

Lalescu T (1911) Mécanique Sociale de Mr. Spiru Harert (in Romanian). Convorbiri Literare XLX. In: Lalescu T (2013), Opere, Ed. II, Editura Academiei Romane, Bucuresti :632–637

Lozada G (1995) Georgescu-Roegen's defense of classical thermodynamics revisited. Ecol Econ 14:31–44

Mayumi K (1995) Nicholas Georgescu-Roegen (1906–1994): an admirable epistemologist. Struct Chang Econ Dyn 6:261–265

Mayumi K, Gowdy JM (1999) Bioeconomics and sustainability: essays in honor of Nicholas Georgescu-Roegen. Edward Elgar, Cheltenham

Mehedinți S (1931) Terra. Introduction in geography as science (in Romanian). Editura Națională S. Ciornei, Bucuresti

Mehedinți S (1946) Premises and conclusions to Terra (in Romanian). Imprimeria Națională, Bucuresti

Mendelsohn E et al (eds) (1977) Sociology of the sciences, vol 1. Reidel, Dordrecht

Moisil GC (1937) Les Etapes de la connaissance mathematique. In: Joja et al (ed) 1971:131–144

Moisil GC (1942) La tendance nouvelle dans les sciences mathematiques. In: Joja et al (ed) 1971:145–154

Moisil GC (1971) Les mathématiques structurelles et leur applications. In: Joja Ath et al (ed) (1971). pp 131–173

Neculcea I (1942) Thought and formalism (in Romanian). Caiete de filosofie 4:181–195

Nemoianu V (1990) Mihai Șora and the traditions of Romanian philosophy. Rev Metaphys 43(3):591–605

Onicescu O (1938) Chaine statistique et déterminisme. Revue de l'université libre de Bruxelles (1):45–59

Onicescu O (ed) (1940) The problem of determinism. Seminar of philosophy of Sciences of the Bucharest University, (in Romanian), Editura Oficiul de Librărie, București (French trans: Joja Ath. et al (eds) (1971))

Onicescu O (1944) Principles of scientific knowledge (in Romanian). Editura Oficul de Librărie, Bucuresti

Onicescu O (1971) Principles de logique et de philosophie mathématique. Ed. de l'Académie Roumaine, Bucharest

Onicescu O (1975) Invariantive mechanics. Springer, Vienna/Berlin

Pinch T (1977) What does a proof do if it does not prove? A study of the social conditions and metaphysical divisions leading to David Bohm and John von Neumann failing to communicate in quantum mechanics. In: Mendelsohn E et al (eds) (1977)

Rickert H (1901) Les Príncipes fondamentaux de l'histoire par Alexandre D. Xenopol. Historische Zeitschrift, München und Leipzig, t. LXXXVI, pp 464–470

Schlegel R, Pfouts R, Hochwald W, Johnson GL Four reviews of Nicholas Georgescu-Roegen, The entropy law and the economic process. J Econ Iss 7(3):475–499

Seppanen L (1982) Bedeutung, Bezeichnung, Sinn: Zur Sprachauffassung Eugenio Coserius. Neuphilosophische Mitteilungen 83:329–338

Stoilow S (1945) Axiomatic thinking in modern mathematics (in Romanian). Caiete de filosofie 10:3–18

Suppes P (1993) The transcendental character of determinism. Midwest Stud Philos Xviii:242–257

Suppes P (1994) In appreciation of the work of Alexandre Froda. Lib Math xiv:1–2

Teodorescu N (1966) Physique théorique et physique mathématique. Bulletin Mathématique de la Société des sciences mathématiques de la République Socialiste de Roumanie 10:3–12

Ward L (1913) Two books on social mechanics. Am J Sociol 18(6):814–819

Xenopol AD (1900) "Les sciences naturelles et l'histoire" I. Rev Philos XXV 50(10):374–387

Xenopol AD (1908) Théorie de l'histoire. Leroux Ed., Paris (2th edition of the Principes fondamentaux de l'histoire. Leroux Ed., Paris, 1899)

Xenopol AD (1909) Zur Logik der Geschichte. Hist Z 102 Bd(3 Folge 6 Bd):473–498
Xenopol AD (1911) L'inférence en histoire. Revue Synth Hist XXII:357–368
Zlatev J (2011) From cognitive to integral linguistics and back again. Intellectica 56:125–147
Zub A (1973) Alexandru D. Xenopol. Biobibliografie. Editura enciclopedică română/Editura militară, Bucuresti

Chapter 2
What Ought to Be Done and What Is Forbidden: Rules of Scientific Research as Categorical or Hypothetical Imperatives

Mircea Flonta

> *Celui qui sème ne peut donc juger ce qui vaut la graine mai il faut qu'il ait foi dans la fécondité de la semence, afin que, sans défaillance, il suive le sillon qu'il a choisi, jetant des idées aux quatre vents du ciel.*
>
> P. Duhem, *L'évolution de la mécanique*

Similar with other activities of large groups or specialized ones, scientific research is a rule following activity. The learning process at the end of which somebody becomes a scientist consists, among other things, in being able to follow certain rules. Most of these rules are assimilated through paradigmatic practices. Future scientists learn how to follow rules. Because they learn the practice of scientific research, they are not asked to elaborate them. The rules of scientific practice, as well as the cognitive values scientists associate with them, are the elements which glue together people who work in a certain scientific field, thus forming a scientific community. These rules represent the shared background which makes consensus possible in a group.

A research practice that produces important results will generate, what might be called, a scientific tradition. As Thomas Kuhn argued convincingly, a scientific tradition is usually inaugurated by certain scientific breakthroughs, such as Newton's *Mathematical Principles of Natural Philosophy*, which provide models for stating and solving problems. The rules of scientific research are implicitly assumed in the way problems are stated and solved, serving as a model for active scientists. These rules are acquired through the practical exercise of scientific research. Those who

M. Flonta (✉)
Department of Theoretical Philosophy, University of Bucharest, Bucharest, Romania
e-mail: mircea.flonta@yahoo.com

© Springer International Publishing Switzerland 2015
I. Pârvu et al. (eds.), *Romanian Studies in Philosophy of Science*, Boston Studies in the Philosophy and History of Science 313, DOI 10.1007/978-3-319-16655-1_2

work in a certain scientific tradition follow rules which express the objectives of scientific research and paths recommended for attaining them. Any decision will be assessed by the group members with reference to the mental abilities acquired in the training process of becoming a scientist. Usually, there is little tolerance for a critical attitude towards the rules the tacit acceptance of which constitutes the foundation of a scientific tradition. Skeptics will be viewed as strangers. And challengers will be easily put aside. The analogies with other activities seem convincing.

The justification of the rules which shape a scientific tradition is usually done by formulating general considerations regarding the object of research, the requirements of scientific description, the objectives and paths for scientific advancements. Important scientific traditions in the exact sciences have received such justification. One aims at providing a justification for the rules of a scientific tradition by pointing out that these rules agree with allegedly universal norms of reason and scientific knowledge. This type of justification is considered important in order to make the case that a scientific tradition is superior to a competing one. The existence of competitors is an excellent opportunity to elaborate such considerations. In general, the explicit formulation of rules and requirements of scientific research made by prestigious scientists is closely tied with their preoccupation to justify them. When a scientific tradition is not challenged by strong competitors, there's no urgent need to underline its excellence.

As a lesson which can be gleaned from the history of natural science, I suggest to distinguish between a *doctrinal justification* and a *pragmatic justification* of the framework of a scientific tradition. The doctrinal justification is achieved through principled considerations. Such justification tends to confer to the rules of scientific research the statute of categorical, unconditional imperatives. In contrast to the doctrinal justification, the pragmatic one justifies certain rules on the basis of their fruitfulness. Therefore, the rules of scientific research are considered adequate decisions in relation with certain objectives. In the case of some prestigious scientists who explicitly formulated principles of scientific research, it seems harder to draw such a distinction. Nevertheless, there are situations where this can be made with enough clarity.

The features of these two ways of justifying the rules of a scientific tradition are presented and analyzed with reference to several historical episodes: the confrontation between the mechanistic and the phenomenological approaches in physics in the last decades of the nineteenth century, the relation between the continuum mechanics approach and approaches based on structural theories, the conflict between the late Einstein and supporters of the Copenhagen interpretation of quantum mechanics. The tension between a doctrinal and a pragmatical justification will be analyzed with reference to the confrontation between representative scientists belonging to competing traditions, as well as to the views adopted by one and the same scientist.

2.1 Doctrinal Justification and Pragmatical Justification of the Rules and Norms of Scientific Research

The justification of a research practice which I call *doctrinal* is usually achieved on the basis of general considerations regarding nature, the part of the world that constitutes the object of research, as well as allegedly universal requirements for scientific description. The rules followed in a certain scientific tradition are justified by pointing out that they are adequate to the nature and peculiarities of the object of research and are in agreement with norms of rational thinking, with requirements of scientific knowledge in general. In other words, a justification of the rules of scientific research is doctrinal if it is achieved by stating principled considerations concerning nature, the capacities and limits of human knowledge, the objectives and requirements of scientific knowledge. This way, the doctrinal justification tends to qualify research-orienting rules as unconditional imperatives. For instance, an imperative of the mechanistic research tradition, settled already in the seventeenth century, prescribed that all natural phenomena ought to be described by mechanical models. Christiaan Huygens, in his treatise on light, described "true philosophy" as that philosophy which conceives of all natural phenomena in terms of mechanical causes. Because Newton's universal attraction did not satisfy this requirement, Huygens perceived it as an "occult quality". In 1690 Huygens wrote to Leibniz that Newton's principle of attraction is absurd. When grounded in a doctrinal manner, a requirement of description becomes an absolute, unconditional requirement. Alternatives will be rejected as inadequate with respect to elementary requirements of scientific knowledge.

The pragmatic justification of a scientific practice offers as well arguments that favor a certain orientation of the scientific endeavour. This orientation is usually justified by appealing to successful strategies which have led to important discoveries. Similar with other cases where we invoke maxims of practical wisdom, the reasons we appeal to are based on what has been learned from experience. Therefore, the plea in favor of following certain rules will not be authoritarian. The rules of scientific research will be promoted not as unconditional imperatives, but as useful decisions justified by their fruitfulness. Following the rules will be presented as a recommendation. The decision to adopt these rules is not constrained by the considerations about the nature of reality or about general requirements of scientific knowledge. The term *pragmatic* expresses well enough this distinct feature.

The contrast between a doctrinal understanding and a pragmatical understanding of the rules of scientific research, often labeled *methodological rules*, can be expressed in the following way. In the case of scientists who favor a doctrinal justification it seems essential that they impose in an authoritarian manner a certain strategic orientation of scientific research and try to convince their peers that there's no alternative. For those scientists who advocate a pragmatic orientation it is important, on the contrary, to acknowledge the fact that there is a variety of options and decisions the authority of which is weakened or strengthened only with reference to failures or achievements, just as in the case of other strategic practices, i.e. financial, managerial, medical, military or sports. Therefore, on the one hand, the proponents of a doctrinal position view the rules of scientific research as categorical imperatives. On the other hand, for pragmatically oriented scientists the rules are decisions under risk conditions.

For a preliminary clarification of the way I see this distinction I will use some examples. Galileo Galilei said that the book of nature is written in mathematical characters and, therefore, only those who know this language can read it. Thus, the practice of using mathematical models for discovering natural laws receives a justification. This justification can be thought of as doctrinal in so far as the mathematical structure of the natural world is attributed to the Creator. So, the search for natural laws by means of mathematical models seems to be a categorical imperative. More than three centuries after Galilei, physicist Paul M. A. Dirac claimed that the mathematical beauty of a physical theory is the most important mark of its excellence. Dirac characterized the belief in mathematical beauty as a creed. He thought his creed was inoculated to physicists such as Albert Einstein, Erwin Schrödinger and himself by professional experiences which had influenced decisively their style of thought. Dirac noted that Schrödinger, like himself, highly appreciated mathematical beauty. This appreciation dominated the work of both. They shared the creed that any equations which describe fundamental laws must exhibit great mathematical beauty. Dirac writes: "It was a very profitable religion to hold and can be considered as the basis of much of our success."[1] No doubt, Dirac would have agreed that certain experiences can shake the most adamant faith. His creed is very different from a categorical, unconditional imperative. Dirac's justification for favoring mathematical beauty is a pragmatical one.

Usually, stating and justifying principles of scientific research succeed important discoveries and serve to legitimate the orientation that made them possible, which most of the time competes against other rival traditions. This becomes clear in the case of the rules of scientific research formulated by Newton.[2] It's worth pointing out that the first three have been formulated as *rules* in the second edition, from

---

[1]P. A. M. Dirac, *Recollections of an Exciting Era*, in (ed.) C. Weiner, *Proceedings of the international School of Physics Enrico Fermi, 1972*, London Academic Press, 1977, p. 136. Taking into consideration certain experiences, such as the provisory disaccord, eventually eliminated by the discovery of the electron spin, between Schrödinger's equation, which described the motion of the electron in an hydrogen atom, and experimental data, Dirac made some general remarks. He believed that in order to have a successful theory it is important to have beautiful equations and scientists with the right intuition. In case the output of the theoretical research and experimental data do not match perfectly, we should not worry because this mismatch might be caused by sharper details, not yet taken into consideration. We can hope that this mismatch will be eliminated by further theoretical developments. Dirac's proposal is not based on a categorical imperative.

[2]Newton introduces them under the title "Rules of reasoning in natural philosophy". The expression "natural philosophy", present also in the title of his main work, designated the research of inorganic nature using mathematical tools and being under the scrutiny of experience. In that period "philosophy" referred to any theoretical research. Newton characterized his "natural philosophy" as "experimental philosophy", in opposition with "speculative philosophy", a philosophy which is not under the scrutiny of experience. Already in the second half of the XVIII century, Madame

1713, of the *Mathematical Principles* and the fourth rule, maybe the most important, not until the third edition, from 1726. The General Scholium, which is closely linked with the fourth rule, also appears in the second edition. It seems to be a consensus among scholars that the main function of the General Scholium was to answer certain critics, notably the Cartesians and other strict adherents of the mechanical philosophy. Newton's intention to emphasize the well-groundedness of his science and to reject the claims of rival traditions becomes especially transparent both in the fourth rule and in the General Scholium, placed at the end of the third book. Here there are two revealing passages: "In experimental philosophy, propositions gathered from phenomena by induction should be considered either exactly or very nearly true notwithstanding any contrary hypotheses, until yet other phenomena make such propositions either more exact or liable to exceptions." "For whatever is not deduced from the phenomena, is to be called an hypothesis; and hypotheses, whether metaphysical or physical, whether of occult qualities or mechanical, have no place in experimental philosophy. In this philosophy particular propositions are inferred from the phenomena, and afterwards rendered general by induction. Thus it was that the impenetrability, the mobility, and the impulsive force of bodies, and the laws of motion and of gravitation, were discovered. And to us it is enough, that gravity does really exist, and act according to the laws which we have explained, and abundantly serves to account for all the motions of the celestial bodies, and of our sea."

For Newton "phenomena" are facts and observable regularities, as well as their generalizations. And the expression "deduction" was used by Newton in a broad sense to refer to any attempt to ground a statement. The meaning of "deduced from phenomena" was, therefore, "grounded with respect to experiential data". By formulating the four rules and the general scholium, Newton clearly wanted to justify a scientific practice and to point out its excellence, in opposition with its competitors. In a new preface for his book, written after the one from 1713, and left unpublished, Newton insisted on his opposition against the Cartesians, stating that the first two books of *Mathematical Principles* deal with forces in general "without investigating the causes of the forces, but only their quantity, direction and efects". Natural philosophy must not be grounded on metaphysical conjectures, but on its own principles "deduced" from phenomena. "In all philosophy we must start with phenomena and reject principles, causes, explanations beyond those deduced from phenomena."[3]

What can be said about Newton's justification of the rules of scientific research? It seems that his conception of the Creator's intentions supplied the warrant for these rules. Let us consider the first rule: "We ought to admit no more causes of natural phenomena than such as are both true and sufficient to explain their appearance."

---

de Châtelet, the French translator of the *Mathematical Principles*, used "physics" for "natural philosophy".

[3] All citation cf. Bernard Cohen, *A Guide to Newton's Principia*, Los Angeles, London, University of California Press, 1999, pp. 52–54.

The assumption supporting these claims seems to be the following: the world is the creation of a superior intelligence and this intelligence represents the supreme warrant of its rational order. Neither does it seem hard to find an answer to the question how Newton grounded the imperative of limiting the scientific research of nature to what can be controlled by experience. This imperative receives different formulations in the fourth rule and in the general scholium. His conviction seems to be that providence limits the human knowledge of nature to what can be "deduced" from phenomena. To ask for the knowledge of ultimate causes, as the Cartesians did for example, means to go against the providential order. It does not seem an exaggeration to attribute such thoughts to Newton, an author who did not hesitate, in talking about the Creator, to start from the phenomena. In this manner, the rules of research seem to receive a doctrinal justification. This justification is similar to that of which Galilei could have provided for the claim that the book of nature is written in mathematical characters and that only those who know this language can read it.

Of course, we can also suppose that the temptation of natural scientists to provide a doctrinal justification for the orientation of their research becomes weaker once they distance themselves gradually from philosophical reflections about ultimate causes. But this is not always the case. Paradigmatic for an authoritarian justification of the orientation of research is the justification given to the mechanistic program, a point of view about the objectives of scientific research in physics, defended by prestigious scientists almost two centuries after Newton published the first edition of his book. The fundamental assumption of this orientation was that the ultimate layer of all natural events is the motion of particles with mass or of a continuous medium, motions which take place in conformity with the laws first formulated by Newton. In a programmatic speech, Emil du Bois Reymond, the president of the Berlin Academy, formulated this assumption by saying that the goal of scientific knowledge is the reduction of changes in the world of bodies to motions of atoms, activated by central forces. In other words, the goal is the dissolution of the elements of nature in the mechanics of atoms.[4] What becomes clear from such formulations is that a certain prescriptive orientation is not endorsed only by results which can

---

[4] *Über die Grenzen des Naturerkennens*, 1872. Other important physicists in those days tried to ground the mechanist approaches by similar considerations. William Thompson ("Steps Towards a Kinetic Theory of Matter", 1884, p. 218, in W. Thompson, *Popular Lectures and Addresses*, vol. I, Cambridge University Press, 2011): "The now well-known kinetic theory of gases is a step so important in the way of explaining seemingly static properties of matter by motion, that it is scarcely possible to help anticipating in idea the arrival at a complete theory of matter, in which all its properties will be seen to be merely attributes of motion." Surprisingly, even James Clerck Maxwell wrote ("On the Dynamical Evidence of the Molecular Constitution of Bodies", in *Nature*, Volume 11, Issue 279, 1875): "When a physical phenomenon can be completely described as a change in the configuration and motion of a material system, the dynamical explanation of that phenomenon is said to be complete. We cannot conceive any further explanation to be either necessary, desirable, or possible, for as soon as we know what is meant by the words *configuration*, *motion*, *mass*, and *force*, we see that the ideas which they represent are so elementary that they cannot be explained by means of anything else."

justify its fruitfulness, but also by definite claims about the ultimate nature of physical reality. These considerations are the basis for categorical, unconditional imperatives because are backed up by conclusions about the essence of the material world. For those who advocated the mechanist approach, the ultimate goal of research was to reduce the apparent diversity of physical phenomena to motions and mechanical interactions of entities which are not open to direct observation.

2.2 Doctrinal Justification and Pragmatical Justification in the Clash Between the Research Program of the Mechanist Physics and the Research Program of the Phenomenological Physics

In this climate of thought, some claims made by the prestigious German physicist Gustav Kirchhoff in the foreword of his *Lectures of mathematical physics*, published in 1876, were provocative. Referring to the definition of mechanics as science of motions and to the characterization of forces as causes which produce or tend to produce motions, the author assert that this definition is heavy-laden with the ambiguity of concepts such as cause and tendency (*Streben*). Kirchhoff believed that such ambiguities can be discarded by limiting the objectives of mechanical science. "For this reason, I set the task of mechanics to describe motions which take place in nature, that is completely and in the most simple way. By this I mean only to indicate the phenomena which take place and not the discovery of their causes."[5] Kirchhoff's claims were perceived as a challenge to the point of view that the physical research must explain natural phenomena by developing mechanical models. Against this point of view, he proposed to limit the goal of science to a much economic and simpler description of natural phenomena, that of, the correlations between facts established by observation and experiments. What did Kirchhoff had in mind when he opposed description to explanation? Apart from what he said explicitly, we can suppose that a reserved attitude was at stake, skepticism about the fruitfulness of much of the mechanist hypotheses proposed in those years. What Kirchhoff suggested was that the development of mechanistic models must be abandoned when the requirement of simplicity in mathematical description of phenomena is not satisfied. Mechanistic explanations must not be sought at all cost in such fields as heat, light, electricity etc.

In the spirit of Kirchhoff's recommendation, Heinrich Hertz, taking into consideration the fact that the explanation of electromagnetic phenomena which occur in ether in terms of mechanical processes raises great difficulties, proposed that these phenomena should be described in simple terms as relations between electric and

---

[5]G. Kirchhoff, *Vorlesungen über die mathematische Physik*, zweite Auflage, Leipzig, B.G.Teubner, 1877, p. III.

magnetic charges. Like Kirchhoff, Hertz believed that a physicist ought to be pleased if he manages to formulate laws that make the calculus and prediction of observable phenomena possible, reducing as much as possible the number of fundamental concepts. Kirchhoff did not accepted force as primary notion in mechanics and Hertz eliminated it from his exposition of the principles of mechanics. This approach seems to spring from their conviction that in relation with an extremely ambitious objective, such as the mechanistic one, whose accomplishment would pose great difficulties, it's more practical to choose more modest, realistic and accessible objectives.

On the other side, physicist Ernst Mach, who even wrote a history of mechanics,[6] promoted a phenomenological, anti-mechanist and anti-atomist orientation from doctrinal premises. This contrast is evident in the manner in which Mach related to both Kirchhoff and Hertz. In his history of mechanics, Mach underlined the fact that he formulated a "more radical" point of view about the objectives of scientific research earlier and independently of Kirchhoff's remarks of the matter, in a book published in 1872, before Kirchhoff's *Lectures*. Mach valued Hertz' proposal to eliminate the concept of force, because he believed that it contained the principle of asserting only what is observable.

In his critique of the Newtonian principles and concepts of mechanics, Mach did not focused on the idea that these principles and concepts were not universally valid, that their domain of validity can be established only by experience, but insisted on the fact that notions such as absolute space and time cannot be grounded on processes which are directly observable and measurable. Mach formulated a theory of knowledge which aimed to lay the foundation for the conclusion that scientific research must be limited to the objective of describing the relations between observable magnitudes and the prediction of these magnitudes. He claimed that explanations in terms of entities inaccessible to direct observation exceed the objectives of science. It is important to keep in mind that in his works on the principles of scientific knowledge, Mach did not intended to pursue general philosophical goals, but to influence the orientation of scientific research. He believed that his works will help scientists to better understand the reason why atomistic hypotheses fail. "I don't wait for the approval of philosophers, but the acknowledgment of the natural scientist." Of course, Mach accepted that atomistic hypotheses can be useful as heuristic means of phenomena representation, for instance in the theory of chemical combinations or in the molecular theory of heat. It can serve the role of auxiliary means in order to discover natural laws. Its role can be compared with that of scaffolding in constructions. In those parts where scientific research is complete, hypotheses which refer to entities inaccessible to direct observation are not useful anymore and ought to be eliminated. For that purpose, Mach claimed that atoms, electrons or quanta are nothing more than "auxiliary concepts". Therefore, he denied the existence of atoms on principled grounds. His argument was that scientific statements are statements which refer to

---

[6]*Die Mechanik in ihrer Entwicklung. Historisch-kritisch dargestellt*, first edition 1883.

correlations between phenomena, and that scientific theories are nothing more than means to express these correlations in an economical manner. This understanding of the function of theoretical concepts and principles implies the conclusion that they does not refer to a reality beyond direct observation. A physical theory describes nothing which cannot be known by a more comprehensive experience.

The chemist and promoter of energetism Wilhelm Ostwald, who considered himself Mach's pupil, also illustrates the attempt to provide a doctrinal justification to the phenomenological orientation of scientific research. In his dispute with Ludwig Boltzmann, who defended the legitimacy of atomistic approaches, at the Congress of German naturalists in Lübeck, 1895, Ostwald gave the following argument. Let's suppose that two handspikes interlock in a closed box and by observation and experiment we can establish a functional relation between their speed quantities from which it can be derived successful predictions. In this case, the explanation of correlations established by a mechanism of interlocking handspikes will not produce any new knowledge. Reacting in this way, Ostwald characterized the kinetic theory of gases as "sterile" and rejected the statistical interpretation of the second law of thermodynamics. In his *Vorlesungen über Naturphilosophie* published at Leipzig in 1902, Ostwald grounded the constraint on scientific research to limit itself to phenomenological approaches on the distinction between "laws" and "hypotheses". The formulas of laws must include only measurable magnitudes. The formula of gas pressure in molecular theory – $pv = \frac{1}{2} m \ n \ c^2$ – includes however three variables, m = the mass of a gas molecule, n = the number of gas molecules and c = the speed of gas molecules, which refer to magnitudes that cannot be measured. Therefore, this formula is a "hypothesis" and not a "law". Ostwald discarded these "hypotheses", as opposed to working hypotheses which are heuristics tools, as being not only useless, but also detrimental. Their elimination from science was an unconditional imperative for him.

The doctrinal justification of the mechanist and phenomenological approaches is in opposition with the pragmatic attitude adopted by Boltzmann. In his interventions which caught the attention of the scientific community, the Austrian physicist opposed the principled rejection of atomistic hypotheses by machists and energetists. Boltzmann claimed that the old mechanist approach and the new phenomenological approach should not be advocated for as dogmatic points of view, which imply categorical imperatives and interdictions. He emphasized that all lines of scientific research must remain open: "We are, thus, in agreement with the idea that any conception ought to be freely developed. Instead, Mister Ostwald's attempts to prove that the old conceptions of theoretical physics are obsolete and that the new energetist ones are preferable seem to me groundless."[7] Against what he called "the new dogmas of the theory of knowledge", Boltzmann claimed that "with all necessary prudence" we ought to defend the right to formulate hypotheses, based on observations, about "what we cannot perceive". It cannot

---

[7]"Ein Wort der Mathematik an die Energetik", in L. Boltzmann, *Populäre Schriften*, Leipzig, Verlag von Johann Ambrosius Barth, 1905, pp. 131–133.

be said "that any mechanical hypothesis has failed".[8] Opposing the exclusivist preference for a certain approach, Boltzmann asked if "it would not be a loss for science if the current idea of atomism would not be developed with the same avidness as those of phenomenology".[9] We can try, for example, to develop a mechanical explanation of heat and electricity, but this is not an imperative. It might be successful or it might not. Neither the unfruitfulness nor the fruitfulness of the mechanistic-atomistic approaches have been proved beyond doubt. Boltzmann was not against the phenomenological orientation of scientific research. He did not deny its utility. He just wanted to warn against the risks of favoring it unconditionally, of accepting the principle that "the only goal of physics is to write equations for every series of processes, without any hypothesis, any intuitive representation or mechanical explanation". Boltzmann defended the right to foster both atomistic and phenomenological approaches. To the dogmatic confrontation between two mutually incompatible schools of thought, he opposed the opportunistic development of alternative approaches.

It's worth pointing out that already at no more than 10 years after Boltzmann's death, the well-known German scientist Wilhelm Voigt compared the results obtained by atomistic and phenomenological approaches. He proposed the following general reformulation of Kirchhoff's characterization of the phenomenological approach in mechanics: the general goal of the theory is to obtain laws of natural processes by stringent reasoning, based on a minimum of presuppositions. The requirement of the phenomenological approach is to derive a diversity of mathematical consequences from a small number of principles based on experience. In opposition with the phenomenological approach, the atomistic approach explains correlations established through observation and experiment by introducing hypotheses about invisible events and processes. The results of the atomistic approach will be considered important whenever the hypotheses allow the derivation and prediction of a variety of correlations between observable magnitudes.[10] Voigt reviewed the results obtained by the two approaches. He determined that the phenomenological approach produced important results in pure thermodynamics, it has a dominant position in rigid body mechanics and ideal fluid mechanics and that it plays an essential role in electricity theory and hydrodynamics. On the other hand, the atomistic approach was very successful in the kinetic theory of matter, in the scientific research of the relations between caloric and elastic phenomena. Great

---

[8]Boltzmann's position can be fully assessed only by those who know that it was an unorthodox position in relation to the point of view shared by leading figures in the community of German physicists. So in 1891, at the Halle Congress of German naturalists and physicians, Max Plank claimed that the development of kinetic theory of gases, advocated by Boltzmann, is "doubtful and conjectural". Attributing an absolute value to the principle of increase of entropy, Planck was skeptic about Boltzmann's probabilistic interpretation. Many outstanding physicists talked at that time about "a crisis of atomism".

[9]"Über die Unentbehrlichkeit der Atomistik in der Naturwissenschaft", in *op. cit.*, p. 142.

[10]See W. Voigt, "Phänomenologische und Atomistische Betrachtungsweise", in *Physik* (editor E. Warburg), Leipzig und Berlin, Verlag von B.G. Teubner, 1915, pp. 715–717.

achievements of the atomistic approach were made by the discovery of the electron and photon. The general conclusion of the review is formulated as follows: "The brilliant successes of the electrons hypothesis, alongside the results of the kinetic theory of matter, contributed decisively in the last decade to the supremacy of the molecular approach over the phenomenological one."[11] If Boltzmann was still alive he would have had the satisfaction to see how much he was right!

Some of Einstein's methodological remarks shed light on Boltzmann's position, which can be considered paradigmatic for the pragmatical approach to the justification of the rules of scientific research. Two decades latter, Einstein formulated some considerations about the relative virtues of the approaches and theories which he named *constructive* and respectively *principle theories*. Constructive theories, such as the kinetic theory of gases, explain correlations which are accessible to observation by introducing hypotheses about entities and correlations situated at a more fundamental level. Principle theories, such as thermodynamics, are based on general features of natural phenomena, which constitute the foundation for mathematical criteria. Einstein pointed out that the advantage of constructive theories is "comprehensiveness, adaptability, and clarity", while principle theories can claim in their favor attributes such as "logical perfection, and the security of their foundation".[12] Like Boltzmann, Einstein did not believe that we are entitled to speak about the superiority of one approach over the other, but only about their relative capacity to indicate the direction of answering questions raised by the scientific research of nature. This is a pragmatic style of thinking, in sharp contrast with the style that lead to the confrontation between the mechanist and phenomenological orientation, which was at its peaks in the last decade of the nineteenth century and in the first decade of the next century. Both positions resemble in their tendency to grant to certain rules of research the statute of unconditional, absolute imperatives. In his book *Science and hypothesis*, Henri Poincaré emphasized that scientists who endorse this kind of imperatives "want to limber nature after a certain form, beyond which their spirit will not be satisfied". And he asked ironically: "Is nature flexible enough for this?"

2.3   The Meaning of the So-Called "Reductionist Program in Physics" from the Perspective of the Distinction Between the Doctrinal and the Pragmatical Approach

From the point of view of the distinction between doctrinal and pragmatical justification of the rules of scientific research, the discussion about the statute of continuum mechanics as a field of scientific research is very instructive. If it

---

[11]*Idem*, p. 730.

[12]See A. Einstein, "My Theory", in *Times*, 28 November, 1919.

aims to discover regularities and laws of phenomena accessible to observation, and does not try to explain them in terms of corpuscular parts of the structure of matter, then it can be said that the scientific research in continuum mechanics has a phenomenological orientation. However, we can ask the question whether the results of such an research ought to be considered its ultimate objective. Divergent points of view have been formulated in order to answer this question. One of these points of view, which was and still is influent among theoretical physicists, is that particle physics is more fundamental than other scientific fields.[13]

There are different ways to understand this assertion. Steven Weinberg, for instance, characterizes his position as that of a reductionist open to compromises. He does not claim that disciplines such as thermodynamics or hydrodynamics ought to be reduced to molecular physics, that properties such as vorticity, turbulence, entropy or temperature ought to be investigated only from this point of view and not from a phenomenological perspective as macroscopic properties. Also, he does not claim that particle physics can help us to make new discoveries in the fields of hydrodynamics or condensed matter physics. For him, reductionism is just an expression for the belief that the natural world is constituted in such a way that the laws of thermodynamics and hydrodynamics are explained by the laws of fundamental physics. To the question why the behavior of certain macroscopic systems, studied in various fields of physics, is governed by certain laws, the reductionist will say that we must always look for the answer in fundamental physics. It this case, reductionism expresses a feature of nature itself.[14] Weinberg wants to demarcate himself from stronger versions of reductionism, from what he called "reductionism without compromise". But since he claims that the constitution of nature itself imposes a regress towards fundamental physics in explaining natural laws, then it can be said that his position had doctrinal shade.

While acknowledging the autonomy of disciplines which study different macroscopic systems, their thorough importance in discovering specific laws, we can still accept the legitimacy of explaining these laws by appealing to universal laws of fundamental physics. From this viewpoint, we can understand the concern of some scientists, such as the American physicist Clifford Truesdell. He emphasized the right to existence of a phenomenological approach of macroscopic systems, contrary to the point of view that continuum mechanics is just an approximation or "a secondary theory" in comparison with explanations based on the research of the discrete structure of these systems. Besides the insurmountable mathematical difficulties incurred by the prediction of macroscopic systems' behavior in terms of entities such as intra-molecular forces, those who favor structural approaches on principled considerations neglect the fundamental fact that materials with very

---

[13]See Steven Weinberg, "Newtonianism, Reductionism and the Art of Congressional Testimony", in *Nature*, vol. 330, dec. 1987.

[14]"Now reductionism, as I've described it in term of the convergence of arrows of explanation, is not a fact about scientific programmes, but is a fact about nature. I suppose if I had to give a name for it, I could call it objective reductionism" (*Idem*, p. 436).

different corpuscular structures can react identically to the same forces. These limits show that the phenomenological approach is not only legitimate but also useful. Nevertheless, to acknowledge this fact does not mean to doubt the utility of a structural approach when certain objectives are aimed at. For example, the properties of a fluid can be examined as the result of motion and interaction of a large number of particles. Fluid properties like pressure, speed and ropiness, as well as the laws of ropy fluids, are presented in this case as derived consequences from a statistical theory of particles motion.

The manner in which Truesdell reacted against the tendency to oppose, on doctrinal reasons, the structural approaches to the phenomenological approaches of continuum mechanics, is paradigmatic for the pragmatical justification of the rules of scientific research: "Widespread is the misconception that those who formulate continuum theories believe matter 'really is' continous, denying the existence of molecules. That is not so. Continuum physics presumes *nothing* regarding the structure of matter. It confines itself to relations among gross phenomena, neglecting the structure of the material on smaller scale. Whether the continuum approach is justified, in any particular case, is a matter, not for the philosophy or methodology of science but for *experimental test*. In order to test a theory intelligently, one must first find out what it predicts." And also: "It should not be thought that the results of the continuum approach are necessarily either less or more accurate than those from a structural approach. The two approaches are *different*, and they have different uses."[15] Scientists ground their decision to follow certain rules on convincing research experiences. The only justification of these rules is their capacity to direct in a fruitful way the scientific research.[16]

2.4 From Pragmatical to Doctrinal Approaches: The Einstein Case

It is not always easy to draw the distinction between a doctrinal and a pragmatical justification with regards to orientations of thought and programmatic declarations made by prestigious scientists. One of Albert Einstein's decisions as well as the justifications provided to support these decisions illustrates this point.

The author of the theory of relativity repeatedly stressed that Mach's historical and philosophical writings had a great influence in shaping that orientation of

---

[15]C. Truesdell, "Purpose, Method, and Programm of Nonlinear Continuum Mechanics", (1965), in C. Truesdell, *An Idiot's Fugitive Essays on Science*, New York, Berlin, Springer Verlag, 1984, p. 54 and p. 57.

[16]"With this sober and critical understanding of what a theory is, we need not see any philosophical conflict between two theories, one of which represents a gas as a plenum, the other as a numerous assembly of punctual masses. According to the physicists, a real gas such as air or hydrogen is neither of these, nothing so simple. Models of either kind represents aspects of real gases; if they represents those properly, they should entail many of the same conclusions, though of course not all" (C. Truesdell, "Statistical Mechanics and Continuum Mechanics", 1973, 1979, in *op. cit.*, p. 73). These are reflections which call to mind Hertz's and Boltzmann's characterizations of "images" and "theories" as descriptions of the physical world.

thought which lead him to his epochal discovery. Many of Einstein's remarks leave the impression that he followed Mach from a pragmatical perspective, not a doctrinal one. In his intellectual *Autobiography*, Einstein said that he was directed towards the formulation of the theory of relativity because he acknowledged the "arbitrary nature" of the principle that time and simultaneity have an absolute status. This orientation, he admitted, was "decisively stimulated" by Mach's writings. The young Einstein was sympathetic to the anti-metaphysical trend that Mach voiced so vigorously. He was influenced especially by Mach's skeptical position against the mechanist conception which was still dominant in those years. At the same time, Einstein admitted that in that period he was also impressed by Mach's phenomenological approach, and by the Machian conception of physical theory.[17] Starting from such declarations, it seems hard to asses if Einstein was influenced only by Mach's critical research of the foundations of physics or he also embraced Mach's epistemology. I especially refer to Mach's claims that notions such as force, electric charge, time element, atom, are nothing more than "auxiliary concepts" that are legitimate only if the propositions which contain them are derivable from propositions which can be verifiable by experience. According to this point of view it wouldn't make sense to see these notions as having a value other than heuristic. Physical science should not set the objective of describing what is beyond the phenomena known by experience. It's plausible to claim that young Einstein was not reticent to such points of view. In 1912, Einstein signed an address, initiated by Mach, to create a society of positivist philosophy. Among those who signed it were notable figures like Felix Klein, David Hilbert and Sigmund Freud.

The appreciation of Mach's ideas from a pragmatical perspective, and not a doctrinal one, stems more clearly from Einstein's article *Ernst Mach*, published in the journal *Physikalische Zeitschrift* in 1916, the year Mach died. I refer to statements like the one that notions which had proven useful to systemize phenomena can gain an authority in as much as to be considered "necessities of thought". This can block the path to scientific progress. Einstein emphasized that it is important to pin point the conditions of justification and utility of these notions in order to undermine their "excessive authority". For instance, in a letter from 1930 to Armin Weiner, Einstein still appreciated that the theory of relativity is in line with the general orientation of Mach's conception. This view was widespread. For instance, it was formulated in the letters of prestigious scientists who recommended the Nobel Prize award to Mach.

---

[17]See A. Einstein, "Autobiographisches", in P. A. Schilpp (ed.), *Albert Einstein: Philosopher-Scientist*, Open Court, La Salle, Illinois, 1949. No doubt, Einstein appreciated Mach's liberating influence on his style of thought. Mach expressed his position by saying that all principles of mechanics rely on experiences regarding the positions and *relative* velocities of bodies. In the field in which they are considered valid, Mach thought they should not be used without proper examination. Therefore, nobody would be justified in extending such principles beyond the borders of experience. (See E. Mach, *Die Mechanik in ihrer Entwicklung. Historisch-kritisch dargestellt*, Berlin, Akademie Verlag, 1988, p. 252).

In his references to Mach from latter years, Einstein adopted an ambivalent attitude towards his ideas, pointing out, on the one hand, the fortunate influence which Mach's historical and critical analysis of the concepts of mechanics had on his orientation of scientific research, and, on the other hand, the doubts he had about Mach's empiricist philosophy of knowledge and science. Now Einstein repeated the claim that the principles of theoretical physics cannot be derived from experience. These principles ought to be considered free inventions of the scientist's mind. In opposition with Mach, which had principled concerns about all atomistic hypotheses, Einstein accepted them without whenever it was proven that such hypotheses can provide a simple and adequate description of phenomena accessible to observation. This is a pragmatical position *par excellence*.

Einstein expressed himself in a clear and direct manner about the evolution of his ideas on the method of science in connection with what he had learned from his experience as a scientist. Here is what he wrote to Cornelius Lanczos in January 1938: "Coming from a skeptical empiricism of somewhat the kind of Mach's, I was made, by the problem of gravitation, into a believing rationalist, that is, one who seeks the only trustworthy source of truth in mathematical simplicity."[18] The change in Einstein's position is very well illustrated by Werner Heisenberg's recording of a discussion he had with Einstein in Berlin, in 1926. With regards to his first formulation of quantum mechanics, Heisenberg told Einstein that it's rational to introduce in a physical theory only observable magnitudes, for example the radiation emitted by atoms. Heisenberg was surprised when his discussion partner objected to this principle. He told Einstein that exactly this principle lead to the theory of relativity. Einstein's reaction indicates clearly the evolution of his ideas on the matter: "Maybe I did used this kind of philosophy, but nevertheless it is a nonsense. Or I can say it in a more careful way: it can be useful from a heuristic point of view to remember what is really observable. But from a principled point of view it is entirely false to ground a theory only on observable magnitudes. In reality it is the other way around. Only the theory decides what is observable."[19]

---

[18]Cf. G. Holton, More on Mach and Einstein, in G. Holton, *Science and Antiscience*, Cambridge, MA, London, Harvard University Press, 1993, pp. 65–66.

[19]W. Heisenberg, *Der Teil und das Ganze. Gespräche im Umkreis der Atomphysik*, München, R. Piper Verlag, 1969, p. 92. Heisenberg's recording about the fact that Einstein had changed his position towards the Machian principle that physical theories ought to describe only observable quantities, is confirmed by physicist Philipp Franck in a chapter of his book on Einstein, entitled suggestively "Einstein's critique of the fruits of trees he himself planted". Frank recalls visiting Einstein at Berlin, in 1932. In their discussions, Einstein made some ironic remarks regarding what he called "a new fashion in physics" – the interdiction to introduce magnitudes which cannot be measured in the language of physics – probably he was referring to those physicists around Niels Bohr. To Frank's observation that he himself used the same principle in 1904, Einstein replied: "A good joke should not be repeated too often." We also find a convincing testimony in Max Born's comments about his correspondence with Einstein. These comments were written in 1965 for the German edition published in 1969. Born claims that Einstein grounded the theory of relativity on the supposition that notions which do not refer to observable facts, such as absolute simultaneity, should be ruled out from physics. Quantum theory was born in the same way. Heisenberg applied

The way Einstein related to the new quantum mechanics, in the standard interpretation of the Copenhagen School, is an excellent illustration of the evolution which took place in his thought. Einstein repeatedly claimed that the general theory of relativity opened the door to a unified field theory, to what he called "a deeper understanding of the connections". Hence, he did not show enthusiasm to those evolutions in scientific research that weren't in agreement with this direction. To his old friend from Bern, Michele Besso, he wrote that his aim was to explain the reality of quanta, the laws established by experiments, without *sacrificium intellectus*, that is, without transgressing the requirements of what he considered an acceptable scientific description. Einstein's belief was based on his deep convictions about the simplicity, harmony and intelligibility of the natural universe, convictions which he did not hesitate to qualify as "religious".[20] In his later years, Einstein expressed his inflexible conviction that the fundamental laws of nature are simple and beautiful in their mathematical expression. B. Hoffmann, one of his assistants at Princeton, noted that when he asked what he thinks about a good physical theory, Einstein answered that in assessing a theory he asks whether God could have created the universe in such a way that the theory could be true. "If the theory does not posses the kind of beauty demanded by God then in the best case scenario it is only provisional."[21] Referring to the fact that Einstein was inclined to asses the scientific evolution in physics only from this perspective, it has been noticed that his most pragmatic and temperate colleagues were inclined to see him more like a philosopher of nature rather than a professional scientist. This is because his scientific creed was based on considerations about the features of the natural universe and on absolute requirements of scientific description. Very important in this respect is Einstein's own characterization of the evolution of his position from that of an "empiricist skeptic" to that of a "rationalist believer".

Being aware that his position regarding the development of physics is in a complete disagreement with the positions which were dominant among the younger generations of physicists, Einstein insisted in his published work, as well as in private conversations and correspondence, to state the principles which ground and justify his position. He wanted to be better understood and to win supporters eventually. With his convictions about the direction and ultimate goals of scientific knowledge in the background, Einstein claimed that although the achievements of the new quantum mechanics cannot be denied, we cannot also consider it a satisfactory physical theory. This claim opposed the dominant position among physicists. Many accepted Paul Dirac's reflection that the physicist is pleased if the prediction of a theory is in agreement with experimental data and that he doesn't want more. Einstein expected more from a theory which represents progress

---

this principle to the structure of the atom. (See Max Born, *The Born-Einstein Letters 1916–1955*, New York, MacMillan, 1971.)

[20]See A. Einstein, *Religiosität der Forschung* and *Science and Religion* (I–II).

[21]See B. Hoffmann, P. Bergmann, "Working with Einstein", în (Ed.) H. Woolf, *Some Strangeness in the Proportion*, Reading, MA, Addison-Wesley, 1980, p. 476.

towards what he marked as the ultimate goal of scientific knowledge. He conceived these requirements as postulates in the sense that they cannot be challenged and negotiated.

Referring to such requirements, Einstein claimed that the goal of any physical theory is to describe a reality which exists independently from the subject and the act of scientific research. This requirement implies the supposition that objects and physical events which are spatially separated have an independent existence, the so-called "principle of separability". Einstein believed that if we do not comply with this requirement then scientific reasoning itself is not possible anymore. So, for him the case was about unconditional imperatives. Observing that the new quantum mechanics does not meet these requirements, Einstein reached the conclusion that that theory didn't give a complete description of the physical reality. Nevertheless, he admitted that physicists are not rationally compelled to accept the principle of separability. Hence, it results that the unconditional adhesion to this principles is an act of faith. Einstein had no doubts that the future evolution of scientific research will give him credit. Especially in his correspondence and particular discussions, he expressed the firm trust that the objects of a future theory will not be probabilities but what we call facts. However, there are some ambiguities in his statements. Thus, in a letter to Erwin Schrödinger from August 1939, after specifying that the core of his disagreement with Bohr is visible in the different answer to the question whether the wave function from quantum mechanics describes a final physical state or just its probability, Einstein added: "Beide Standpunkte sind logisch einwandfrei; aber ich bin nicht imstande zu glauben, das einer dieser Standpunkte sich schliesslich bewähren wird." ("Both points of view are logically impeccable, but I'm not able to believe that one of these two points of view will be confirmed")[22] This formulation can suggest an oscillation between a doctrinal approach and a pragmatical one. On the one hand, Einstein's attachment to requirements of description established by classical physics is categorical. On the other side, the expression "confirmed" suggests that the future evolution of scientific research will decide the matter. But what distinguishes a categorical imperative from a hypothetical one is precisely the fact that it cannot be "confirmed" or "infirmed". Maybe Einstein's usage of the expression "confirmed" was just a turn of phrase for his conviction that future developments will be give him credit.

However, the categorical formulations of his refusal to accept theoretical descriptions which do not meet his requirements remain dominant. It is important to point out that this position is in sharp contrast with the position held by physicists from Bohr's group. They were skeptical about universal requirements of theoretical description in physics. Wolfgang Pauli formulated very clearly this point of view writing, in an editorial for a collection of papers which were published in an issue of the journal *Dialectica*, from 1948, that quantum mechanics marks a new era in physics. An era in which there will be other revisions of the so-called

---

[22]K. Prizibram (ed.), *Schrödinger, Planck, Einstein, Lorentz: Briefe zur Wellenmechanik*, Wien, Springer Verlag, 1963, p. 33.

"classical" requirement of natural description. The increasing isolation in the scientific community did not determined Einstein to soften his position regarding the acceptance of quantum mechanics as a complete theory. Until the end of his life, he constantly attempted to elaborate a general field theory from which the laws of quantum mechanics can be derived as consequences. In his paper *Quantenmechanik und Wirklichkeit*, published in that issue of *Dialectica*, Einstein expressed his hope that such a theory will contain quantum mechanics "just as the optics of radiation is contained by ondulatory optics: the relations remain the same, but the basis will be deeper, namely it will be replaced with a more comprehensive one."[23]

The next testimony also shows how categorical and inflexible was Einstein's position. Physicist Arthur Komar recalled that in a lecture held at the Palmer Physical Laboratory Einstein claimed that the fundamental laws of nature ought to be mathematically simple. The speaker called the lecture "his final exam". To the question from the audience "What if the fundamental laws are not simple?", Einstein replied: "Then, I'm not interested."[24] For Einstein, to abandon this requirement and accept compromises amounted to giving up theoretical science altogether. All this seems to indicate how far Einstein had gone in his departure from a pragmatical approach in his assessment of new evolutions in physics.

At the same time, Einstein's attitude draws attention upon a significant difference between doctrinal and pragmatical approaches. Theoretical physicists who take a pragmatical approach are willing to revise the rules and norms of scientific research in the light of those experiences which show that these rules and norms have become a burden for the advancement of knowledge. Things are the other way around when a doctrinal approach is assumed. The unconditional commitment to certain ideal requirements of scientific description and explanation is not touched by continuous failures to advance the prescribed direction. Eventually, Einstein perceived his ideal requirements of theoretical excellence in physics as "necessities of thought", namely, in the same way as he described, in his article in memory of Mach, the attitude of many physicists from the old generation towards the foundations of their science. Confronted with constant failures to finish his project on the general field theory, Einstein did not show any signs of willingness to take a step back from convictions which endorsed his scientific stand. His reaction was typical. Einstein used to say: "I've lost a battle, but not the war." What can one reply? Well, we can eventually ask when the war is expected to end.

---

[23] See *Dialectica*, vol. II, no. 3–4, 1948, p. 320.

[24] Cf. John Archibald Wheeler, "Mercer Street und andere Erinnerungen", in (eds.) P.C. Aichelborg, R.U. Sexl, *Albert Einstein. Sein Einfluss auf Physik, Philosophie und Politik*, Braunnschweig, Vieweg, 1979, p. 214.

Part II
Mind, Language, and Technology

Chapter 3
Memory as Window on the Mind

Radu J. Bogdan

This paper argues for two propositions. The first is that memory is not only indispensable to a mind but also constitutionally implicated in shaping its operation. As a result, a study of memory systems that dominate a kind of mind opens a unique explanatory window on what that kind of mind can and cannot do. This angle on the memory-mind relation has not been widely adopted in cognitive science so far; it should be.

Adopting this angle in a major instance as illustration, the second proposition is that autobiographical memory, which is unique to humans and emerges late in childhood, operates in ways and with resources that reveal an entirely new kind of mind that only older children develop and adults inherit – a mind unknown in the rest of the animal world.

3.1 Memory Systems

The list of memory systems being familiar, I will limit myself only to those details that are relevant to the two propositions that frame the argument of this paper.

I begin with the distinction between procedural and declarative memory systems. The work of the former, which we may call *procedural recall*, is a sort of stored and retrievable know-how. How to tie one's shoes or write or swim or drive or salute or handle a fork or a telephone (to take examples of what was once learned) is a skill stored and retrieved, hence remembered, in a procedural mode. Procedural memory is the most widespread in the animal world, probably the oldest historically, and most standard outcome of learning. Even though almost

R.J. Bogdan (✉)
Department of Philosophy, Tulane University, New Orleans, LA, USA
e-mail: bogdan@wave.tulane.edu

© Springer International Publishing Switzerland 2015
I. Pârvu et al. (eds.), *Romanian Studies in Philosophy of Science*, Boston Studies
in the Philosophy and History of Science 313, DOI 10.1007/978-3-319-16655-1_3

all instances of learning begin with explicit encodings of information through observation, instruction or imitation, most of them end up as implicit and procedural memories. Even though learning to ties one's shoes or write or handle a fork begins with laborious and attentive intake of mostly perceptual and often verbal information, explicitly encoded, one ends up remembering how to do it, and do it, unthinkingly; indeed, intrusions of thinking in the retrieval or deployment of procedural memories are likely to interfere with and slow down the activity in question. This memorial "proceduralization" (as we may call it) of initially explicit information is worth noting because, among other things, it may explain the loss of many explicit memories of early childhood (early childhood amnesia), which is a period of intense acquisition of procedural skills in many domains – behavioral, social or interpersonal, communicational and cultural.

Childhood is also a period of intense acquisition of *facts* and *experiences* in those very same domains. Memory of facts and experiences belongs to another type of memory – the *declarative* type, also known as explicit or descriptive. The remembered facts and experiences are encoded, stored and retrieved in the form of data structures, as images (broadly construed as visual, motor, tactile), feelings, emotions, words and other mental or artificial symbols. Several versions of declarative memory systems dominate animal and human minds – semantic, episodic and (only in humans) autobiographical. Despite their declarative format, these memory systems are vastly different in what they do and how they do it.

Semantic memory encodes, stores and retrieves information that was once represented in the form of some data or signal structure through some sensory channel – usually something seen or heard – without a record of the actual experience of how and when that information was registered in the past and from what sources. My semantic memory reminds me that Nairobi is the capital of Kenya or that WWII ended formally on May 9, 1945, but I have no recollection of the initial experiences and sources that introduced me to these items of information or the time of the introduction. Many if not most of our mundane beliefs are stored in this fashion in semantic memory and lack experiential associations and a sense of their sources or origins. Semantic memory is memory of facts, lots of them.

Episodic memory, another and much richer version of declarative memory, operates under two principal constraints: it is about *specific* episodes (events, situations) in one's past as sources of information; and those episodes are registered, encoded, stored and recalled in *experiential* detail, typically perceptual and often including recollections of once vivid reactions to and emotions about those episodes. To this very day, for example, I recall the chloroform experience and its smell, and visualize the clinic bed and room, when and where I had surgery around the age of 7.

Two pieces of neuropsychological evidence suggest a tight link between perceptual experience (in various modalities) and episodic memory. One is that episodic memories are represented in the same brain areas as actual perceptual experiences, particularly visual. The other piece of evidence is that episodic memories evoke relatively short-lived experiences and the access to such memories tends to degrade rather quickly (Conway 2001). The implication is that the primary function of episodic retrieval is to *reenact* and *recapitulate* an initial perceptual or action experience and its associated reactions and emotions in their immediacy and vividness (Smith and Kosslyn 2007, pp. 216, 221).

Why, then, do I remember rather vividly my surgery episode of late childhood? The main reason is that now (and then too) I am (and was then) capable of *autobiographical* recall. The latter is uniquely able to reframe and as a result consolidate and retrieve episodic memories (Conway 2001). More on this later. Since many animal species are also credited with some sort of episodic memories, which were said to be relatively short-lived, how come that those animal minds, while not being autobiographical, manage to remember events and experiences in their distant past? A plausible answer is that (like humans) they retain, encode and, when appropriate, retrieve their past physiological, emotional and behavioral *reactions* to such events and experiences but not necessarily the specific details of the initially registered episodes. In other words, such reactions constitute a sort of episodic traces or cues that facilitate the retrieval of the past experiences. An indelicate example, familiar to many people and surely many animals, is the long-term adverse reaction to some food with which they once had a bad experience: one need not, and usually does not, remember the initial episode of eating the bad food or the first adverse reaction in order to retain the bad-food susceptibility for many years. Even though episodic memories are rather short-lived, the reactions they generate can and usually are durable, even more so when reframed and consolidated by autobiographical memory.

Autobiographical memory is the declarative system that is capable of consciously, deliberately and reflectively projecting one in the past, with the one's past mental states, experiences, reactions (e.g., emotions and feelings) and actions as well as their relations to specific targets (things, persons, events, situations) encountered in that remembered past and usually retrieved or reconstructed in terms of some intelligible context, normally defined by some script, scenario or narrative. Autobiographical recall need not always be deliberately initiated and explorative, for it can be and often is occasioned by some unbidden input or association. What matters, though, is that one has the capability to engage consciously, deliberately and reflectively in autobiographical recall even when the initial memory trigger is not of one's initiative.

To see why dominant memory systems characterize distinct kinds of minds and why in particular autobiographical memory is uniquely human and shapes – as well as reveals – a new kind of mind that develops after the age of 4, I propose to broaden the explanatory picture and look at the intimate and intricate relation between memory and mind.

3.2 Mind as Memory

Memory is central to a mind in several respects, all of which are crucial in understanding the unique specificity of autobiographical memory and its role in the new kind of mind it animates.

First, the memory systems that dominate an organism's mind can be expected to shape the operation of the faculties that use those systems. If an organism draws mostly on procedural memory, has a relatively poor declarative memory and no significant if any working memory, we can reasonably predict that it has no significant thinking mind, for there is no mental buffer for thinking and no significant internal source of stored and retrieved data structures as thoughts. We can also reasonably predict that the temporal range of its cognitive activities is narrow if not entirely limited to fast changing stimuli that activate its know-how as habits and routine. Likewise, a mind dominated by an episodic memory that stores perceptual and action experiences is likely to be limited to displaying fragmented snapshots of a fast-moving present.

In contrast, an organism's capacious declarative memory, both semantic and episodic, suggests an ample storage of data structures and hence an ability to learn and retain facts about its environment and to experience and recognize distinct patterns behind sensory stimuli. If, furthermore, an organism's has a working memory of some significant size, we can reasonably predict some commensurate ability to hold steady and process rather complex strings of mental representations and perhaps thinking.

The overall point, in short, is that a dominant memory system or a dominant mix of such systems can be a reliable guide to the sort of cognitive mind these systems are likely to service. As noted next, the memory-mind connection is even tighter since memory systems not only store know-how and data but also configure and even anticipate the world to which a mind has access.

Second, a memory system is vital in making accessible a world beyond the sensory inputs. The reason is familiar: sensory interactions with the world are fragmentary, unstable and short-lived, yet organisms perceive the world around them as fairly stable, well organized, continuous, and orderly. This is because past facts and experiences, stored by memory systems, are recruited to fill in the gaps. Declarative memory systems are fillers of sensory gaps. Whereas sensation samples information from the world, perception stabilizes the shaky and transient sensory inputs into durable and actionable encodings of information with the help of memory systems. Organisms exercise their cognitive faculties and act in and on the world on the basis of what they perceive distally, not what they sense proximally. On the procedural side, activating skills, habits and routines, upon receiving sensory stimuli, is an organism's adaptive way of regularizing and engaging the world in the light of past successes in action, thus again filing in the sensory gaps with behavioral expectations about the world regularities that matter to the organism.

To sum up so far, it is a dominant memory system or a set of such, whether procedural or declarative, that tells an organism what to expect in the world, in the present and the future, beyond the transient and fragmentary surface of sensory stimuli. This familiar idea has been recently fortified and expanded dramatically by the neuropsychological discovery that the brain is an intrinsically projective engine that constantly anticipates and predicts actions, experiences and states of the world (Bar 2011; Clark 2013).

This is the third respect in which memory makes a major difference to the mind it services. The brain cannot help but project – this is its "default mode." It has been thought for some time that this default mode is that of spontaneous conscious activity, reflected in daydreaming and hallucinations as well as focused mentation. But it has been recently discovered that the default-mode projection is present even under general anesthesia, during sleep and other unconscious states. The brain is always working.

The constant projectivity of the brain is beginning to reshape the understanding of *memory* as well. Instead of being primarily a repository of information about the past, memory is increasingly seen in neuropsychology as a database for predicting the future and a forward-looking facilitator of responses to stimuli (Bar 2011). Some experts have suspected this future-orientation of memory for some time (Nelson 1996) but now we have a deeper neurological rationale for this. Memory is for the future, not the past: what is deposited in memory is primarily material for projections about the future or the possible. The past, in other words, is just a springboard for the future or, in French, *reculer pour mieux sauter.*

Instead of generalizing from past experience, learning itself may be seen as proceeding from projections adjusted to incoming experiences. Indeed, one model of projective learning gaining currency is Bayesian in spirit: beginning with a set of advanced predictions that form a sort of prior "mindset", the brain then generates best guesses about the environment, to be revised in the light of further experience (Bar 2011; Clark 2013).

In both anticipation and learning, the projective brain is already equipped with a version of the world it is about to engage (its Bayesian "priors"), allowing inputs from perception as well as further thinking and other available data to adjust or modify this prior version. It matters a good deal, then, what sort of memory systems are the ones that store, maintain and access that prior version, at what level of complexity and with what range of domains – physical, biological, communicational, cultural and so on.

In the three respects noted so far, autobiographical memory is radically different from the other memory systems: it animates cognitive and executive capabilities that are quite unlike those of minds lacking this memory system; it stores, proposes and anticipates (prior) versions of the world that are very different from those supplied by the other memory systems; and it enables projections, both in the past and future, that are vastly more distant in time, probing in details and durably stored than those delivered by the other memory systems. In short, autobiographical memory is at the vital center of a mind – the autobiographical mind – that operates in ways and according to "readings" of the world that are not accessible to the minds – of nonhuman animals or young children – serviced by other memory systems. Since, in various degrees, both animal and young human minds are dominated by procedural and episodic memory systems, and only the latter are declarative, as is autobiographical memory, a useful way to mark the unique contours of the autobiographical mind is to contrast it with the episodic mind, based on some key differences between the respective memory systems.

3.3 Episodic Versus Autobiographical Memory

Let us begin on the memory side of the divide. Episodic and autobiographical memory systems have often been conflated in the psychological literature, mostly in the sense that episodic memories are also autobiographical as well (Bauer 2007; Smith and Kosslyn 2007; even Tulving 2002, perhaps the most influential memory theorist of recent decades – for a critical review of the debate, see Fivush 2011).

The assumption behind this conflation seems to be that that by reenacting consciously a past experience and perhaps one's reaction to it one necessarily remembers autobiographically the source of one's experience – its what-it-is-about, so to speak – where it occurred and when. It turns out, upon careful research, that none of these parameters – the what, when and where of remembered experiences – are available to minds dominated by an episodic memory system or episodic minds (as we may call them). Children develop slowly a sense of past time only after the age of 4 and the same is true of their sense of the sources or causes of their experiences (Perner 1991, 2000). Eminently episodic, animal minds are not likely to fare better.

It takes however much more than a sense of what, when and where of a past experience to remember that experience autobiographically. Exploiting terminology, I will just focus on the two dimensions of 'autobiography' – 'auto' or self-regarding and 'biography.' I begin with the 'auto' part.

To remember something in one's past, *a sense of selfhood* must minimally be in place, in the triple sense of (a) being distinct from the surrounding world and somehow aware of this distinction, (b) the owner of one's mental states and actions and somehow aware of this ownership, and (c) the initiator of one's mental and physical actions and somehow aware of this fact. It is in implicit or procedural ways, through specialized self-regulatory mechanisms, that animal minds and those of young human children manage all these vital dimensions of selfhood (Bogdan 2010). In remembering autobiographically, the 'auto' or 'self' component indicates a fairly *explicit* sense of a past (or future or possible) self with its mental and behavioral relations to sundry targets (situations, events, persons). It is an explicit sense because, first, the autobiographical mind is aware of the distinction between the current and the projected self, and second, it can think (and talk) about the projected self, embellish or revise its properties, move it in time backward or forward, and so on.

I am not offering here an analysis of the projected self at the heart of autobiographical memories (but see Bogdan 2010, Chap. 7). I will only make two remarks, amply documented and defended elsewhere (Bogdan 1997, 2000, 2010, 2013). First and crucially, to project oneself in the past or future or some possible scenario, with one's mental states and actions, one's mental activity must go entirely offline and display introvert or mind-oriented consciousness. This is something that animal and young human minds cannot do and children gradually develop only after the age of 4. Assuming they are conscious at all, the episodic minds of animals and young humans are conscious only extrovertly or in a world-oriented direction. Their episodic memories are as extrovertly conscious as their ongoing perceptual and actions experiences, both materialized in the same brain centers, as noted earlier. By definition, such an extrovert or world-oriented consciousness cannot deliver a sense of projected self with its projected mental states and actions.

Second, animal and young human minds do not recognize and introspect their own mental states and attitudes in general (even though they recognize their perceptual and action experiences) and particularly the relations of these states and attitudes to their targets. This is to say that these episodic minds do not have a self-directed theory of mind. Such a competence, together with introspection and an introvert consciousness, develops in children gradually only after the age of 4.

Let us pause for a moment and reflect on this late development in the light of the puzzling phenomenon of early childhood amnesia. Older children and adults do not normally remember much if anything from the first 3 years of life, despite the fact that young children have reasonably effective procedural, semantic and episodic memory systems (Bauer 2007; Conway 2001). Why aren't the semantic and episodic memories of the first 3 years sufficiently durable, anchored and stable – or indeed sufficiently useful – to be later retrieved and employed in thinking and action? Revealingly, once children begin to develop an autobiographical memory after the age of 4, their experienced past is no longer lost to memory and a personal history begins to take shape and endure. Is there a link between the end of early childhood amnesia and the onset of autobiographical memory?

I think the answer should be sought in the late development of a self-directed theory of mind and introvert consciousness, jointly resulting in introspection. David Foulkes (1999) claims that people do not remember the first 3 years of life because their original experiences as young children were not conscious, to begin with. As a dream expert, Foulkes bases his claim on research showing that young children do not dream or dream very poorly and on the assumption that to be subject of a dream an event must first be consciously experienced. Peter Carruthers (2005) reaches the same conclusion from a different direction: to be conscious one must have the capacity to form thoughts about thoughts, which originates in one's theory of mind and which young children (and animals) do not possess.

I think (and argued elsewhere) that Foulkes and Carruthers are half right and half wrong (Bogdan 2010). They are right to the extent that remembering autobiographically (and apparently dreaming) requires consciousness and a theory of mind. But they are wrong in assuming that young children lack both altogether. Consciousness and theory of mind are not unitary competencies. What young children seem to lack is *introvert* consciousness and a *self-directed* theory of mind, and hence introspection; but young children do possess an extrovert, perceptual and action consciousness and other-directed theory of mind. It is just that these latter capacities do not seem able to deliver autobiographical memories or indeed any durable memories of early childhood. The implication, left open here, is that to remember durably one must remember autobiographically and, to do that, one must initially experience and later recall some event or situation in consciously introspective ways, centered on an explicitly represented self.

I turn now, and finally, to the 'biography' part of the analysis. The opposite of somewhat disjointed episodic snapshots held together by an ephemeral context, which is how young children and animals tend to experience the world, autobiographical recall normally has (from an 'auto' or 'self' perspective) a theme and, as part of personal history, an intelligible contextualization of the past experience and some continuity in time, often reaching to the present. Required for this biographical part of recall is an immersion not just in culture and its basic scripts (which already begins in early childhood) but also an immersion in the complicated and fast-changing sequences of cultural patterns. Also required is some intuitive mastery of narrative thinking and communication (gossip, thematic conversation), needed to represent, track and manage such cultural patterns. Both requirements are met only after the age of 4 (Nelson 1996; also Fivush 2011). Biography thus appears to matter in autobiographical recall because – and to the extent that – it consolidates memories around independently accessible scripts, routines and narratives that act as anchors of the initial experiences. But biography does more than just consolidate memories.

Recall the fact, noted a few paragraphs ago, that young children do not compute the sources or causes of their ongoing experiences until around the age of 4, and they are even less able to do so in memory recall, when the challenge is to connect memories of experiences with the initial sources or causes and surrounding contexts of those experiences. This, I suggest, is where biography is critical: it provides a coherent and often reliable matrix for the mental reconstruction of those sources, causes and contexts, while also inviting confabulation. Such are the costs and benefits of biography in memory. Strictly episodic memories may be more accurate experientially but they are not solidly anchored and hence do not last.

The first moral of the story barely sketched here is that we – actually our selves – are what and how we remember: procedural selves, episodic selves and autobiographical selves. These kinds of selves have very little if anything in common, and that is true even of the selves of children, before and after the age of 4. The second moral is that the kinds of memory systems that dominate a mind explain a good deal of what that mind can think about and how, by way of various sorts of projections both current and across times and possibilities.

References

Bar M (2011) Predictions in the brain. Oxford University Press, Oxford
Bauer PJ (2007) Remembering the times of our lives. Erlbaum, Mahwah
Bogdan RJ (1997) Interpreting minds. MIT Press, Cambridge, MA
Bogdan RJ (2000) Minding minds. MIT Press, Cambridge, MA
Bogdan RJ (2010) Our own minds. MIT Press, Cambridge, MA
Bogdan RJ (2013) Mindvaults. MIT Press, Cambridge, MA
Carruthers P (2005) Consciousness. Oxford University Press, Oxford
Clark A (2013) Whatever next? Predictive brains, situated agents and the future of cognitive science. Behav Brain Sci 36:181–253

Conway MA (2001) Sensory-perceptual episodic memory and its context: autobiographical memory. Philos Trans R Soc Lond 356:1375–1384

Fivush R (2011) The development of autobiographical memory. Annu Rev Psychol 62:559–582

Foulkes D (1999) Children's dreaming and the development of consciousness. Harvard University Press, Cambridge, MA

Nelson K (1996) Language in cognitive development. Cambridge University Press, Cambridge, MA

Perner J (1991) Understanding the representational mind. The MIT Press, Cambridge, MA

Perner J (2000) Memory and theory of mind. In: Tulving E et al (eds) The Oxford handbook of memory. Oxford University Press, Oxford

Smith EE, Kosslyn SM (2007) Cognitive psychology. Pearson/Prentice Hall, Upper Saddle River

Tulving E (2002) Episodic memory. Annu Rev Psychol 53:1–25

Chapter 4
A Momentous Triangle: Ontology, Methodology and Phenomenology in the Philosophy of Language

Manuela L. Ungureanu

I am interested in examining some core elements of the understated relationship between two theses which occupy center stage in the Chomskyan approach to the study of language. Typically, the theses I focus on are presented as follows:

(A) investigations in the theory of language (ought to) adhere to methods of empirical inquiry already established in the natural sciences, i.e., they subscribe to what Chomsky calls Methodological Naturalism, and
(B) a language is best understood as an individual, internal mental system, i.e., an idiolect, or an I-language, again, to use Chomsky's terminology.

More specifically, while Chomsky characterizes thesis (A) in terms of his commitment to Methodological Naturalism (1993, 1995a, 2000), in the works of his supporters (A) has been construed and defended as the view that linguistics is a branch of psychology (Laurence 2003) or that psychological research is relevant to (philosophical) theories of language (Antony 2003; Smith 2009; Stainton 2012). The formulations we find in the literature for (B) occupy a much wider range. (B) is presented either as the view that there is no such a thing as a public language (Chomsky 1993; Stainton 2011), or that the study of what we commonly call a social language cannot be the target of a feasible scientific project (Chomsky 1993; Collins 2010; Stainton 2011), or alternatively, that an ontology of idiolects suffices for the diverse purposes of a theory of language (Chomsky 1997; Heck 2006; Isac and Reiss 2008).

Taken individually, theses (A) and (B) have been vigorously defended by Chomsky in many of his works (1965, 1986, 2000) and also by his, by now, numerous supporters among philosophers of language (Jackendoff 2002; Antony 2003). But in much less explicit ways, Chomsky and his followers have subscribed

M.L. Ungureanu (✉)
Department of Philosophy, University of British Columbia, Okanagan, Kelowna, BC, Canada
e-mail: manuela.ungureanu@ubc.ca

© Springer International Publishing Switzerland 2015   55
I. Pârvu et al. (eds.), *Romanian Studies in Philosophy of Science*, Boston Studies in the Philosophy and History of Science 313, DOI 10.1007/978-3-319-16655-1_4

to there being a natural, or even a privileged relation between (A) and (B). Despite its various versions, Chomskyans often take (B) as intuitively warranted by one's commitment to Methodological Naturalism, and regard (B) as the *only* sound ontological position about public languages consistent with (A) (Stainton 2006; Smith 2009; Collins 2010).[1]

Like Chomsky's followers, I submit that, when keeping in line with the exigencies of Methodological Naturalism –outlined only roughly in (A)–, we are well advised to take some version of the thesis concerning the I-language as part and parcel of an empirical account of linguistic abilities and of their development (Margolis and Laurence 2001; Crain and Pietroski 2001, 2002). I question, however, the Chomskyan's restrictive view of the I-language as the *only* sound ontology afforded by (A). More specifically, I examine the weight of the ontological commitment in (B) if and when a theory of linguistic competence subscribes to (A).[2] I ask, for instance: is (B) itself an empirical thesis for a Chomskyan theory of linguistic abilities, i.e., re-assessed in order to accommodate the relevant data? Or is it interpreted rather as a desideratum, or even a criterion of success on such an account? If the latter, can (B) remain consistent with the broad methodology of an empirical account of language as sketched in (A)?

Briefly, my focus is on the epistemological status of (B) within the paradigm of investigation established by Chomsky in philosophy of language. But since Chomsky's arguments offer rather scarce information on this theme, I apply my questions to the works of John Collins and Barry C. Smith, two main representatives of Chomsky's newer generation of followers, whose recent work supports in stimulating, novel ways a strong relation between the two theses which concern us here. While Collins (2010) attempts to show that, for an account of speakers' grammatical intuitions, a version of (B) follows from thesis (A), Smith commits to including a phenomenology of understanding speech as part of the empirical theory of language, and requires this more elaborate account to be consistent with both (A) and (B) (2007, 2008, 2009). I provide reconstructions of some central arguments in these recent works, and argue that the status Collins and Smith bestow on (B), albeit for different reasons, is much stronger than that of an empirical thesis. But, while I examine the weight their reconstructed arguments place on thesis (B), I treat them

---

[1] Stainton (2011) represents an important exception on this.

[2] As introduced by Chomsky, the central term of the thesis (B), 'I-language', refers to an individual mental structure whose postulation does not depend on it being manifested in speakers' ability to engage in social practices (1997). Chomsky uses the term 'E-language' to refer to external, social, or pragmatic aspects of speakers' abilities, especially when he emphasizes their alleged intractable nature, e.g., idioms indicating social status, linguistic fashions, political debates surrounding linguistic rights, etc. (1993: 18). Here I will follow Stainton (2011) and take public languages as external social entities usually denoted by our common-sense terms 'language' 'English,' etc., thus objects such as Armenian, Cantonese, French. As Stainton claims, despite their vague boundaries and various materializations, at least intuitively, public languages have a variety of features such as the following: "[they] have not just a morpho-syntax, but also phonology; some are spoken, some signed, and some are no longer spoken nor signed.[. . .] They have a history, and belong to language families [. . .] Some have corresponding writing systems, but not all do . . . " (2011: 480).

as representative samples of the Chomskyan approach to the two theses. To be clear, the intent of my present investigations is not exegetical. As the themes surrounding the theses (A) and (B) require, I draw not only from arguments by Collins and Smith, but also from a variety of works by Chomsky and his other supporters. But if my diagnosis of the status of (B) in Chomskyan philosophy of language is correct, then their position on the relation between (A) and (B) faces additional challenges. Thus my submission is the more general idea that in order to remain faithful to some sound core principles defended by Methodological Naturalism, thesis (B) ought to be revised, and by extension, the received wisdom which construes its restrictive reading as originated in (A) needs to be modified.[3]

(1) In some of his explicit presentations of Methodological Naturalism (MN), Chomsky describes it briefly as a standpoint which extends to the study of human cognition a particular style of inquiry elaborated on in other domains of the natural sciences.[4] At least initially, this new approach is outlined in terms of three core methodological commitments. First, MN is introduced as supporting *explanatory* accounts of linguistic capacities, stressing the need for hypotheses which posit underpinning mental structures (e.g., entities, events or processes internal to the mind) in order systematically to accommodate the phenomena under investigation (Chomsky 1986, 1992). Second, while MN does not imply the reduction of linguistics and cognitive psychology to neurology or biology, it does view language and our knowledge of it as a part of human biology, or broader, as *a part of the natural world* (Chomsky 1980, 1991). Indeed, supporters of MN take it as a regulative ideal that the science of linguistic abilities does not remain isolated from other empirical scientific accounts of human nature (Collins 2010). Third, in keeping to the practices of the natural sciences, MN eschews speakers' own understanding of their cognitive capacities as a source of explanatory hypotheses. Thus, Chomskyans commit to a constant *re-evaluation of the common-sense views* we, as speaker-hearers, tend to construct about our linguistic behaviour and capacities (Chomsky 1995b; Collins 2010).

In the supportive exegetical literature, the three methodological tenets are typically presented as converging towards the two-fold requirement for (a) explanations based on non-demonstrative arguments, and (b) which are expected to draw from a very generous evidence base (Chomsky 2000; Margolis and Laurence 2001;

---

[3]The repudiation of the notion of a public language is shared by Chomsky and Davidson, although in response to different explanatory tasks. But Davidson does not support key components of Methodological Naturalism, such as the approach to the mind as a natural object. Thus Davidson's position provides a remarkable counter-example to the correlation under study here between support for Methodological Naturalism and an internalist, individualistic ontology of language.

[4]Chomsky outlines this methodological stance, for instance, in the following passage: "[. . .]the study of the mind is [for the methodological naturalist] an inquiry into certain aspects of the natural world [. . .] and that we should investigate these aspects of the world as we do any others, attempting to construct intelligible explanatory theories that provide insight and understanding the phenomena that are selected to advance the search into deeper principles" (1993: 41).

Crain and Pietroski 2002). To illustrate, thesis (A) has been taken to express the prerequisite that the theory is at least open, if not required, to take into account an ever expansive set of empirical findings, from data concerning speakers' linguistic intuitions surrounding various properties of linguistic expressions, to their processing errors, language development, language breakdown, cross-cultural and cross-linguistic data about the course of language development, or language change over long periods of time, etc. (Laurence 2003; Antony 2003). More recently, Chomsky defends the inclusion of a wide evidence base in the theory as a general norm of scientific methodology, readily observed in the practices of mature natural sciences, and thus also regarded as symptomatic of an empirical investigation in the theory of language (2000).[5]

But what appears to Chomsky as a mere observation about how empirical sciences operate has received a more principled defense by philosophers interested in what warrants these methods as applied to the study of language. Of great importance here are the exchanges between Antony and Soames with regards to whether, in principle, psychological data are germane or not for conclusions about linguistic properties. Soames (1984) argues that psychological data are not relevant for identifying linguistic properties as they do not provide information on properties or relations which are constitutive of languages, e.g., grammaticality, ambiguity, synonymy, contradiction, etc. In her powerful rejoinder to Soames, Antony questions his view that only data constitutive of or (already) nomologically connected to linguistically significant properties are directly relevant to the theory of language (2003).

For our purposes it is useful to recap briefly one of Antony's central arguments in which she focuses on the reasons for taking various empirical findings as appropriate for assessing core hypotheses in (other) natural sciences, such as paleontology or biology. As illustrated in her examples from paleontology, theorists do not know in advance which properties of the data reviewed in their hypotheses concerning species are nomologically connected or not, and particularly to which other phenomena. But if so, in a natural science like paleontology and biology, we do not exclude the possibility that data about, for instance, geographical distribution of animal kinds, may serve to predict core theoretical theses, such as those concerning species membership. Antony also reminds us that highlighting empirically regular connections between the area in which an animal lives and the kind of thing it is counts as evidence for a theory concerned to identify animal conspecifics (2003: 60). Thus, more generally, in natural sciences facts that are only contingently related to the main properties of the domain can still constrain theorizing in that domain, namely based on their reliability. Antony then argues by analogy that,

---

[5]In his early elaborations of (A), Chomsky sees it as motivated by the need to ensure that the general notion of language applied in the theory is not merely borrowed from the study of symbolic, formal languages but rather particular to our biological endowment, i.e., our species-specific language faculty (1965/1985).

as theorists of language, we cannot simply assume we are already familiar with what is empirically significant (contingent or nomological) about the conditions in which speakers acquire or manifest their linguistic capacities. Thus Antony disputes Soames' position on the relevance of psychological data for the theory of language.

While her argument also reminds us that the investigation ought not to be constrained by our existing preconceptions about language or thought, her defense of (A) does not outline explicitly any implications MN may have for thesis (B). But her view seems to suggest at least a broad constraint on an ontology of language, namely that the commitment to (B) itself ought to be grounded on the same two-fold methodological standard, i.e., the inclusion of a wide variety of empirical data concerning speakers' capacities as well as the search for the most plausible explanations of our empirical generalizations about them. I suggest thus that if and when this two-fold constraint is met, we can legitimately call (B) an empirical thesis, or analogously, consistent with the methodological commitments of (A).

Let us turn now to the main elements of Chomsky's own arguments for thesis (B) – which I indicate below as (i)–(iv).[6] One of the starting points of Chomsky's arguments towards (B) is (i) the hypothesis that the mind benefits from an innately structured and modular endowment. Indeed, while emphasizing the poverty of the stimuli when contrasted to properties of speakers' capacities, Chomsky has persuasively defended the idea that the human mind possesses a substantial, language-specific, innate endowment, i.e., enjoys specific cognitive structures which help account for core features of language acquisition and processing (Chomsky 1986; Margolis and Laurence 2001). However, Chomsky's move from the innateness and modularity of language to the strong reading of (B) seems to assume some further theses, which also remain unqualified. For instance, when arguing in favour of the idea that a language is *only* internal/individual, Chomsky and his supporters presuppose that (ii) for explanatory purposes, a language can *only* be construed as either internal or external to speaker-hearers, but not both, and (iii) that any external language can *only* be understood either as a Platonic entity, similar to a logico-mathematic system, or as a set of observable behaviours manifested in speech (Chomsky 1997; McGilvray 1999; Jackendoff 2002; Bezuidenhout 2008; Stainton 2011).

Chomsky has argued extensively that, since neither a Platonic language, nor a cluster of observable speech behaviours play any role in explanations of the empirical findings, the two interpretations of external languages in premise (iii) above are both flawed. To be specific, when he refers to empirical findings, Chomsky directs us to the classical arguments from poverty of the stimuli, or some other arguments inspired by MN, and he suggests, on this basis, that thesis (B) is rooted

---

[6]On the one hand, this excursion helps us identify whether Chomsky's own position provides sufficient information for assessing whether and, if so, where his position departs from the constraint on ontology already afforded by Antony's defense of MN. On the other hand, the outline functions as the background against which I examine the contributions of Chomsky's newer generation of supporters.

in MN. However, when defending the strong reading of thesis (B) as the only sound conception of language, by and large, his main preoccupation is to stress that the rationale for the alternative position – that public languages exist – lies only in some obscure need to preserve common-sense intuitions about such social objects. Thus, Chomsky and many of his supporters have defended the stronger reading of (B) with help from the additional claim that (iv) the notion of a public language is a mere relic of common-sense.[7]

While representative of Chomskyan arguments against the notion of a public language and in support of thesis (B), the reasoning outlined in (i)–(iv) is still presented rather sketchily by many of Chomsky's supporters.[8] It is thus unclear whether it provides sufficient resources for addressing the questions concerning the empirical status of the thesis (B) in the architecture of their program.[9] But, albeit indirectly, the recent works by Collins and Smith provide us with new insight into the Chomskyan motivations for the view that public languages cannot *in principle* be included in a theory of linguistic phenomena. While their position supports theses (ii) – (iv) and continues to limit the ways in which one can conceive of a (public) language for explanatory purposes, they introduce novel lines of defense for the strong commitment to an I-language, which help with the query concerning the received epistemological status of the thesis (B), or so I argue.

(2) Unlike Chomsky's early followers, Collins is overt about the need for two different lines of reasoning against public languages, given their dual role as either target or as *explanans* for speakers' abilities. In his 2010, Collins defends the idea that there are no extant explanations of linguistic phenomena which require appeal to an external language. He then argues in favour of the stronger thesis that reference to linguistic externalia, as he calls them, will never be useful for any explanations of linguistic phenomena. But there are also three additional features which distinguish from the start Collins' position on the relationship between (A) and (B) from the Chomskyan stance on public languages (e.g., as reconstructed in Stainton 2011). First, Collins does not challenge the intuitive, common-sense

---

[7]See Stainton (2011) for two more variants of Chomskyan arguments undermining the notion of a public language, none of which are at work in the lines of reasoning introduced by Collins and Smith. In contrast to my focus here, Stainton's defense of the notion of public languages questions Chomskyan arguments which support the idea that in principle they cannot become the *object* of empirical, rigorous science (Stainton 2011).

[8]For instance, as Stainton (2011) stresses, despite their additional canvassing of the theses (ii) – (iv) in this cluster, Chomskyans do not provide an elaborate defense of the idea that a public language as an object of study can only be either a set of concrete behaviours or Platonic abstracta (Stainton 2011: 483).

[9]Chomsky's descriptions of the I-language as "the real object of inquiry" (1991: 12) constrain the inquiry to an internal language understood as an unchanging natural object, while research in neuropsychology of writing/reading suggests it is more appropriate to take it as a natural object able to change within its own innate parameters, under cultural influences it helps create (Deheane 2009). See also (Harris 1986).

ontological commitment which takes public languages as real. Second, he attempts to defend versions of the thesis (B) concerning the proper ontology of language by showing how it can be derived explicitly from a version of (A) he introduces and dubs the Naturalistic Reality Principle. Third, he carefully defends MN as a position about language and mind which follows the model of natural sciences in that it is "independent of any materialist commitments" at least in that it makes no presumption about "a fixed naturalizing base or a fixed conception of what is to be naturalized" when it comes to explaining speakers' abilities (2010: 41–6).

I focus here on his central argument for the idea that linguistic externalia cannot function as *explanans* in a theory of linguistic abilities faithful to MN. For this purpose it is useful to introduce two additional quotations from Collins (2010):

> NRP [Naturalistic Reality Principle] at a given stage of inquiry a category is taken to be (naturalistically) real iff it is either successfully targeted by naturalistic inquiry or essentially enters into the explanations of such inquiry (47).[10]
> [When it comes to an account of speakers' grammaticality judgements] [a]n externalism of grammatical properties [...] looks to be explanatorily supererogatory; the externalism confuses what the language faculty [...] enables – the projection of structure onto sounds/marks – with the target of the explanation itself – the capacity to project, inter alia. (50).

What Collins calls NRP functions as the starting premise of Collins' main line of defense for what he takes as the proper ontological commitment of the theory of linguistic abilities. The remaining premises of his argument introduce criteria for when and how external, public linguistic entities can be taken to play the role of *explanans* in the theory. Then Collins uses these criteria to argue that a public language is "essential," as he puts it, to *no* extant explanation of any linguistic phenomena, and to support the strong, exclusive reading of thesis (B). I argue that his defense of this reading of (B) includes requirements inconsistent with taking it as an empirical thesis of the Chomskyan program.

Here is a reconstruction of Collins' main argument for the idea that no external language works as *explanans* in the theory of language (48):

P1: The naturalistic principle of reality NRP is correct.

P2: A true *explanans* ought to provide constitutive conditions for the properties of the phenomena it explains, i.e., ought to play a role stronger than being merely causally necessary for the acquisition of (the various features of) the language faculty.

P3: To play the stronger role, properties of the entities in the external language (e.g., properties of "sound waves, hand gestures, inscriptions") ought to be either

---

[10]Immediately after introducing this principle, Collins illustrates it as follows: "So, in classical (Newtonian) mechanics, absolute velocity is not 'real', its measurement being relative to a particular inertial frame of reference. After the maturation of electromagnetic field theory, lines of force and potentials ceased to be 'real'; for neither essentially enters into the field-theoretic explanations and their measurement, again, is arbitrary" (2010: 47–48).

necessary or sufficient for the characterization of the linguistic structures which enter into explanations of empirical findings.

P4: But linguistic externalia are neither necessary, nor sufficient for the characterization of the linguistic structures.

Hence, no notion of an external language or linguistic externalia can ever play the role of *explanans* and, by NRP, cannot be (naturalistically) real.[11]

To spell out in detail what has gone wrong in Collins' reasoning would require greater space than is presently available. I question whether his application of NRP here is consistent with MN as he defends it, i.e., as a position about language which makes no presumption about "a fixed naturalizing base or a fixed conception of what is to be naturalized" (41–6). But I also examine the motivations behind premises such as P2 and P3 of his argument. For, in contrast to Antony's defense of MN, Collins' argument introduces here very stringent criteria for what counts as a sound *explanans* for purposes of empirical inquiry, and supports them only with a limited use of examples. For instance, in his analysis of data such as

(1) Mary shot the elephant from Africa

which is ambiguous in precisely two ways, Collins insists that additional readings of the sentence are excluded "not merely contingently, but due to constraints on how we can interpret the string" and presents this as his sample non-contingent relation between the result and the underpinning syntactic structure (53)

I agree with Collins that for this type of judgements the explanation has to point to underpinning structural/syntactic features which determine speakers' interpretations, e.g., that *from Africa* can modify the DP (*the elephant*) or the VP (*shot the elephant*) of (1). But his examples fail to motivate the stronger, more restrictive criteria he introduces in P2 and P3 and the intended generality of his reasoning. Ironically, somewhat like Soames, Collins insists on accounting for data about speakers' abilities by reference only to what he describes as non-contingent features of their grammaticality or ambiguity judgements. Limited as they are to our judgements of ambiguity and grammaticality, and to their (alleged) non-contingent features, Collins' examples do not rule out the possibility that the *explanans* for other findings concerning speakers' abilities may be provided in terms of features they enjoy contingently, just as Antony's interpretation of MN requires.

Collins could reply here that the stringent criteria in P2 and P3 are not restricted to an account of grammaticality or ambiguity judgments, where it is natural to identify non-contingent features of the syntax. For, explanations of phonological changes, such as what is described as the Great Vowel Shift in the pronunciation of English vowels, have also been accounted for in terms consistent with his

---

[11]The argument extends into considerations about the inexistence of an internal *language* through

P5: Only speaker-hearers having mental structures with these generating contents can perform the explanatory job (50).

P6: The target of the explanation is the language faculty, i.e., capacity to project structure onto sound, and not the objects generated by the capacity (51).

requirement for constitutive features of the data. The radical alteration in the pronunciation of English vowels that occurred during the fifteenth century has been explained in terms of how sounds are articulated in the mouths of individuals, e.g., long vowels being articulated with the tongue higher up in the mouth, determined a change in the place of articulation of other sounds. Now, when Chomskyans appeal to phonological properties of individual speakers in the explanation of the Great Vowel Shift, they show no interest in exploring contingent, reliable correlations between speakers and their environments, which may have prompted the initial change to how long vowels are articulated. Rather, they introduce here a constraint to include only individual-based properties, i.e., not to appeal to speakers' external and/or social features as potential *explanans* (Bezuidenhout 2008).

Collins' own position on such an (implicit) individualist constraint pulls us into opposite directions. On the one hand, the individualist restriction conflicts with his characterization of MN as a position where there is no presumption about "a fixed naturalizing base or a fixed conception of what is to be naturalized" when it comes to explaining speakers' abilities (41–6). On the other hand, the same restriction to intrinsic properties of individual speakers is implied by Collins' argument, which rules out contingent correlations with linguistic externalia. But such a restriction is left without justification when we expand the evidential basis for the theory of language and introduce accounts of some other types of linguistic phenomena. Of particular interest here is data concerning speakers' developing an understanding of the sound structures of speech or, broader still, their meta-linguistic awareness of words and phonemes. In the 1990s, the main debates on children's development of meta-linguistic awareness have shifted from identifying the influence of language acquisition (given normal cognitive development) on meta-linguistic abilities to investigating whether exposure to reading and writing contributes to children's or adults' more abstract understanding of speech (Homer 2009). Cross-linguistic evidence indicating that conventional notions of word are not necessarily employed by adult speakers of all languages has led to an interest in designing novel types of experiments and tasks which contrast the performances of pre-literate and literate children and/or adults (Hoosain 1992).[12] Most of these experiments introduce a variety of segmentation and processing tasks and highlight some remarkable correlations between exposure to literacy and children's or adults' grasp of linguistic categories such as words and phonemes (Veldhuis and Kurvers 2012).

---

[12] Another contributing factor to the interest in the new types of experiments has been the appraisal of the spectrum of approaches when it comes to children's meta-linguistic awareness. While Piagetian and neo-Piagetian psycholinguists explained children's difficulty with meta-linguistic tasks in terms of their inability to think abstractly about language, their emphasis on the advent of reflected abstraction as explanans for development provided too general an account of the data. Other approaches to the development of meta-linguistic awareness stress that recording of representations of language in more abstract formats allows for conscious reflection. Those who highlight the contribution of literacy take the abstractness of writing itself, as an external representational system, to influence meta-linguistic awareness (Homer 2009).

To illustrate, some studies demonstrate the effect of literacy on *phonemic* awareness, such as the one by Read et al. in which Chinese adults were asked to add or delete consonants in spoken Chinese words. The results showed that only participants with prior exposure to alphabetic writing (Pinyin) were able to segment words in phonemes. This finding, which has been replicated for Chinese children exposed to alphabetic script, obtains even for subjects who could no longer read and write using Pinyin (Read et al. 1986; Homer and Olson 1999).[13] More generally, research in cognitive psycholinguistics on the influence of writing on meta-linguistic awareness implies that speakers' experiences of word meaning are moulded by linguistic determinants both internally *and* externally, and can be traced back to individual intrinsic constraints as well to cultural ones. This is, of course, in contrast to the assumption in Chomsky's initial argument that for the purposes of explanatory theory a language is construed only as either internal or external to speaker-hearers.[14]

Similarly, and more importantly for our purposes, when confronted with the findings in the psycholinguistics of literacy, the commitment to I-language in Collins' argument appears to be much stronger than that of an empirical hypothesis. In contrast to Antony's support for including in explanatory hypotheses about linguistic phenomena facts that are contingently and reliably connected to speakers' linguistic properties, Collins' P2 and P3 require an *explanans* which is, as he puts it, more than causally necessary for an account of the findings. Thus his position rules out the inclusion of contingent, albeit empirically regular, relations between speakers' engaging with literate artifacts and specific aspects of their meta-linguistic awareness.

But while cognitive psycholinguistics brings to the forefront such empirically regular connections between speakers' exposure to literate artifacts and various features of their meta-linguistic awareness concerning phonemes and words, its research program seems consistent with MN: it offers explanatory hypotheses, without taking common-sense intuitions as criteria of their evaluation. Moreover, just as Antony insists, hypotheses introduced to accommodate the new findings aim to discover the most *plausible* explanations of the empirical generalizations they bring to the forefront (Smith and Tager-Flusberg 1982). Specifically, they introduce linguistic externalia as part of the *explanans* on the basis of an inference to the best

---

[13] Homer and Olson, among others, conclude that children's meta-linguistic understanding of word develops as they attempt to relate written language to speech (Olson 1977, 1996; Homer 2009; Rosado et al. 2013).

[14] With its emphasis on cultural and technological determinants of meta-linguistic awareness, research on the influence of literacy on linguistic intuitions has also moved away from the third element of the Chomskyan position, namely (iii) the exclusive understanding of an external language as either a Platonic entity or as a set of observable behaviours. On the other hand, it is still too early to say whether and, if so, how the reference to regular use of literate artifacts as *explanans* is in tension with the innateness and/or modularity theses. Tolchinsky (2003) provides evidence for children's very early ability to discriminate among notational domains, such as drawing, letter-like or numerical notation.

explanation of the contrasts the experiments help identify (Karmiloff-Smith et al. 1996; Olson 2001; Carreiras 2009).

But if so, contrary to Collins' view, MN does not entail the (ideological) preference for individual intrinsic properties as *explanans*, i.e., does not imply the exclusive reading of thesis (B). Rather, Collins requirement for such properties as *explanans* for the linguistic phenomena comes into tension with the MN desideratum to include a wide variety of empirical data concerning speakers' capacities, and the various types of *explanans* that may be needed to account for them.[15] By extension, in Collins' position, the ontological commitment to (B) a purely individual linguistic structure functions as a criterion of success, and not a mere part of on a sound *explanans* in the theory of speakers' abilities.[16]

(3) While coming in conflict with the requirement to allow for ever expansive empirical data and for their most plausible explanations, Collins' position may still be supported by Chomskyans who take the focus on grammaticality and ambiguity judgments as pivotal to the theory of speakers' linguistic abilities. This type of reply, however, is not available to those interested in the phenomenology of speech and in accommodating data concerning our experiences of word meaning within the broad umbrella of MN. It is thus useful to look at the role thesis (B) plays in Smith's work, as it is representative of a renewed interest in the phenomenology of understanding speech and in highlighting its implications for the Chomskyan research program.[17]

Like Collins, Smith explicitly subscribes to Chomsky's overall theory of language. He identifies it as a commitment to the two main theses (A) and (B) which he outlines as:

(A) a theory of language embedded in psycholinguistics and developmental psychology, and
(B) a conception of language as internally represented (Smith 1992).

But, Smith also focuses on what he characterizes as our first-personal, authoritative knowledge of *word* meaning (Dummett 1978/1993; Dummett 1994). In his

---

[15]Collins' formulation of the NRP principle itself seems to allow, for instance, for an explanation of data concerning phonological awareness, and moreover, one which appeals to linguistic externalia in order to accommodate such data. Indeed, when we give a different reading of the "essential" contribution of the *explanans*, his position is open not only to an account of speakers' identifying words in the sound stream, but also for explaining why this type of phonological awareness is strongly correlated with their having acquired the ability to read and write (Homer 2009). Thus, without the addition of **P2** and **P3**, nothing precludes the supporter of MN/NRP from appealing to linguistic externalia if and when the account of linguistic abilities requires them.

[16]Another way of putting this point is to argue that the ontological thesis concerning public languages the Chomskyans may defend as consistent to MN is not the strong thesis targeted by Collins (and Chomsky), but rather the weaker one that, e.g., *some* linguistic externalia do not play the role of *explanans* for *some* empirical findings. Thus, even Collins' goal of defending the stronger conclusion that *no* notion of an external language can ever play the role of *explanans* is itself in tension with MN.

[17]See, for instance, McDowell 1998; Fricker 2003; Pettit 2010.

2008 paper, he stresses the need for an account of our experiences related to the meaning of words, such as the following "in a language we understand we hear people's words as meaningful, and cannot help but hear them that way when words are familiar" (942). Analogously, as he states, "we hear what the speaker says to us as there in the words uttered" (2009: 190).

In a nutshell, his proposed account for the phenomenological data concerning word meaning centers on the idea that, from the beginning, children invest speech sounds with word meaning, while interacting with their caregivers (2008: 979).[18] To what extent is this (neo-Davidsonian) proposal consistent with Smith's commitment to MN? What does it imply as to Smith's interpretation of thesis (B)? At least *prima facie*, Smith's proposed account of word meaning comes into conflict with his explicit endorsement of MN, for it appears to be in tension with some core empirical results surrounding meta-linguistic awareness of words and/or phonemes. On the one hand, psychological research indicates that children do not begin by investing the speech sounds with *word* meaning (Arnon and Snider 2010). On the other hand, as reviewed in the previous section, cognitive psycholinguists support the view that meta-linguistic awareness of the more explicit varieties develops only by the time children reach a certain age and depends on a variety of factors, such as being brought up in a bilingual environment, or being exposed to symbolic communication (Olson 1996).

By itself, Smith's proposed (neo-Davidsonian) account of word meaning is also unclear on the role I-languages play in children's acquisition of word meaning. On the one hand, it seems to leave open the possibility that linguistic properties of the external language can play a role in accounting for the phenomenology of speech. On the other hand, just as it appears characterized in the works by Davidson which inspire Smith, it is unclear whether the interaction with caregivers is taken to determine word meaning in a contingent, albeit reliable fashion or rather constitutively (Davidson 1992, 1994, 1995).

However, Smith's position on thesis (B) emerges quite clearly in his recent reply to McDowell's take on the phenomenology of speech (McDowell 1998; Smith 2009). Specifically, in his 2009, Smith attacks McDowell's position concerning the implications of phenomenological data for the ontological commitment to public languages, a line of argument which brings us a step closer to determining the weight Smith's own account places on the thesis (B) that only internal languages exist. In his paper, Smith agrees with McDowell on the need to account for the datum that "we hear meaning in people's speech" (184). But he disputes McDowell's (quick)

---

[18]As Smith puts it "[e]xperiences with meaning – in the sense of word meaning, are authoritative and objective *because* we learn to have experiences with words in the context of learning words from others [. . .] such that when word meanings are introduced the experience of two subjects is co-ordinated and involvement with an object and another person are not negligible" (2008: 978, my italics). Smith also states that "[t]here is such an experience as the meaning of a *word* being all there at once" and moreover "[t]here is such an experience of bringing the meaning of a word to mind as when one decides whether the use of a particular word is more apt than another [. . .]" (2008: 978).

move from the phenomenological datum that we hear what the speaker says to us "as there in the words uttered" to the idea that word meanings are indeed to be found in speech sounds (190–1). In the reconstruction of Smith's argument against McDowell's position below, I include his earlier claims about an account of the phenomenology of understanding speech:

P1: Any theory of language subscribing to MN can and need to accommodate phenomenological data such as experiences of word meaning (Smith 2008, 2009).

P2: Such data can be accounted for if we think of word meaning as acquired by children early on in their interactions with other speakers, e.g., caregivers (Smith 2007, 2008).

P3: No account of the phenomenological data about experiences of understanding word meaning can depart from thesis (B) and make reference to elements of a public language (2009).

Thus, when (B) conflicts with a position concerning phenomenological data, we do not question the exclusive commitment to thesis (B). Rather, we correct the (phenomenological) data accounted by the theory, e.g., about where speakers hear meaning. By extension, McDowell's account of phenomenological data by reference to a public language is mistaken, while Smith's corrective phenomenology is, arguably, sound.

To indicate in some detail what is doubtful about this argument and especially about the status it bestows on thesis (B), it is useful to begin with a reminder about the particular phenomenological datum both McDowell and Smith attempt to explain, i.e., we hear what the speaker says to us "as there in the words uttered." But, as highlighted in the reconstructed argument, Smith's disagreement with McDowell appears to be not simply about *where* one may locate meaning, given the phenomenological data, but also about *how* and *to what extent* to accommodate such data about our ordinary experiences with word meaning.[19] Crucially for our purposes, in case of a conflict between the phenomenological datum and the stance taken in (B) on the location of linguistic meaning, Smith suggests leaving (B) unchallenged while correcting the phenomenological datum. Specifically, Smith's alternative is that the real object of perception of speech is the speaker's voice, and not speech as sound stream.[20] By extension, Smith implies that our phenomenology of speech can be taken at face value and explained *only* within the broader metaphysical background that (B) speakers' knowledge of language is internally

---

[19]To bar McDowell's inference from the phenomenological datum (that we hear meaning in people's speech "as there in the words uttered") to the thesis locating word meaning *in* speech sounds, Smith also marshals research in psycholinguistics in favour of "specialized speech processing mechanisms, rather than just general auditory" ones, and which help speakers discriminate, for instance, between ambiguous speech and non-speech stimuli (Dehaene-Lambertz et al. 2005).

[20]Arguably, this hypothesis is grounded on empirical research on the psychology of auditory experience, in that it follows Nudds' position that auditory perception tells us about the sources of sounds (Smith 2009: 204–5, 208–9; Nudds 2009).

represented. In turn, this suggests that he is taking (B) as a desideratum on any successful phenomenology of understanding speech.

However, supporting thesis (B) as a criterion here also entails that his account fails to explain the particular phenomenological datum on which both he and McDowell focus. At least when it comes to the phenomenological finding that "we experience hearing what you say as there in the words uttered" Smith's hypothesis that we actually perceive the voice of the person does not explain why we *do* experience words as there in the sounds uttered. More importantly for our purposes, given the weight Smith's argument puts on (B), empirical hypotheses which point to our engaging in literate practices as part of the *explanans* for the phenomenology of understanding speech (such as those introduced in the previous section) cannot even be evaluated for their plausibility.

But given his commitment to MN, this suggestion has more drastic consequences for Smith's position than for McDowell's. While taking the commitment to an I-language (B) as a constraint on any future account of phenomenological data., Smith's view moves away from a core implication of MN, i.e., that the position concerning (B) ought to be grounded on what we may discover as most plausible explanatory hypotheses (nomological or contingent) concerning our empirical generalizations about speakers' linguistic capacities. Thus, despite Smith's commitment to the inclusion of a wide variety of empirical data concerning speakers' capacities and the search for the most plausible explanations of our empirical generalizations about them, his proposed phenomenology also appears to support thesis (B) without appeal to empirical and/or explanatory considerations standardly supported by, or at least consistent to, MN.

To conclude, my examination of the status of the thesis (B) in recent Chomskyan arguments has highlighted that, either implicitly or explicitly, they interpret (B) as a desideratum on an account of speakers' grammaticality judgments, or respectively, phenomenology of speech. But in both Smith's and Collin's arguments, this strong reading of (B) comes into tension with the broad methodology of an empirical account of language as sketched in (A). Thus, even if phenomenology of word experiences is not at the centre of one's philosophy of language, these analyses of the Chomskyan view undermine its typical construal of the exclusive commitment to an internal, individual language as implied by Methodological Naturalism.

References

Antony LM (2003) Rabbit-pots and supernovas. In: Barber A (ed) Epistemology of language. Oxford University Press, Oxford, pp 47–68

Arnon I, Snider N (2010) More than words: frequency effects for multi-word phrases. J Mem Lang 62:67–82

Bezuidenhout A (2008) Language as internal. In: Lepore E, Smith BC (eds) The Oxford handbook of philosophy of language. Oxford University Press, Oxford, pp 127–148

Carreiras M et al (2009) An anatomical signature of literacy. Nature 461(7266):983–986

Chomsky N (1965) Methodological preliminaries. In: Chomsky N (ed) Aspects of the theory of syntax. The MIT Press, Cambridge, MA, pp 3–47. Reprinted in Katz J (ed) (1985) Philosophy of linguistics. Oxford University Press, Oxford, pp 80–125

Chomsky N (1980) Rules and representations. Columbia University Press: New York

Chomsky N (1986) Knowledge of language. Praeger, New York

Chomsky N (1991) Linguistics and adjacent fields: a personal view. In: Kasher A (ed) The Chomskian turn. Blackwell, Cambridge, MA, pp 3–25

Chomsky N (1992) Language and interpretation: philosophical reflections and empirical inquiry. In: Earman J (ed) Inference, explanation and other philosophical frustrations: essays in the philosophy of science. University of California Press, Berkeley, pp 99–128

Chomsky N (1993) Language and thought. Moyer Bell, London

Chomsky N (1995a) Naturalism and dualism in the study of language and mind. Int J Philos Stud 2:181–209

Chomsky N (1995b) Language and nature. Mind 104(413):1–61

Chomsky N (1997) Language from an internalist perspective. In: Johnson DM, Erneling CE (eds) The future of the cognitive revolution. Oxford University Press, Oxford, pp 118–135

Chomsky N (2000) New horizons in the study of language and mind. Cambridge University Press, Cambridge, UK

Collins J (2010) Naturalism in the philosophy of language; or why there is no such a thing as language. In: Sawyer S (ed) New waves in philosophy of language. Palgrave Macmillan, New York, pp 41–58

Crain S, Pietroski P (2001) Nature, nurture and universal grammar. Linguist Philos 24:139–186

Crain S, Pietroski P (2002) Why language acquisition is a snap. Linguist Rev 19:163–183

Davidson D (1992) The second person. Midwest Stud Philos 17:255–267

Davidson D (1994) The Social character of language. In: McGuiness B, Oliveri G (eds) The philosophy of Michael Dummett. Kluwer, Dordrecht, pp 1–16

Davidson D (1995) Could there be a science of rationality? Int J Philos Stud 3:1–16

Dehaene S (2009) Reading in the brain: the science and evolution of a human invention. Viking, New York

Dehaene-Lambertz G, Pallier C, Serniclaes W, Sprenger-Charolles L, Jabert A, Dehaene S (2005) Neural correlates of switching from auditory to speech perception. Neuroimage 24:21–33

Dummett M (1978/1993) What do I know when I know a language? In: Dummett M (ed) Seas of language. Oxford University Press, Oxford, pp 94–105

Dummett M (1994) Reply to Davidson. In: McGuiness B, Olivieri G (eds) The philosophy of Michael Dummett. Kluwer, Dordrecht, pp 257–267

Fricker E (2003) Understanding and knowledge of what is said. In: Barber A (ed) Epistemology of language. Oxford University Press, Oxford, pp 325–366

Harris R (1986) The origin of writing. Duckworth, London

Heck R (2006) Idiolects. In: Thomson JJ, Byrne A (eds) Content and modality: themes from the philosophy of Robert Stalnaker. Oxford University Press, Oxford, pp 61–92

Homer B (2009) Literacy and metalinguistic development. In: Olson DR, Torrance N (eds) The Cambridge handbook of literacy. Cambridge University Press, Cambridge, UK, pp 487–500

Homer B, Olson D (1999) The role of literacy in children's concept of word. Writ Lang Lit 2:113–140

Hoosain R (1992) Psychological reality of the word in Chinese. Adv Psychol 90:111–130

Isac D, Reiss C (2008) I-language: an introduction to linguistics as cognitive science. Oxford University Press, Oxford

Jackendoff R (2002) Foundations of language. Oxford University Press, Oxford

Karmiloff-Smith A, Grant J, Sims K, Jones M-C, Cuckle P (1996) Rethinking metalinguistic awareness: representing and accessing knowledge about what counts as a word. Cognition 58:197–219

Laurence S (2003) Is linguistics a branch of psychology? In: Barber A (ed) Epistemology of language. Oxford University Press, Oxford, pp 69–106

Margolis E, Laurence S (2001) The poverty of the stimulus argument. Br J Philos Sci 52:217–276

McDowell J (1998) Anti-realist semantic and the epistemology of understanding. In: McDowell J (ed) Meaning, knowledge and reality. Harvard University Press, Cambridge, MA, pp 314–343

McGilvray J (1999) Chomsky: language, mind and politics. Polity, Oxford

Nudds M (2009) Sounds and space. In: Nudds M, O'Callaghan C (eds) Sounds and perception: new philosophical essays. Oxford University Press, Oxford, pp 69–94

Olson DR (1977) From utterance to text: the bias of language in speech and writing. Harv Educ Rev 47:257–281

Olson DR (1996) Towards a psychology of literacy: on the relations between speech and writing. Cognition 60:83–104

Olson DR (2001) What writing is. Pragmat Cogn 9:239–258

Pettit D (2010) On the epistemology and psychology of speech comprehension. The Baltic Int Yearb Cogn, Log Commun 5:1–43

Read CA et al (1986) The ability to manipulate speech sounds depends on knowing alphabetic reading. Cognition 24:31–44

Rosado E et al (2013) Production and judgment of linguistic devices for attaining a detached stance in Spanish and Catalan. J Pragmat 60:36–53

Smith BC (1992) Understanding language. Proc Aristot Soc 92:109–141

Smith BC (2007) Davidson, interpretation and first person constraints on meaning. Int J Philos Stud 14(3):385–406

Smith BC (2008) What I know when I know a language. In: LePore E, Smith BC (eds) The Oxford handbook of philosophy of language. Oxford University Press, Oxford, pp 941–998

Smith BC (2009) Speech sounds and the direct meeting of minds. In: Nudds M, O'Callaghan C (eds) Sounds and perception: new philosophical essays. Oxford University Press, Oxford, pp 183–210

Smith CL, Tager-Flusberg H (1982) Metalinguistic awareness and language development. J Exp Child Psychol 34:449–468

Soames S (1984) Linguistics and psychology. Linguist Philos 7:155–179

Stainton R (2006) Meaning and reference: some Chomskian themes. In: LePore E, Smith B (eds) The Oxford handbook of philosophy of language. Oxford University Press, New York, pp 913–940

Stainton R (2011) In defense of public languages. Linguist Philos 34:479–488

Stainton R (2012) The role of psychology for philosophy of language. In: Russell G, Graff D (eds) The Routledge companion to philosophy of language. Routledge, New York/London, pp 525–532

Tolchinsky L (2003) The Cradle of culture and what children know about writing and numbers without being taught. Lawrence Erlbaum, Mahwah

Veldhuis D, Kurvers J (2012) Offline segmentation and online language processing units. Writ Lang Lit 15(2):165–184

Chapter 5
On Rule Embedding Artifacts

Gheorghe Ştefanov

Philosophy of technology has a busy agenda. Figuring out what is a technological artifact and how the nature of such an object can depend both on physical properties and human intentions occupies a top spot on the to-do list of the discipline.[1] Some pressure to deal with this task comes from the technological changes we are involved with at present.

To take just a case, it seems that we will soon be able to make our environment similar to a computer simulated reality in at least one respect. A virtual world is one in which most of the rules governing the occurring practices can be embedded in the simulated environment. For instance, one cannot infringe your property rights in *Order and Chaos Online.*[2] The rule is incorporated in the game. To take another example, a few lines of code can turn the rule that you ought to wash your hands before eating into a detail of a virtual setting (in an adventure game, perhaps), according to which you simply cannot eat without washing your hands. In a similar way, a rule saying that you are forbidden to honk while driving in a city could actually be embedded in the workings of a GPS-enabled car by a few lines of code (with the result that your car horn will be disabled when you are inside a city). Since

---

[1]See Kroes and Meijers 2006, for instance.

[2]For a general description of the game, see http://orderchaosonline.com/. A player cannot steal items from another player in the game. Identity thefts may of course happen, but they take place in real life, not within the game.

G. Ştefanov (✉)
Department of Theoretical Philosophy, University of Bucharest,
Bucharest 030018, Romania
e-mail: gstefanov@gmail.com

© Springer International Publishing Switzerland 2015
I. Pârvu et al. (eds.), *Romanian Studies in Philosophy of Science*, Boston Studies
in the Philosophy and History of Science 313, DOI 10.1007/978-3-319-16655-1_5

we are able to use various types of sensors and digitally control the functioning of our artifacts, it seems that we will be more and more able to incorporate rules in our technological surroundings.[3]

This brings up a few problems. What rules should we decide to embed in our environment? Who is to assume responsibility, depending on the type of rule in question, for the production and use of rule embedding artifacts? How should we establish the risks of such technologies? Besides risks and costs, what other criteria could we use when faced with the choice of embedding the same rule in different artifacts? Also, how can an artifact embed a rule? The list could contain other questions as well, but in what follows I will focus only on the last one and propose a way in which we could conceive a rule embedding artifact. At the end of my paper, some suggestions inspired by my conceptual proposal for a general approach on technological artifacts will be presented.

1 How can an artifact embed a rule? In order to better understand the problem, let us begin with a contrastive example. A signpost signaling that one should slow down seems to convey the rule that one ought to slow down at some particular point while driving on the road. Does the signpost also embed the aforementioned rule? On one hand, we feel inclined to answer in the negative, since the signpost only expresses the rule and, no matter what a rule is, it should not be identified with its expression. On the other hand, since the signpost is at least part of the constraint that one should slow down at some point while on the road (since one could not get a fine for breaking the rule if the signpost were not in place), we might be inclined to say that the rule is at least partly embedded in the signpost.

Of course, the practice of slowing down before marked crosswalks could function without a special signpost being part of it, so the signpost might be regarded as a nonessential part of the rule. Also, if we restate the rule to make the signpost essential – "When seeing a signpost looking so-and-so you should slow down" – it becomes obvious that the signpost does not embed the rule, but the rule regulates its intended use. Thus, speaking in general, we would want to say that sign occurrences do not embed their intended uses.

So much for signposts. Speed bumps, however, seem to be in a different situation. When placed before crossings, for instance, they seem to enforce the rule that one should slow down before a crossing. The problem, in this case, is that in doing so, speed bumps seem to *cause* a slow down, so we might be reluctant to talk about

---

[3]The literature on the normativity of artifacts seems to focus either on the way in which the physical structure of an artifact can incorporate rules for its use, or on evaluations regarding how well does an artifact fulfill its function as an appropriate means for achieving a purpose. See, for instance, Akrich 1992; Radder 2009; Franssen 2009; Vries et al. 2013: 101–169. However, the case of rule embedding artifacts, which regulate actions not pertaining to their use, is a different one. One example given in Latour 1999: 186–190 is closer to the topic discussed here and will be presented in what follows.

rules anymore in such cases. A rule is supposed to say what one must do or refrain from doing under certain conditions, but speed bumps seem to just make something happen.

Perhaps we could pause for a moment at this point and note that what we say about speed bumps depends on the vocabulary we use in order to talk about the entire situation (the phrase "situation" being used here in a neutral way).

We could talk about speed bumps slowing down cars in a naturalist vocabulary. In this respect, what distinguishes a speed bump from naturally formed bumps on the road is the fact that the speed bump was produced and placed on the road by some human agents. The speed bump is a technological artifact, but on the naturalist view it does not need to be recognized as such in order to function properly. In fact, from a naturalist perspective, being made by humans is only an accidental trait. Humans are, after all, part of the nature. Any artifact, from this point of view, is a natural object.[4] The conclusion, in this case, would be that rule embedding artifacts cannot exist.

Were we to talk about speed bumps in a non-naturalist vocabulary, we would be saying something different, however. First, we would say perhaps that speed bumps are produced and placed on roads not in order to cause some behaviour, but with the open intention to determine human agents perform some *action* – that of slowing down – under certain circumstances. The open intention, we would say, includes not only the intention that human agents should slow down in the presence of speed bumps, but also the intention that human agents should recognize the intention that they should slow down.[5] In addition, when speaking about the way in which speed bumps would get someone to perform a certain action we might avoid causal terms in favor of a phrase like "being responsible for".

Here an objection can be raised, namely that we are not supposed to talk about objects in terms of responsibility, particularly when we explicitly distinguish between "being the cause of" and "being responsible for".

Responsibility cannot be simply reduced to causal relations, as we can easily acknowledge from the example of a parent being responsible for the actions of her child without being causally connected with the changes those actions might effect. However, it might seem that we attribute responsibility only to persons and not to

---

[4]See, for instance, Dipert 1995: 119 for the idea that artifacts "lose their character or essential nature when considered only as physical objects". Since my investigation is focused on rule embedding artifacts, I do not wish to enter into the ontological debate about the relation between artifacts and physical objects here (although I tend to agree with Kroes and Vermaaas 2008 and disagree with Thomasson (2007) and Baker (2008) for reasons that should become obvious at the end of my paper).

[5]These are similar to the first two conditions figuring in the analysis of speaker meaning in Grice 1957. The third condition is missing, since adding it would make speed bumps indistinguishable from signposts.

objects (or rules, for that matter).[6] This is perhaps due to the fact that we take being able to admit responsibility for A as a necessary condition for being responsible for A.[7]

There is, nevertheless, another way in which one could stick to a non-naturalist vocabulary while talking about speed bumps as rule embedding artifacts. The premises of a practical syllogism are not, of course, responsible either for the conclusion of the syllogism, or for my actions, but they can be reasons for a conclusion or an action. The fact that a person accepting the premises might be caused by her acceptance to adhere to the conclusion or even to behave in a certain way does not falsify the previous description. In a similar way, rules can be said to be reasons for actions performed according to them. What the non-naturalist needs at this point is to specify the way in which a grounding relation could hold between an artifact and an action. The problem is that the grounding relation must be conceived as a logico-semantic one and, as such, for it to hold between two terms the respective terms must have a conceptual content. Here we reach a serious difficulty. Intentional actions can be said to have a conceptual content,[8] but it is difficult to see in which way one could say that an object, even an artifact, has a conceptual content.

It seems, therefore, that we have got, although in a cumbersome manner, to the core of our problem. For an artifact to embed a rule it must have conceptual content. Its conceptual content, however, must not be its meaning, so now our question becomes: "How can an artifact have conceptual content without being a sign?"

2 I suggest that we proceed by considering another contrast, between two types of properties. To take an example, let us consider two terms used to express such properties, "heavy" and "fireproof".

A heavy gambler is in a way similar to a heavy weights lifter. The second lifts heavy weights, while the first gambles large amounts of money, which, considered as material artifacts, are supposed to be heavy. Among the uses of "heavy" we can identify one according to which the word "heavy" stands for a natural property. Vague as it can be, the term applies to all things, while being used like this, in the

---

[6]In discussing the case of speed bumps, Latour talks about "nonhumans" being able to perform actions (see Latour 1999: 188–190). Being responsible for A seems, however, to be a necessary condition for doing A (as an intended action).

[7]I am not saying that someone can assume responsibility only for an action. A doctor, for instance, can assume responsibility for an event – the death of a patient – without being able to indicate any action which she assumes responsibility for. She does not have to believe that omissions are also some sort of actions in order to do that.

[8]See Anscombe 1963 for the idea that only for an action under a description the question whether it was intentional or not can be raised. An interesting suggestion along the same line (actions have conceptual content) can be found in Austin 1962: 19–20: "a great many of the acts which fall within the province of Ethics are not, as philosophers are too prone to assume, simply in the last resort physical movements: very many of them have the general character, in whole or part, of conventional or ritual acts [...]". In order to be the object of moral evaluations, our actions have to occur in a conventional space. If we take Austin's conventional space to be a conceptual space and extend the remark to the province of practical reason, we get to a similar conclusion.

same way. An object is heavy among other objects in terms of the relative weight it has compared to the other objects. A heavy pen weights more than most pens, but it is of course lighter than a piano. Still, for both the pen and the piano, their relative heaviness depends on their weight, which, in turn, depends on their mass and the gravitational field they are in. In order to be heavy an object does not have to somehow incorporate a concept of heaviness.

Being fireproof, by comparison, does not simply amount to having a natural property. Water cannot burn, for instance, but we do not usually say about water that it is fireproof. One could pour water on some clothing in order to make it fireproof for a while, but were we to consider that clothing article we would not simply say that it is fireproof because it cannot burn. We would rather mean that it can be used to protect us from fire. When speaking of fireproof artifacts, then, we speak of a way in which they can be used – namely to protect someone or something from being burned. A fireproof safe can be used to protect important papers from disappearing in a fire, for instance. If the practice of protecting things from disappearing in a fire did not exist, there would be no fireproof safes. Were we not interested to protect ourselves and other things from being burned, there would be no fireproof things at all.[9]

What can we conclude from this? Perhaps artifacts can, indeed, have a conceptual content due to the fact that they have not only natural properties like being heavy, but also properties like being fireproof.

We might call the second type of properties "functional properties", but in order to do this we need to clarify the meaning of "functional" in this context. A function can be characterized by the causal role an object plays within a system. Saying that something has a function or an use is ambiguous in this respect. Let us look at the following cases:

(a) One of the functions of a tree in an ecosystem is to produce oxygen.
(b) A tree can be used to produce wood.
(c) A tree can be used for climbing.
(d) A ladder is used for climbing and descending.

Now, it is obvious that (a) states something about trees that amounts to a natural description of trees as a part of our ecosystem. Sentence (b) expresses more than the natural fact that trees are made of wood. It says, in addition, that in describing trees as being made of wood we also regard trees as a raw material, as the starting point for some productive actions. This is, however, different from saying that trees can be used as means for a particular purpose in a practical action. By contrast, (c) states that trees can be used precisely like this, if by "climbing" we mean a specific practice. A tree prepared for climbing activities in an adventure park is, in fact, quite

---

[9]It could also be noted that the existence of artifacts to be used for the protection of people and objects from burning is another condition for the existence of a "being fireproof" property. Keeping a distance of 1 mile from a fire can protect someone from being burned, but we do not call 1 mile "a fireproof distance".

similar to an artifact, with the notable difference that, unlike a climbing wall, the tree was not "artificially produced for this purpose". Finally, (d) states that ladders are used for the practical purpose of reaching to a higher or a lower area by the actions of climbing or descending. It is in this sense that we take "being climbable" as a functional property of ladders and "being fireproof" as a functional property of fireproof artifacts.[10]

In order to conclude that artifacts can have a conceptual content in a different way from the one in which a special type of artifacts – sign tokens – have conceptual content, we still need to understand where could such a conceptual content come from.

3 How can a functional property have conceptual content? Two strategies for answering this question are at hand. The first is to apply a restricted version of an intention-based semantics[11] to artifacts. According to such a view, the conceptual content of a functional property comes from the conceptual content of the intentions accompanying the production of the artifact bearing it.[12] For something to be fireproof, then, to recall our previous example, it seems important that it was produced with the intention to be used in a certain way. A wet coat lying on the ground could be used by someone as a protection from fire, but since it was not made for this purpose the intentional theorist might be reluctant to call it fireproof.[13]

This strategy was implicitly used to talk in a non-naturalist vocabulary about speed bumps as rule embedding artifacts in the first section of this paper. Let us revise it here, starting with an attempt at a general definition. Thus, given a rule R, of the form "Under condition(s) C, all rule addressees ought to do (or refrain from doing) A", a rule embedding artifact X is such that:

(1) X was produced with the intention that X has the functional property F.[14]
(2) F is the property of being useful to achieve the purpose that if C obtains, the rule addressees do (or refrain from doing) A.

---

[10]I am reluctant to talk about "fireproof materials", since raw materials are not directly used in practical actions, but in productive ones. For instance, I would say that asbestos is resistant to heat (which is a natural property), but not fireproof. This does suggest an analysis of the relation between natural and functional properties, but developing such an analysis would be beyond the aim of my paper.

[11]See Grice 1957, 1969.

[12]I think Dipert 1993, 1995; Bloom 1996; Hilpinen 2011 are representative for this strategy.

[13]In fact, the intentional theorist does not need to talk about functional properties in order to specify the content of a productive intention. Dipert simply talks about "properties", while Hilpinen 2011 replaces functional predicates by sortal terms. I suspect this is due to their interest in works of art as artifacts. Also, if we make it a necessary condition for being an artifact that the object in case satisfies a sortal, we might have to exclude some respectable artifacts like glue and whisky (the second being Hilpinen's own example).

[14]Besides production, we could also include modification in our analysis, as Dipert 1993, 1995 suggests. It should also be noted that (1) is weaker than Hilpinen's (DEP) condition ("The existence and some of the properties of an artifact depend on an author's intention to make an object of certain kind.").

(3) X was produced with the intention that the users of X recognize that (1).[15]

(4) X's production was successful.[16]

Several problems can be raised with respect to this attempted definition. One could produce artifacts unintentionally.[17] We can even conceive of an artifact coming into existence due to a random arrangement of atoms. Other difficulties raised with respect to Grice's theory of meaning could perhaps apply here as well. Also, if X's possession of F depends on X's author's intention that X has F and the success of X's production depends on X having F, our definition seems to involve a suspect circularity. However, the most important problem seems to be that X, thus defined, cannot be a rule embedding artifact. In order to see this we need to recall the old distinction between doing something in accordance to a rule and obeying a rule.[18] Based on the attempted definition, X is produced to make the rule addressees act according to the rule, but not to obey the rule. It seems, then, that we need to add another condition to our definition:

(5) X was produced with the intention that X being F is considered by the addressees of R as a reason for doing A under C.

This, however, makes X indistinguishable from a sign used to communicate the rule. To return to our first example, the definition (1)–(5) would be only satisfied by a road hump with the message "Slow down before crossing!" written on it. The message would not embed the rule, but only express it, while the road hump would make drivers act according to the rule, but not obey it. It is disputable that the combination of the message and the street hump would embed the rule, but we do not need to enter into a dispute right now. After all, we were trying to understand how an artifact could embed a rule *without being used to communicate it*.

4 We are still facing the same problem. How can a functional property have conceptual content in such a way that having certain functional properties would make an artifact embed a rule? Perhaps we need to turn to a different strategy at

---

[15]This condition is due to Dipert. Hilpinen 2011 also accepts it as a "plausible condition, since an F-object can presumably be a good F-object only if its potential users recognize it as such". His critique that "recognizability should not be taken to mean general recognizability" can be answered by my distinction between the users of X and the addressees of X.

[16]Here we can define successful production by combining Hilpinen's conditions (SUC) ("An object is an artifact made by an author only if it satisfies some sortal description included in the author's productive intention.") and (ACC) ("An object is an artifact made by an author only if the author accepts it as satisfying some sortal description included in his productive intention."): X's production was successful IFF either X has F, or the author(s) of X accept that X has F.

[17]Suppose I automatically fold an origami bird while talking on the phone and leave it on a table. Someone else takes it and uses it as a bookmark. The origami bird was not produced with the intention to be used in a certain way and it is not used according to the way in which it is regularly used (as a decoration). Proper intentions (at least as psychological states) seem to be missing from this picture, but my origami bird would still be an artifact (here "my" does not mean that I am the author of the origami bird, but only that I am the cause of its existence).

[18]See Wittgenstein 1953: §§201–2.

this point, one which can be related to contemporary use theories of meaning.[19] According to such a strategy, in order to specify the conceptual content of a functional property we do not need to talk about the intentions accompanying the production of the property bearing artifact, such a content being provided by the ways in which the artifact in case is used within series of goal-oriented actions.[20] After all, if words can have a conceptual content by being used in relation to perception, reasoning and action, why could not artifacts get a conceptual content in a similar way[21]?

Let us see, then, how a rule embedding artifact could be defined within this approach. We start as before. Given a rule R, of the form "Under condition(s) C, all rule addressees ought to do (or refrain from doing) A", a rule embedding artifact X is such that:

(6) X is a designed structure.
(7) X realizes F (given a goal G, X can be used to achieve G).
(8) R is a means to achieve G (that the rule addressees do (or refrain from doing) A under C).
(9) R is involved in the design of X.

To take an example, suppose that one of my goals is that my son does not use the computer after 10 P.M. The rule which could be a means to achieve this goal would be "Răzvan Florea-Ştefanov ought to refrain from using the computer after 10 P.M.". The rule can be embodied in the functioning of the computer by being translated into a script to be automatically executed at 10 P.M.[22]

---

[19] I believe it is safe to assume that conceptual role semantics (see Harman 1982, 1999; Horwich 1994) and inferentialism (see Brandom 1994, 2000, 2007) are such theories.

[20] This functional approach to artifacts can be traced back to Skolimowski (1966) and Heidegger (1927), Chap. 3, recent representatives being involved in the *Dual nature of Artifacts* research project (Kroes and Meijers 2006; also: http://www.dualnature.tudelft.nl/). This last claim is disputable, since Kroes and Meijers invoke Dipert and talk about intentions, although they do not seem to consider intentions as psychological states, so I tend to agree with Vaesen 2011 in this respect. For other representative positions for the functional approach, see Houkes 2006; Vermaas 2006; Vermas and Houkes 2006; Preston 2009. The distinction between intentional and functional approaches to artifacts was suggested to me by Verbeek and Vermaas 2009 (although they put Dipert in a distinct category).

[21] We do not want to say that artifacts are like words in all respects, of course. The distinction could be made by pointing out that artifacts are only used in relation to perception and action (and perhaps they can be also involved, in a sense, in practical reasoning, in a different way from the one in which words are involved in practical reasoning).

[22] On the Linux operation system, for instance, a *bash* script would be executed (by the *cron* daemon) by including the following line in the *crontab* file:

```
00 22 * * * /usr/bin/somedirectory/script.sh
```
and it would look like this:
```
#!/bin/bash
if [[ `whoami` == "razvan" ]]; then
halt
fi
```

In particular, the digital control of artifacts seems to make the way in which rules can be involved in their design quite transparent. But is this really so? Even if we ignore the distinction between rules and commands, we still have to note that the initial rule was a *rule of action*, while the best translation of it into a script can only be a *rule of criticism*[23]: "at 10 P.M. the computer ought to halt if the logged in user is Răzvan".

The relation between the two rules can be analyzed along the following lines:

(RA) All rule addressees (D) ought to refrain from doing A under C.

(RC) The artifact X ought to remove E (where E is a necessary condition for doing A) or to produce E' (where A can be done only in the absence of E') under C and C' (where C' is the additional condition that the person interacting with X is identified as D).

If we take into account not only interdictions, but also permissions, the analysis would be:

(RA) All rule addressees (D) may do A only if C.

(RC) The artifact X ought to prevent A from being done (either by removing E, which is a necessary condition for doing A, or by producing E', such that A can be done only in the absence of E') and stop preventing A from being done only if C and the person interacting with X is identified as D.

However, such an analysis does not answer our main problem. How can we say that an artifact incorporates a rule of action if the only rule involved in its design is a rule of criticism? Even if we stick to definition (6)–(9), it seems that our "R" in that definition stands for a rule of criticism. In addition, the goal that can be achieved through the functioning of an artifact embedding a rule of criticism is not an action (that my son refrains from using the computer[24]), but an event (that the computer halts). In this respect, the situation seems to be worse than before. If my son stops using the computer thinking that the computer is malfunctioning, then the script I wrote does not embed any rule.

From a practical point of view, the functional analysis of rule embedding artifacts is more useful than the intentional analysis, since it allows us to talk about the ways in which rules can be involved in the design of rule embedding artifacts. The result of such an analysis is the idea that the respective artifacts can include means to identify the rule addressee and also means to form causal links between events which are necessary conditions for the regulated actions (E's) and events which are the normative conditions (C's) of the same actions.

---

[23] The distinction between rules of action and rules of criticism comes from Sellars 1969. The form of a rule of action is familiar: "If one is in C, one ought to do A" (Sellars 1969: 507; C stands for the particular circumstances under which the rule applies). An example of a rule of criticism is "Clock chimes ought to strike on the quarter hour" (Sellars 1969: 508).

[24] I consider 'refraining from A' as designating an action.

There are also conceptual reasons to prefer the functional approach. Talk about intentions as psychological states leads to naturalization. However, if there is one lesson to be learned from the debate fostered by Wittgenstein's rule-following considerations,[25] it is precisely that rules cannot be naturalized.[26] This is not to say that one could not talk about intentions in an attempt to understand rule embedding artifacts. We could do this without considering intentions as a type of mental objects, since intentions are already embedded in our actions.[27] After all, it would not be preposterous to assume that artifacts have a dual nature, because our actions have a dual nature. Talk about actions involved in the design of an artifact can capture all that can be said in an attempt to define artifacts by talking about the artifact authors' intentions.

At this point we might pause for another remark. It might be that the functional approach cannot account for rule embedding artifacts because such objects fall outside its scope. Another moral from the rule-following debate, regardless of the theoretical differences between its participants, is that rules are embedded in social practices. If we agree to this, then rule embedding artifacts must have a social character.[28] They must be, so to speak, like money,[29] but not quite like words. Nevertheless, it is precisely artifacts like money or words that the functional theorists are not interested in.

However, since such an interest does not seem incompatible with the functional approach, perhaps we could hold on to the functional approach and try to improve it in such a way as to cover the case of rule embedding artifacts.

5 Now, let me introduce the hypothesis that rule embedding artifacts come into being by being *inserted* into pre-existent rule-following practices. The question of how could a practice embed a rule is too remote from our present interests to be answered here. It is sufficient to note, for instance, that if the practice of issuing fines for drivers honking their horns within cites was missing and cars were manufactured by default such that their horns would be disabled within city limits, we could not say that cars embedded the rule that one should not honk while in a city. In a similar way, the practice of asking my son to stop using the computer after 10 P.M., warning him that it is close to 10 P.M. when he is still using the computer, turning off the computer if he does not comply and so on must be in place for the script in my previous example (conceived, perhaps, as a virtual artifact) to embed the rule that my son ought to refrain from using the computer after 10 P.M.

---

[25]For the starting point of this debate, see Wittgenstein 1953: §§142–202; Kripke 1982; Hacker and Baker 1985.

[26]My own arguments for this claim can be found in Ştefanov 2004.

[27]I take this suggestion from Anscombe 1963.

[28]This is different from saying that rule embedding artifacts are socially constructed (as in Bijker 1995) or "constructed and constructing" (as in Latour 1993), since according to either SCOT or ANT all artifacts are like this.

[29]See Searle 2007, for instance.

The problem is to understand how can the rule embedding artifact be inserted into such a practice. If the artifact gets only a place in the pre-existent practice, then why should we say that it embeds the rule? The rule is embedded in the entire practice, of which our artifact is just a component. Locks, for instance, are only a part of the practices embedding the rule that one ought to refrain from stealing, so it could hardly be said that they embed the rule.

Perhaps an artifact can be said to embed a rule when it comes to replace the entire practice in question. To this it could be replied that when something like that happens, the rule should rather disappear than be embedded in an artifact. If our environment would be modified in such a way that stealing would become impossible, why keep thinking that the rule that one ought to refrain from stealing is embedded in it? There is no need to refrain from something which you cannot do.

We can concede to this reply. Embedding rules into our technological environment might ultimately lead to the disappearance of said explicit rules, although not immediately. Embedded rules can be hacked[30] but even if they could not, some time should pass until no one sees the rule embedding artifacts on the background of the pre-existent practices anymore.

It seems that we can finally return to our main concern. A rule-following practice within which some actions are either permitted or forbidden under certain circumstances includes rule-invocations in providing reasons for the rule addressees doing A only if C or refraining from doing A if C. Due to this, the relation between such a practice and an artifact replacing it must be conceived in such a way that the reasons-providing part of the practice is somehow preserved. The problem is that the reason-providing part of the practice falls under the heading of communication, while the functioning of the artifact replacing the practice did not fall under such a heading.

Allow me, then, to propose a solution to our problem. The first step towards the solution consists in pointing out the distinction between invoking a rule to provide *a reason why* one ought to do or refrain from doing something and invoking a rule to provide a *reason to* do or refrain from doing something. Another idea we can get from Wittgenstein's rule-following considerations is that by invoking rules we can never properly provide *reasons why* one ought to do or refrain from doing something, since by using the appropriate interpretation we can put any action apparently breaking the rule in accordance with the rule, but only *reasons to* do or refrain from doing something. So "you ought to wash your hands before eating" or "you may eat only if you have washed your hands" can be invoked only as reasons to wash your hands before eating.

---

[30]It could also be noted that in the speed bumps example the artifact does not simply reduce the speed of the car. A driver could maintain a high speed when encountering a speed bump, at the risk of damaging her car suspensions. The opponent could of course reply that in this case speed bumps should not be considered "proper rule embedding artifacts".

Now, since actions can have conceptual content and enter into logico-semantic relations,[31] my actions opposed to your action of eating, performed when you did not wash your hands[32] can also provide you with reasons to wash your hands before eating without communicating them to you.[33]

Finally, it is precisely by replacing this part that the functioning of a rule embedding artifact provides one with reasons to do or refrain from doing something. In short, a rule embedding artifact can embed a rule by replacing in its functioning the rule-invoking actions performed in the corresponding rule-following practice.

6 To conclude, we may attempt to complete our analysis of a rule embedding artifact as follows. Let R be a rule either of the form "Under condition(s) C, all rule addressees ought to do (or refrain from doing) A" or of the form "All rule addressees may do (or refrain from doing) A only if C.". Also, let I stand for the set of rule-invoking non-communicative actions belonging to the R-following practice P. X is an R-embedding artifact IFF:

(10) X is designed to replace the practice P (in other words, replacing P is the function of X).

(11) R is not related to the use of X.

To this we need to add a specification of what is meant by an artifact replacing a practice. So X replaces P IFF:

(12) The successful functioning of X produces the same result as the successful performance of the actions in I.

A few observations might be useful here. Condition (11) is necessary in order to distinguish rule embedding artifacts from regular artifacts which may incorporate some rules pertaining to their proper utilization. Utilization rules, it can be easily noticed, can be considered constitutive, while rule embedding artifacts enforce only regulative rules.[34] The phrase "produces the same result" in (12) may seem ambiguous, but it does not cover any hidden glitch in my conceptual proposal. Human actions can both provide reasons for other actions and have effects. The functioning of a rule embedding artifact, which we distinguish from an action, preserves this dual nature. A speed bump placed before a crosswalk can be said

---

[31] I have tried to give support to this idea in Ştefanov 2013.

[32] To these could be added actions implied (i. e. necessary conditions in a routine) by your action of washing your hands (opening the bathroom door, turning on the water etc.).

[33] Someone could be taught to play Hashiwokakero (See http://en.wikipedia.org/wiki/Hashiwokakero) only through such actions. It would be at least awkward to say that the rules of the game were communicated to such a learner.

[34] A rule embedding artifact can, of course, incorporate some rules pertaining to its proper utilization as well. The addressees of these constitutive rules are its users (the people who place bumper speeds on roads, for instance), which are not the addressees of the regulative rules which the artifact embeds. The distinction between constitutive and regulative rules comes from Searle 1965, although its source can be traced back to Austin 1962.

to provide a reason to slow down before the crosswalk in case and also to make drivers slow down. My computer script provides my son with a reason to refrain from using the computer and also makes him do it. Condition (12) only states that X's functioning produces the same double result as the actions by which we were both enforcing the rule and invoking the rule without communicating it. Further complications can of course appear on the course of an attempt to get a clearer view of the relation between the two aspects, but they are not specific to the case of rule embedding artifacts.

7 To sum up, then, our problem was to conceptually accommodate the view according to which we produce and use some artifacts in order to embed certain rules in our environment. We have started by noting that we cannot simply resort to an entirely naturalist vocabulary to account for such cases, since the vocabulary in question would not allow us to distinguish between artifacts and natural objects.[35] We have thus reached the conclusion that in order to conceive something as a rule embedding artifact we must be able to conceive it as having a conceptual content. In its turn, this thought led us to the idea that we must conceive rule embedding artifacts as having functional properties. The next step was to find a way in which we could conceive a functional property as having conceptual content. Two possible strategies were suggested by two contemporary semantic approaches – intention-based semantics and inferentialism. Having traced the existent attempts to analyze the concept of a generic artifact to these semantic approaches, we have concluded that the second strategy might be better for our purposes, the first strategy being abandoned mainly due to its failure to distinguish a rule embedding artifact from a regular sign.

By focusing on the functional approach to artifacts we have reached a new problem, for it seemed that this approach only allowed us to conceive rule embedding artifacts as incorporating *rules of criticism*, while we wanted to say that the artifacts in question also incorporate *rules of action*. In order to overcome this difficulty we took a suggestion from Wittgenstein's rule-following considerations and came to think that in order to incorporate rules of action the rule embedding artifacts must be inserted into a pre-existent rule-following practice.[36]

Finally we had to face one last problem, namely that it was not very clear how could a rule embedding artifact replace the reason-providing part of a rule-following practice. Here a distinction between *reasons why* one ought to do A and *reasons for*

---

[35]To this one could add that unless we have a convenient way to naturalize concepts like 'responsibility' and 'justification', talking about rule embedding artifacts in a naturalist vocabulary would prevent us from asking questions like 'Who is to assume responsibility for the production and use of rule embedding artifacts?' or 'How should we justify the rules to be embedded in some artifacts?'.

[36]One could of course wonder what should be said about an artifact the practical use of which produces an entirely now practice. I think that such a case would fall outside the scope of what I have called 'rule embedding artifacts' here, since the artifact in question could only embed rules *for its own use*.

someone *to* do A came in handy. Thus, it seemed conceivable that an artifact could replace the rule-invoking actions performed in a rule-following practice, since such actions were providing agents with *reasons to* do or refrain from doing something. The rule-invoking actions being replaced by rule embedding artifacts, we noted, had conceptual content without being performed as part of a communication.

One problem remained. Rule embedding artifacts, according to this conceptual proposal, appeared to have inherited the dual nature of our actions. We could conceive our actions as events and talk (in a naturalist vocabulary) about their causes and effects, but we could also talk about the same actions as having agents, which were responsible for them, as being justified or unjustified and so forth. The same remained true of the rule embedding artifacts' functioning. This problem, however, was not considered specific to the case of rule embedding artifacts, since it required a separate attempt to go beyond the surface of what was called 'the dual nature of our actions'.

We have recently moved forward from the old muddled conception that a technological artifact is to be distinguished from a natural object by the fact that it was made by a human being to some more refined views. The treatment of rule embedding artifacts seems to suggest that in talking about technological artifacts it would be more useful to adopt an action oriented approach.

According to such an approach to artifacts it might be less important to distinguish artifacts from natural objects and more important to distinguish them from other objects figuring at the intertwinement of practical actions like design, production, use, adjustment, repair, hacking, disposal, recycling etc. In particular, it could be important to distinguish artifacts from raw materials – the starting point of any productive actions but also the result of recycling actions – and waste – the starting point of disposal and recycling actions. This, however, might be a topic for another research.

Bibliography

Akrich M (1992) The De-scription of technical objects. In: Wiebe E. Bijker, John Law (eds) Shaping technology/building society: studies in sociotechnical change. MIT Press, Cambridge, MA, pp 205–224

Anscombe GEM (1963) Intention, 2nd edn. Harvard University Press, Cambridge, MA/London

Austin JL (1962) How to do things with words. Clarendon Press, Oxford

Baker LR (2008) The shrinking difference between artifacts and natural objects. Am Philos Assoc Newsl Philoso Comput 7(2):2–5

Bijker WE (1995) Of bicycles, bakelites and bulbs: toward a theory of sociotechnical change. MIT Press, Cambridge, MA/London

Bloom P (1996) Intention, history, and artifact concepts. Cognition 60:1–29

Brandom RB (1994) Making it explicit. Harvard University Press, Cambridge, MA

Brandom RB (2000) Articulating reasons. Harvard University Press, Cambridge, MA

Brandom RB (2007) Inferentialism and some of its challenges. Philos Phenomen Res 74(3):651–676

Dipert RR (1993) Artifacts, art works, and agency. Temple University Press, Philadelphia

Dipert RR (1995) Some issues in the theory of artifacts: defining "artifact" and related notions. Monist 78:119–135

Franssen M (2009) Artefacts and normativity. In: Meijers A (ed) Philosophy of technology and engineering sciences. Elsevier, Amsterdam, pp 923–952

Grice HP (1957) Meaning. Philos Rev 66:377–388

Grice HP (1969) Utterer's meaning and intention. Philos Rev 78(2):147–177

Hacker PMS, Baker GP (1985) Rules, grammar and necessity: an analytical commentary on the philosophical investigations. Blackwell, Oxford

Harman G (1982) Conceptual role semantics. Notre Dame J Formal Logic 23:242–257

Harman G (1999) Reasoning, meaning and mind. Oxford University Press, Oxford

Heidegger M (1927) Sein und Zeit. Max Niemeyer Verlag, Tubingen

Hilpinen R (2011) Artifact. In: Zalta EN (ed) The Stanford encyclopedia of philosophy (Winter 2011 Edition). http://plato.stanford.edu/archives/win2011/entries/artifact/

Horwich P (1994) What It is like to be a deflationary theory of meaning. In: Villanueva E (ed) Philosophical issues 5: truth and rationality. Ridgeview, Atascadero, pp 133–154

Houkes W (2006) Knowledge of artefact functions. Stud Hist Philos Sci 37(1):102–113

Kripke S (1982) Wittgenstein on rules and private language. Harvard University Press, Cambridge, MA

Kroes P, Meijers A (2006) The dual nature of technical artifacts. Stud Hist Philos Sci 37:1–4

Kroes P, Vermaaas PE (2008) Interesting differences between artifacts and natural objects. Am Philos Assoc Newsl Philos Comput 08(1):28–31

Latour B (1993) We have never been modern (trans: Porter C). Harvard University Press, Cambridge, MA

Latour B (1999) Pandora's hope: essays on the reality of science studies. Harvard University Press, Cambridge, MA

Preston B (2009) Philosophical theories of artifact function. In: Meijers A (ed) Philosophy of technology and engineering sciences. Elsevier, Amsterdam, pp 213–233

Radder H (2009) Why technologies are inherently normative. In: Meijers AWM (ed) Philosophy of technology and engineering sciences. Elsevier, Amsterdam, pp 887–921

Searle JR (1965) What is a speech act? In: Black M (ed) Philosophy in America. Cornell University Press, Ithaca, pp 221–239

Searle JR (2007) Social ontology and the philosophy of society. In: Margolis E, Laurence S (eds) Creations of the mind. Theories of artifacts and their representation. Oxford University Press, Oxford/New York, pp 3–17

Sellars W (1969) Language as thought and as communication. Philos Phenomen Res 29(4):506–527

Skolimowski H (1966) The structure of thinking in technology. Technol Cult 7(3):371–383

Ştefanov G (2013) Opposing actions. Transylvanian Rev 22(2):121–131

Ştefanov G (2004) How to be a good philosopher? Are there any rules?. Analele Universitatii Bucureşti – Filosofie 53:81–96

Thomasson AL (2007) Artifacts and human concepts. In: Margolis E, Laurence S (eds) Creations of the mind. Theories of artifacts and their representation. Oxford University Press, Oxford/New York, pp 52–73

Vaesen K (2011) The functional bias of the dual nature of technical artefacts program. Stud Hist Philos Sci 42(1):190–197

Verbeek P-P, Vermaas PE (2009) Technological artifacts. In: Jan Kyrre Berg Olsen Friis, Pedersen SA, Hendricks VF (eds) A companion to the philosophy of technology. Wiley-Blackwell, Oxford, pp 165–171

Vermaas PE (2006) The physical connection: engineering function ascriptions to technical artefacts and their components. Stud Hist Philos Sci 37(1):62–75

Vermaas PE, Houkes W (2006) Technical functions: a drawbridge between the intentional and structural natures of technical artefacts. Stud Hist Philos Sci 37(1):5–18

Vries MJ, Hansson SO, Meijers A (eds) (2013) Norms in technology. Springer, Dordrecht/Heidelberg/New York/London

Wittgenstein L (1953) Philosophische Untersuchungen. Blackwell, Oxford

Chapter 6
Issues in Modeling Open-Ended Evolution

Andreea Eșanu

Using computational models in science is a very widespread practice. It proves effective in exploring mathematical properties of systems for which there are no standard analytical methods available. But this widespread application of computational models has also generated some misunderstandings about computer modeling. Computer models are not only means to calculate intractable equations as an extension to the analytical methods, but in fact they possess enough expressive power to actually represent phenomena of arbitrary complexity. Complex processes can be simulated exactly by concrete computational means (Vichniac 1984). This opens the perspective to use computer models for entirely different purposes than strictly exploring mathematical properties of systems. They can be used to reproduce, quite realistically, very intricate processes in nature and to explore hypotheses about what is happening, more like in an experimental activity (Morrisson 2009).

The issue of developing proper computational models in order to explore complex phenomena has become a significant topic in evolutionary biology, especially since the process of evolution has been acknowledged as a process that generates complexity (McShea 1996, 2005), for instance in the form of biodiversity. The fruitfulness of applying concrete computer models in order to study and explain complex evolutionary processes of large or small scale has been widely agreed upon (Bedau 1998a, 2009). Yet the best modeling strategies are still prone to numerous difficulties.

In the following, I intend to discuss some of the issues facing concrete bottom-up modeling in evolutionary biology. My contention is that some of these issues could be resolved if certain constraints on concrete model construction were developed. The idea to place constraints on computational models in order to maximize their

A. Eșanu (✉)
Department of Theoretical Philosophy, University of Bucharest, Bucharest, Romania
e-mail: aesanu2@gmail.com

© Springer International Publishing Switzerland 2015
I. Pârvu et al. (eds.), *Romanian Studies in Philosophy of Science*, Boston Studies
in the Philosophy and History of Science 313, DOI 10.1007/978-3-319-16655-1_6

explanatory power is not entirely new in biology (see Levins 1966) and it represents an intrinsic part of model construction, but what I believe is specific to this proposal is that it seeks to develop such constraints for concrete bottom-up models, which traditionally have no constraints.

Realistic bottom-up modeling, which is expected to simulate open-ended processes, is by definition experimental in the sense that one can hardly anticipate what one is going to get once the assumptions, the input and the basic local rules of the model are set and the simulations start (Ray 1994). However, some find it important to distinguish between unpredictability *within* a predictable model and unpredictable models. In spite of many researchers' emphasis on predictable models, the main tenet of the present paper is that, in order to maximize the explanatory power of concrete bottom-up models, their architecture should rather be globally unpredictable, while only locally predictable under the constraints. Global unpredictability could be, in fact, a key feature in order to define concrete bottom-up models as a class of models different from standard equation solving models, significant especially in experimental research which deals with hypothetical outcomes.

6.1 Concrete Computational Models in Evolutionary Biology

There is a well-established class of concrete computational models of biological evolution called genetic algorithms. One could exemplify with Richard Dawkins' well-known BIOMORPH (1986) or with Chris Adami's AVIDA (2000). In spite of their concrete architecture, these computational models belong in a specific class of analytical models. Their explanatory effectiveness resides, in fact, in the ability to simulate evolutionary equilibriums at gene level (Beatty 1980). In this respect, concrete equilibriums in BIOMORPH or AVIDA are very similar to the frequency-dependent equilibriums found in the standard equation-based models from population genetics (see Fisher 1958). They all are fitness-maximization equilibriums (see Lensky 2003).

From the point of view of concrete modeling, the major problem with genetic algorithms is that very few evolutionary processes in the natural world are equilibrium driven processes. At the natural scale, evolutionary equilibriums are only local and not fully stable – otherwise the evolution of life on Earth would be an already closed process. Therefore, at least in principle, in order to model "open" realistic evolutionary processes, it is preferable to develop models that do not evolve towards external or global equilibriums. Obviously, standard algorithms cannot do the task.

During the last few decades (see Ray 1991, 1994; Bedau 2009), open-ended processes turned into a real corner-stone for modelers in evolutionary biology. A large diversity of concrete models shifted focus from fitness-maximization processes to local evolutionary dynamics, inherent mechanisms of variation, selection and adaptation, evolution of biodiversity (which is obviously a far from equilibrium process) and so on, in order to get a virtual grasp of what evolution of life on Earth might actually involve. In this respect, "replaying the tape of life" (Gould

1989) on computational machines has become a real research paradigm in computer modeling. One can check, for instance, Tim Taylor's large scale artificial life system COSMOS released in 1999, the latest version of Thomas Ray's TIERRA released in 2004 or Ken Stauffer's EVOLVE 4.0 released in 2007.

COSMOS, TIERRA, EVOLVE are synthetic bottom-up computational models that work in an entirely different manner from genetic algorithms. The basic design of such models is neither placed at the gene, nor at the population level – but at the cell/organism level. Following the theoretical principles of a von Neumann constructor, digital cells/organisms are designed as executable DNA programs with an algorithm for self-replication. The programs replicate in a computer environment – i.e. a computer's RAM or a cellular automata grid – and, given a finite design of the environment, they soon enter a competition for resources. Organisms that manage to get to the resources survive longer and succeed to reproduce in larger numbers, transmitting their genetic traits to the next generation of programs. Natural selection occurs spontaneously and fitter organisms tend to emerge from one generation to the other – without the need of an external fitness function and an optimization algorithm. In this sense, artificial life models are said to offer quite realistic bottom-up simulations of evolutionary processes taking place in the natural world.

The development of concrete bottom-up modeling was set on track by an important advancement in cellular automata research (see Langton 1986). John Conway's Game of Life (Gardner 1970), which is basically an algorithmic model of emergence and self-organization, offered the computational proof that a bottom-up design of complex evolutionary processes is actually possible. In the Game of Life one can successfully see how evolution occurs from scratch.

The bottom-up design behind cellular automata is not complicated. A cellular automata environment is a two-dimensional orthogonal grid of cells. Each cell is in one of two possible states, *alive* or *dead*. Every cell interacts with its neighbors and at each step in time several transitions take place based on a set of simple rules.[1] The initial pattern in the grid is named the *seed* of the system. A first generation of cells is created by applying the transition rules to every cell in the seed, so that births and deaths occur simultaneously. Transition rules continue to be applied repeatedly and further generations of cells are created.

What is peculiar to cellular automata is that they *require just as much computational power as our current computers can provide* in order to spontaneously generate, from the seed and by applying the transition rules repeatedly, diverse and intricate cell configurations that might resemble (up to a certain point) complex organisms living presently on Earth.

---

[1] Survivals: every cell with two or three neighboring cells survives for the next generation. Deaths: every cell with at least four neighboring cells dies from overpopulation; every cell with at most one neighboring cell dies from isolation. Births: every empty cell adjacent to exactly three cells is a birth cell. A cell is placed on it at the next move (see Gardner 1970).

Fig. 6.1 The Mitchell parasite (a gilder) on a back-rake in the Game of Life.[2] The figure presents two stable types of patterns in a Game of Life simulation. The complex large pattern in the *upper* part of the figure is an orthogonally moving pattern, called a back-rake. The two simpler patterns at the *bottom* of the figure, resembling small spaceships, are gliders moving diagonally. One of the two small gliders (the one on the *left*) emerges in the simulation as debris liberated from the backward part of the back-rake. The other glider (the one on the *right*) is already near the back-rake when the second glider (the one on the *left*) emerges. The gliders are said to display parasitic behavior because the glider on the *right* reproduces only when a host, i.e. a back-rake, moves nearby (see also Fig. 6.2)

Fig. 6.2 Reproduction of the Mitchell parasite in the presence of two back-rakes in a game of life simulation.[3] In this figure, another back-rake approaches the already stable patterns (see Fig. 6.1) coming from the *bottom* part of the figure. It moves orthogonally and gets closer to the two gliders from the *upper* part. Once the upward moving back-rake is close enough to them, a new small glider gets liberated from its backward part. The new glider appears to the *left* of the other two gliders (see Fig. 6.1). The unfolding of the simulation pertinently shows that the small gliders multiply only when new back-rakes populate their neighborhood

Let us look, for instance, at Figs. 6.1 and 6.2.

What we see here might be simply called digital parasitism. The Mitchell parasites emerge as complex and stable self-reproducing cell patterns given a very simple setting of local conditions and transitions in a cellular automata environment.

---

[2]Source: http://pentadecathlon.com/lifeNews/2011/01/sprouts_and_parasites.html

[3]Source: http://pentadecathlon.com/lifeNews/2011/01/sprouts_and_parasites.html

The example showcases sophisticated reproductive innovation evolving all by itself in a concrete computational model. The important advancement and departure from standard genetic algorithms is that cellular automata designs manage to generate complex evolutionary outcomes, while still presenting them as bottom-up processes.

Nevertheless, cellular automata principles seem not fully sufficient for realistic models of open-ended processes. In spite of the ability to engineer emergence and self-organization, cellular automata are not far from equilibrium models. In the Game of Life simulations, the evolving structures get in fact too robust too fast due, for instance, to their little sensitivity to the local environment. If one takes the time to look at the evolution of the Mitchell parasite in the Game of Life, one can notice that nothing else happens in the simulation after the parasite's reproduction. The behavior of the entire Game is, in fact, an exploratory behavior towards some evolutionary equilibrium. Once the equilibrium is discovered, the process stalls because no external pressures intervene. In this sense, cellular-automata processes are not open-ended – a Game of Life grid is, ultimately, a lattice of finite state machines that sometimes reach equilibrium, while some other times don't and simply stall.

In this respect, the Game of Life behaves more like an evolutionary niche than a diverse ecosystem. Synthetic models of evolution, like the complex artificial ecosystems TIERRA or COSMOS, rely on the bottom-up design of cellular automata, but also manage to set in motion an evolutionary process with considerable sensitivity to contingent factors of evolution. Such models display less idealization than the Game of Life and, maybe, they have to pay the inevitable price of making certain complex evolutionary outcomes intractable. Yet, in spite of all the risks, the main goal of synthetic modeling is to "replay the tape of life" – i.e. to work in far from equilibrium conditions and see how evolutionary diversity adds up, more like in some engineering experiment that in a mathematical model.

The following table resumes briefly the large variety of computational models available at present time in the field of evolutionary biology. My following discussion on epistemological constraints will focus on the last category (Table 6.1).

Table 6.1 Computational models in evolutionary biology

	Equation-based frequency models (Fisher)	Genetic algorithms (Dawkins)	Cellular Automata (The Game of Life)	Artificial life systems (TIERRA, COSMOS)
Computational profile	Numeric	**Concrete**	**Concrete**	**Concrete**
Architecture	Analytic (top-down)	Analytic (top-down)	**Synthetic** (bottom-up)	**Synthetic** (bottom-up)
Behavior	Equilibrium driven (closed)	Equilibrium driven (closed)	Exploring for equilibrium (closed)	**Far from equilibrium** (open)

6.2 Concrete Synthetic Models: Limitations and Constrains

The synthetic approach to computer modeling is not yet very widespread in evolutionary biology, in spite of the arguments speaking in its favor.

To count only a few of them, in artificial life ecosystems like COSMOS or TIERRA, virtual idealized cells and grids from cellular automata designs are developed into fully engineered and fully functional digital cells that possess both a genotype[4] and a phenotype.[5] Such a design feature is very important, for instance, when tackling the sources and mechanisms of variation in real cells and organisms. Also, the computer environment is not just a uniform orthogonal grid, but it exhibits local differences and peculiarities. This design feature favors rich interactions in the model and promotes coevolution – as organisms react not only to their "natural" environment, but also to the other organisms in their neighborhood. Mechanisms for regulating the genome of cells are systematically implemented. The regulatory routines are expected to facilitate the evolution of differentiated programs and, thus, promote biodiversity (Taylor 1999b). Last but not least, the rules that govern transitions between cell states are correlated with elaborate mechanisms for resources exploitation on the grid (for instance, programs find and convert resources into energy) generating real competition between programs and conditioning survival.

Nevertheless, in spite of such interesting developments, the main reason for reticence against synthetic models is the fact that the modeler faces huge difficulties when it comes to identify the proper ways of carrying out the modeling. These models still lack solid theoretical and methodological grounding (Taylor 1999a; Bedau 1998b). Therefore, skepticism still appears justified.

For example, in Thomas Ray's famous TIERRA, Tierran organisms develop interesting adaptive traits only at very high rates of mutation and at very fine-tuned population sizes. Small changes in these parameters modify the evolutionary patterns entirely. Repeated experiments conducted with Tierran populations (see Yedid and Bell 2002) showed that increased *contingency* is a general feature of evolution in TIERRA. Also, the huge sensitivity to model parameters was mainly explained by the fact that all variation is unbounded and nearly neutral in the system (Yedid and Bell 2002, 811). If all variation in some population is random, each simulation would need a large population of cells and a large fraction of possible combinations of mutations in any generation of cells, in order to observe, possibly, the emergence of some interesting adaptive traits. Given increased contingency in the system, TIERRA lacks the capability to generate significant evolutionary outcomes in computable time – except, perhaps, for several forms of emergent parasitism (see Fig. 6.3), which is observable only in large populations of cells and at very high rates of mutation.

---

[4]Genotype is the source code of the program.

[5]Phenotype is the behavior of the program when its source code is executed.

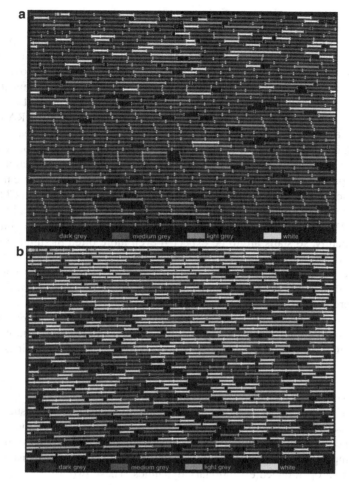

Fig. 6.3 The evolutionary race between hosts and parasites in Tierra.[6] Images are made using the Artificial Life Monitor (ALmond) program developed by Marc Cygnus. Each image represents a memory space of 60,000 bytes which can hold the same number of Tierran machine instructions. Each organism occupies a block of memory in this space and is represented by a color bar. Colors correspond to genome sizes (e.g., *medium grey* = 80 Tierran machine instructions, *white* = 45, *dark grey* = 79). In the first image, hosts with the standard 80 genotype size, depicted in *medium grey*, are very common. They dominate the 60,000 bytes memory space. Parasites, depicted in *white*, have appeared but are still rare. In the second imagine, after the simulation has run for a while, hosts decrease in number, as parasites become increasingly common and take over the memory space. Usually, a digital organism in a Tierran space needs a genome of at least 60 machine instructions to be able to reproduce. Parasites have only 45 machine instructions. In order to reproduce, parasites find the portion in memory where larger organisms' replication instructions are run and take that portion of memory over. Some scattered immune hosts, depicted in *dark grey*, also appear. These are emergent 79 machine instructions genomes resistant to parasitic attacks

---

[6]Photo credit: Marc Cygnus.

Another complicated issue concerns the design of the system. In spite of its bottom-up architecture, TIERRA permits very little local competition between digital programs. The main competition is actually indirect, as each program seeks to occupy as much computer memory space as it can in order to run its code, irrespective of what the other programs do (see again Fig. 6.3). The obvious downside of such a design approach is that most simulations in TIERRA end due to premature convergence (Adami and Brown 1994) – i.e. the system reaches stable states with ease, but it does so without developing much adaptation and diversity. Most stable states are blunt cases of stalling due to full memory occupation.

In fact, premature convergence is a very problematic issue in synthetic modeling even in cases where local competition is implemented better, as it is in COSMOS, because all models running on finite computers are ultimately finite and they stall (Sober 1992). Even the Game of Life, which is a far less contingent simulation, stalls frequently, so stalls are to be expected even more often in simulations with increased contingency.

Discovering how to make evolutionary patterns more robust, but not equilibrium driven, so that synthetic life forms evolve to increasingly more diverse states is a crucial problem in synthetic modeling. The case of TIERRA indicates that it is still very unclear under what modeling conditions executable DNA evolutionary software would continue to produce novel life forms and develop complex adaptations, while keeping the evolutionary process tractable. In short, it is still an issue how open-ended evolution as a "continual production of adaptively significant innovations" (Taylor 2011, 1) "in rich ecosystems of complex organisms" (*ibidem*) could be synthetized by successful computer engineered models.

As Tim Taylor points out (2011), there are two sides to the issue. On the mathematical side, it is important to develop a proper model for the dynamics of open-ended evolution, while on the engineering side, it is crucial to develop a proper computational design, so that artificial evolutionary systems display *all by themselves* an open-ended evolutionary dynamics. These two aspects need not be treated separately as many synthetic modelers, including Taylor, seem to believe. The synthetic modeler has to figure out from the very beginning what model of open-ended dynamics is adequate in respect to their modeling goals (e.g., the evolution of biodiversity), and only then convert the mathematical model into an engineering puzzle. In this sense, mathematical instruments might still be needed in synthetic modeling besides experimental engineering.

Traditionally, synthetic models are said to have no constraints, but only limitations like: (a) a finite space of potential phenotypes (finite memory space); (b) limited mutational pathways (finite mutations); (c) static adaptive landscapes (organisms have limited interactions with each other, as seen in TIERRA). Many argue (see Taylor 2011) that once these limitations are overcome, sophisticated synthetic systems will display open-ended evolution as a continual production of diverse adaptive innovations, all by themselves. However, as we have seen in TIERRA's case, there is also a different kind of problem with such systems. The evolution in TERRA relied on unbounded neutral variation of digital cells, which led to huge instability or contingency in the model.

Instability is a mathematical problem, if we remember Taylor's distinction introduced earlier. It concerns evolutionary dynamics. Even if all limitations of computer design such as (a) to (c) were resolved by implementing the system on a universal Turing machine, the system would still display *no trend* towards the production and accumulation of diverse adaptive innovations *because of that problem*. This means that at least some constraints are needed in synthetic modeling. Otherwise, one would have real difficulties in holding that synthetic models really promote open-ended evolution, as previously characterized.

The proposition I will make in the following section addresses instability or increased contingency in synthetic models. Concrete computational models of open-ended evolution should be developed on a set of well-designed boundary conditions that would constrain their behavior into less unstable patterns. Contrary to Taylor (2011), I maintain that design limitations are only secondary to boundary condition specifications in successful synthetic modeling. The main suggestion will be that, in a proper specification of boundary conditions, the dynamics simulated by concrete synthetic models, such as populations of digital organisms evolving in a computer environment, will emerge as a *special case* of open-ended evolutionary dynamics.

6.3 Boundary Condition Designs

First, let me clarify what I mean by boundary conditions with a simple example from classical mechanics. If we are to pinch suddenly the middle of a string of a swinging string-and-weight pendulum, we would see that it starts oscillating faster than by simply pushing the weight. One could question the nature of this new oscillation, and try to describe it. However, if attention is paid only to the swinging dynamics, one would fail to notice that the important mathematical event that takes place when pitching the string of the pendulum is the introduction of a new boundary condition in the description of the oscillation – i.e. the pendulum system changes its condition from rigid to elastic oscillation. Instead of thinking "there must be a different force that is pushing the weight along faster" and go looking for that force, one could see that rigid oscillation is mathematically a simplification of elastic oscillation. Basically, the *same* phenomenon is being observed, but under two different conditions which make the first observation a special case of the second.

In short, boundary conditions may be seen as simplification tools. Because of that, in computational modeling, their role could be very important. A process may be tractable or intractable *depending on how such conditions are defined*; in other words, depending on the degree of mathematical simplification one is willing to accept in their descriptions of the processes they are interested in modeling. In Richard Levins' terminology (1966) that is an epistemological trade-off. Knowing more about something means, sometimes, simplifying it substantially. In concrete bottom-up modeling, mathematical simplification could serve the same epistemological purpose as it does everywhere. Boundary conditions could help manage increased instability by constraining the behavior of the system into less random patterns and, thus, facilitating the bottom-up tractability of interesting and diverse evolutionary innovations. In this manner, one might get to know more about how evolution generates biodiversity, but through a simpler model.

As said already, a major cause of increased contingency or instability (if we favor the dynamic term) in artificial life systems like TIERRA was found by research to be unbounded, neutral variation of genes and phenotypes in digital cells and organisms. A pertinent question is therefore: what would happen if boundary conditions are introduced in synthetic models constraining cellular/organismal variation?

In order to see how boundary conditions could influence model dynamics, at least two alternative scenarios can be imagined:

(i) A scenario with bounded variation – variation is bounded at the scale of the entire evolving system: for each organism x ever to appear in the system X, the variation $V(x)$ will be smaller or equal to M, where M is the *total* variation boundary.

This case presents a restrictive boundary condition. Each cell or organism in each generation would tolerate the same degree of internal variation. However, if mutation rates are thus bounded, the evolutionary process towards the emergence of diverse and complex adaptive outcomes will have to be biased. Most mutations will have to be *directed mutations* because low-rate random variation in a finite population cannot generate enough internal diversity to see complex adaptive traits emerge all by themselves through blind selection. The obvious drawback of specifying a boundary condition like that is that it imposes a strong idealization on the evolutionary model, forcing biased selection. This implies that the corresponding computational design would probably be highly unrealistic.

So a less restrictive condition should be in view:

(ii) *Locally* bounded variation – variation is bounded around every class (e.g. species) of organisms appearing in the evolving system: for any organism x from the class of organisms A and living in the system X, the variation $V(x)$ will be smaller or equal to M, where M is a *local* variation boundary for class A.

Organisms evolving under this constraint could display what in evolutionary biology is currently called *evolvability* (Wagner and Altenberg 1996; Kirschner and Gerhard 1998) – i.e. an inherent capacity of each organism to diversify in an adaptive manner by eliminating spurious, lethal and deleterious internal variation. Due to "the locking" of basic gene control mechanisms that reduce the chances of undesired mutations across different species (classes), variation is locally bounded as each organism displays internal variation only in a given "beneficial" range. Yet, at the scale of the entire ecosystem, variation is not bounded so mutation rates can vary across populations, as they do not need to be kept low. There is a rich literature on evolvability in cellular, computational and evolutionary biology (Kirschner and Gerhard 2005; Klingenbert 2005; Parter et al. 2008) – but this concept was fairly ignored in the literature on concrete synthetic models of biodiversity. An explanation might be found in the interest of most artificial life modelers for pure bottom-up modeling without constraints that might bias variation, as it is the case here, or bias selection as it is the case in the previous scenario. Yet, as biological theory tends to show, variation might actually be biased in cells and organisms. So, to specify a boundary condition like that would not mean *ipso facto* an explanatory loss. Quite the contrary.

The situation might be illustrated further with an intuitive example found in Kirschner and Gerhard (2005) who elaborate on the WEASEL example from Dawkins (1986). Let us imagine that we want to create a computer program that would generate an intricate line from one of William Shakespeare's plays – for instance, "Methinks it is like a weasel" from *Hamlet*. A program might compute the 28 characters string by permuting all the letters in the English alphabet and space (27 characters in total) simultaneously. This is a case of unbounded variation, as the mutation rate tends to 100 %. Yet, the probability to hit the right combination of characters in the string is so small that the program will probably have to run indefinitely.[7] If no competition between mutating strings is introduced, the increased contingency of the model will make the line "Methinks it is like a weasel" practically intractable.

Alternatively, a different program might be designed with boundary conditions on permutations (letters and space), and so we may see the two scenarios unfold. Let us consider that character permutations are an intuitive representation of internal variation in a string. The program might bind the internal variation of the 28 character string by a total variation boundary (a low mutation rate) and approximate the desired outcome stepwise by means of an optimization algorithm. Taking from Dawkins (1986) the procedure could be the following:

1. Begin with a random string of 28 characters.
2. Make 100 copies of the string.
3. Replace each character in each of the 100 copies with a new random character, at a probability (mutation rate) of 5 %.
4. Compare each resulting string with the target string "METHINKS IT IS LIKE A WEASEL". Count the number of characters in the string that are correct and in the correct position.
5. If any of the resulting strings has a perfect score (28), halt. Else, keep the highest scoring string and go back to step 2.

Given the little amount of variation in the string (only a 5 % mutation/substitution rate in each generation of strings) it is implausible to expect the program to deliver a specific line from Shakespeare's in a reasonable time without biased selection. Here, directional selection is introduced in the model through the scoring algorithm – in each generation of strings only one string survives (the string that mostly resembles "METHINKS IT IS LIKE A WEASEL" from all 100 copies). The new string

---

[7]Any random string has the probability $\approx 1/1{,}040$ of being correct. If a program generating ten million strings per second had been running since the beginning of the universe ($\approx 1{,}017$ s), it would have generated by now only $\approx 1{,}024$ strings.

reproduces at a 5 % mutation rate (which now is low enough to preserve the acquired resemblance to the desired outcome) and the selection process starts again until the goal is met.

The example illustrates briefly how bounded neutral variation introduces strong idealization in the model, and why synthetics modelers are so reticent to boundary conditions: on one hand, low mutation rates require biased (e.g., directional) selection which is exceptional in the natural world; on the other hand, low mutation rates artificially stabilize strings that are partially successful. Bottom line, the total variation boundary generates a very unrealistic description of how interesting adaptive traits emerge in the natural world.

The last scenario, however, requires less idealization, although it involves a higher degree of simplification. In that modeling approach, total variation can be kept unbounded (so no artificially low mutation rates will be designed). But, borrowing from the theory of evolvability in evolutionary biology, variation will be characterized as *non-neutral* – i.e. it will locally facilitate certain "beneficial" permutations of characters in the string. The consequence will be that, beside low mutation rates, the model will eliminate the necessity for biased selection.

Nevertheless, one must be careful at this point because "Methinks it is like weasel" might be a too complicated outcome for a bottom-up computer model that does not optimize heavily on its search strategies. The epistemic trade-off for the synthetic modeler would be, in fact, to settle for a process with a simpler outcome – for instance, the line "To be or not to be" from the same Shakespearian play. This string is easier to compute, with local constraints on internal variation and without directional selection, than "Methinks it is like a weasel". From this perspective, accepting the epistemic trade-off would not necessarily entail that the resulting evolutionary model is highly unrealistic – as in the previous case; it would only entail that evolvability is not always tractable and some simplification (or boundary conditions design) is needed in order to succeed.

But let us finish the example. Kirschner and Gerhard (2005) propose implementing two control mechanisms on character permutations in the Shakespearian string "To be or not to be": (a) discard the combinations of letters which are not English words (this would eliminate "spurious" variation), and (b) discard the words that are more than three letters long (this would eliminate "lethal" and "deleterious" variation).

The algorithmic procedure might be the following:

1. Start with a random word from the English dictionary.
2. If a word is two or three letters long, keep it (natural selection occurs) and add it to the string. Else, go back to step 1.
3. When the resulting string amounts to 18 characters (counting spaces),[8] move to step 4. If the string is less than 18 characters long, move to step 2. If it is more than 18 characters long, go back to step 1.

---

[8]Spaces need not be introduced separately as words are by default introduced with spaces.

4. Assess if the resulting string is "To be or not to be". If yes, halt. Else, go back to step 1.

In steps 1 and 2, new character strings are generated under two simple constrains on internal variation. Here, variation is said to be *non-neutral* because: (i) there is a meaning constraint on one large class of sub-strings (i.e. the ones which do not contain spaces); and (ii) there is a length constraint on the class of meaningful sub-strings. It is hardly the case to maintain that in this example mutations/changes in the string are directed. They are only locally constrained.

The algorithm facilitates the emergence of Shakespearian strings in English language by biasing the manner in which characters in a string vary. What is obvious is that, for different classes of sub-strings, variation is biased, but selection is not. What one can see here is a very simple bottom-up model of how interesting complex innovations can emerge in an artificial life system, with just non-neutral variation and natural selection modeled as a simple preserving mechanism for survival promoting traits.

This is a very intuitive example. The key observation, however, is that in this specific design of boundary conditions, optimization takes place only at the local scale in the system – as each organism in the population seeks to secure its own survival. Basically, if one organism does not survive, it cannot evolve. Yet, once basic survival mechanisms are in place, mutations can lead practically anywhere. Moreover, in any stage of the evolutionary process, each organism has to maintain its survival, so, in each step of the evolutionary process, certain mutations are *facilitated* by non-neutral variation. Under these conditions, the process modeled could be said to be locally stable, but globally open-ended.

Based on this argument, the design hypothesis I am trying to suggest is the following. If one could encapsulate more of the concept of evolvability in concrete artificial life systems, one may gain some significant insight into open-ended evolution and the dynamics of biodiversities through the experimental study of virtual organisms and populations. Such a modeling approach would present a fair amount of realism (by not biasing interactions with the environment, or biasing selection, or engineering low mutation rates), but also a fair amount of *simplification* – as boundary conditions may be easily manipulated to make evolutionary processes more or less tractable in the computer environment (Table 6.2).

Table 6.2 Real biodiversities vs. artificial life systems

Real biodiversities	Artificial life systems
Display evolutionary dynamics	Display discrete state transitions (embedding survival rules)
Are open-ended	Are contingent (or promote contingent outcomes)
→ Design boundary conditions to model their evolutionary dynamics	→ **Engineer evolvability** to generate tractable evolutionary outcomes?

6.4 Final Remarks

In this final section, I will only mention that several steps towards engineering evolvability have been taken already, in recent research from computational biology (Parter et al. 2008) and artificial life systems (Taylor 2013). Both these engineering perspectives stress heavily on the importance of rich interactions with the local environment, so that digital organisms develop internal variation mechanisms to optimize local survival and, thus, promote adaptability. As far as actual engineering is concerned, low-level design to favor rich interactions (Taylor 2013), locally embedded survival goals (Parter et al. 2008) and modular regulator systems developed in digital cells and organisms (Parter et al. 2008; Taylor 2013) are considered first hand tools in synthetizing evolvability.

Building from that and in the longer run, concrete bottom-up modeling might turn into an effective epistemological tool when seeking to gain insight beyond the restrictive high-level cellular automata, artificially low mutation rates or directional selection models, and they might supplement successfully the set of evolutionary explanations based on genetic algorithms. In an adequate setting, synthetic models might provide some bottom-up clues to the problem of how organismal diversification accumulates in an evolutionary system and what the systemic dependencies between variation, mutations and diversity might involve.

Another thing to stress in the end is that designing simpler boundary conditions for synthetic evolutionary systems does not alter the significance of evolvability from biological theory – it alters only the scale at which evolvability is studied and reproduced. In concrete synthetic environments, one seeks to engineer special, yet interesting, cases of evolvability keeping in mind that one has to work only with computable processes.

References

Adami C, Brown CT (1994) Evolutionary learning in the 2D artificial life system Avida. Artif Life 4:377–381

Adami C et al (2000) Evolution of biological complexity. Proc Natl Acad Sci U S A 97:4463–4468

Beatty J (1980) Optimal-design models and the strategy of model building in evolutionary biology. Philos Sci 47(4):532–561

Bedau MA (1998a) Four puzzles about life. Artif Life 4(2):125–140

Bedau MA (1998b) Philosophical content and method of artificial life. In: Bynam TW, Moor JH (eds) The digital phoenix: How computers are changing philosophy. Basil Blackwell, Oxford, pp 135–152

Bedau MA (2009) The evolution of complexity. In: Barberousse A, Morange M, Pradeu T (eds) Mapping the future of biology, Evolving concepts and theories. Springer, Dordrecht, pp 111–132

Dawkins R (1986) The blind watchmaker: why the evidence of evolution reveals a universe without design. Norton, New York

Fisher R (1958) The genetical theory of natural selection, 2nd edn. Dover, New York

Gardner M (1970) Mathematical games: the fantastic combinations of John Conway's new solitaire game 'life'. Am Sci 23:120–123

Gould S (1989) Wonderful life. Norton, New York

Kirschner MW, Gerhard JC (1998) Evolvability. Proc Natl Acad Sci U S A 95(15):8420–8427. doi:10.1073/pnas.95.15.8420

Kirschner MW, Gerhard JC (2005) The plausibility of life, Resolving Darwin's dilemma. Yale University Press, New Haven

Klingenberg CP (2005) Developmental constraints, modules, and evolvability. In: Hallgrimson B, Hall BK (eds) Variation: a central concept in biology. Elsevier Academic Press, Amsterdam, pp 219–248

Langton C (1986) Studying artificial life with cellular automata. Phys D 22:120–149

Lensky RE et al (2003) The evolutionary origin of complex features. Nature 423:139–144

Levins R (1966) The strategy of model building in population biology. Am Sci 54:421–431

McShea D (1996) Metazoan complexity and evolution: is there a trend? Evolution 50(2):477–492

McShea D (2005) A universal generative tendency toward increased organismal complexity. In: Hallgrimson B, Hall BK (eds) Variation: a central concept in biology. Elsevier Academic Press, Amsterdam, pp 435–454

Morrison M (2009) Models, measurements and computer simulation: the changing face of experimentation. Philos Stud 143(1):33–57

Parter M, Kashtan N, Alon U (2008) Facilitated variation: how evolution learns from past environments to generalize to new environments. PLoS Comput Biol 4(11):e1000206. doi:10.1371/journal.pcbi.1000206

Ray T (1991) An approach to the synthesis of life. In: Langton CG et al (eds) Artificial life II. Addison Wesley, Redwood City, pp 371–408

Ray T (1994) An evolutionary approach to synthetic biology: Zen and the art of creating life. Artif Life 1(1/2):195–226

Sober E (1992) Learning from functionalism – prospects for strong artificial life. In: Langton CG et al (eds) Artificial life II. Addison Wesley, Redwood City, pp 749–765

Taylor T (1999a) From artificial evolution to artificial life. PhD thesis. http://www.timtaylor.com/papers/thesis/

Taylor T (1999b) The Cosmos artificial life system. http://www.tim-taylor.com/papers/cosmos-tech-v2n0.pdf

Taylor T (2011) Exploring the concept of open-ended evolution. http://www.tim-taylor.com/papers/taylor-alife13-oee-abstract.pdf

Taylor T (2013) "Evolution in virtual worlds", to appear. In: Grimshaw M (ed) The Oxford handbook of virtuality. Oxford University Press, Oxford. https://www.academia.edu/4254804/Evolution_in_virtual_worlds

Vichniac G (1984) Simulating physics with cellular automata. Phys D 10:96–115

Wagner GP, Altenberg L (1996) Complex adaptations and the evolution of evolvability. Evolution 50(3):967–976. doi:10.2307/2410639. JSTOR 2410639

Yedid G, Bell G (2002) Macroevolution simulated with autonomously replicating computer programs. Nature 420:810–812

Part III
Logic, Semantics, and Social Choice

Chapter 7
On a Combination of Truth and Probability: Probabilistic Independence-Friendly Logic

Gabriel Sandu

7.1 Two Interpretations of First-Order Languages: Game-Theoretical Semantics and Skolem Semantics

One of the main tasks of first-order logic is to express interdependencies of quantifiers. For instance, we express the sentence

> For all x there is a y which depends only on x such that $R(x, y)$

by

$$\forall x \exists y R(x, y)$$

and the statement

> For all x for all z there is a y which depends only on x and a w which depends on x, y and z such that $Q(x, y, z, w)$

by

$$\forall x \exists y \forall z \exists w Q(x, y, z, w).$$

The Skolem semantics for first-order formulas expresses such dependencies in an explicit way. Consider the first sentence above and a model $\mathbb{M} = (M, I)$ which assigns an extension $I(R) \subseteq M^2$ to the relation symbol R. According to the Skolem semantics, the sentence is true in $\mathbb{M}$, $\mathbb{M} \models_{Sk}^{+} \forall x \exists y R(x, y)$, if there is a method (function) $f : M \to M$ which gives an individual $f(a) \in M$ for every $a \in M$

G. Sandu (✉)
Department of Philosophy, History, Culture and Art Studies, University of Helsinki,
Helsinki, Finland
e-mail: sandu@mappi.helsinki.fi

© Springer International Publishing Switzerland 2015
I. Pârvu et al. (eds.), *Romanian Studies in Philosophy of Science*, Boston Studies
in the Philosophy and History of Science 313, DOI 10.1007/978-3-319-16655-1_7

so that $(a, f(a)) \in I(R)$. f is called a *Skolem function*. Symmetrically, we declare the sentence to be false in $\mathbb{M}$, $\mathbb{M} \models^-_{Sk} \forall x \exists y R(x, y)$ if there is a (0-place function) individual $a \in M$ such that for every $b \in M$ we have $(a, b) \notin I(R)$. We call such an a a (Kreisel) *counter-example*.

The definition of the truth (falsity) of a sentence in terms of the existence of appropriate Skolem functions (counter-examples) has an intuitive appeal if game-theoretically motivated. Henkin (1961) and Hintikka (1974), define truth (falsity) of a sentence φ in a model $\mathbb{M}$ as the existence of a winning strategy for Eloise (Abelard) in a certain semantical game $G(\mathbb{M}, \varphi)$. In such a game there are two players, Eloise (the verifier) who tries to show that φ is true in $\mathbb{M}$ and her opponent Abelard (the falsifier) who tries to show that φ is false. Keeping things simple, Eloise chooses values for the occurrences of existential quantifiers and Abelard does the same thing for the occurrences of the universal quantifiers.

A strategy for either player is a function which gives a player the choice to be made for each possible move of the opponent. In our example, a strategy for Eloise is any function $f : M \rightarrow M$; and a strategy for Abelard is any individual $a \in M$. The strategy f is winning for Eloise if $(a, f(a)) \in I(R)$ for every choice a of her opponent. And a strategy a is winning for Abelard if $(a, b) \notin I(R)$ for every choice b of his opponent. This simple example nicely illustrates the correspondence between Skolem functions (counter-examples) and winning strategies for Eloise (Abelard) in semantical games.

When the interdependence of quantifiers in a sentence is more complex, like in the second example, we need a Skolem function (counterexample) of appropriate arity for each existential (universal) quantifier. In such a case, in the definition of truth (falsity) the arguments of the Skolem functions (Kreisel counterexamples) are appropriately embedded.

Thus

- $\mathbb{M} \models^+ \forall x \exists y \forall z \exists w Q(x, y, z, w)$ if and only if there are Skolem functions $f_y : M \rightarrow M$ and $g_w : M^3 \rightarrow M$ such that for all $a, c \in M$: $(a, f_y(a), c, g_w(a, f_y(a), c)) \in I(R)$.

Symmetrically

- $M \models^- \forall x \exists y \forall z \exists w Q(x, y, z, w)$ if and only if there are counterexamples $a \in M$ and $h_z : M^2 \rightarrow M$ such that for every $b, d \in M$: $(a, b, h_z(a, b), c) \notin I(R)$.

For the general case, we define by a double induction the Skolem form $Sk(\varphi)$ and the Kreisel form $Kr(\varphi)$ of an arbitrary first-order formula φ:

Sk1. $Sk(\psi) = \psi$; $Kr(\psi) = \neg\psi$, for ψ an atomic formula
Sk2. $Sk(\neg\psi) = Kr(\psi)$; $Kr(\neg\psi) = Sk(\psi)$,
Sk3. $Sk(\psi \wedge \theta) = Sk(\psi) \wedge Sk(\theta)$; $Kr(\psi \vee \theta) = Kr(\psi) \wedge Kr(\theta)$
Sk4. $Sk(\exists x \psi) = Sub(Sk(\psi), x, f(y_1, \ldots, y_n))$; $Kr(\exists x \psi) = \forall x Kr(\psi)$
Sk5. $Sk(\forall x \psi) = \forall x Sk(\psi)$; $Kr(\forall x \psi) = Sub(Kr(\psi), x, g(y_1, \ldots, y_m))$.

In (Sk4), $y_1, \ldots, y_n$ are all the free variables of $\exists x \psi$ and f is a new function symbol. $Sub(Sk(\psi), x, f(y_1, \ldots, y_n))$ denotes the formula which is the result of the substitution of the Skolem term $f(y_1, \ldots, y_n)$ for the variable x in the formula ψ. Analogously for (Sk5).

The content of the above clauses should be quite clear. To verify (falsify) an atomic sentence, do nothing: just see that it is true (false). To verify a negated sentence, falsify the sentence itself, etc. All the action takes place when quantifiers occur. To verify an existential sentence, produce a witness for the existential quantifier, taking into account, the values which have been introduced before. Etc. Hence the definitions below should come as no surprise:

- $\mathbb{M}, s \models_{Sk}^{+} \varphi$ if and only if there exist functions $g_1, \ldots, g_n$ of appropriate arity in $\mathbb{M}$ to be the interpretations of the new function symbols in $Sk(\varphi)$ such that

$$\mathbb{M}, g_1, \ldots, g_n, s \models Sk(\varphi)$$

- $\mathbb{M}, s \models_{Sk}^{-} \varphi$ if and only if there exist functions $h_1, \ldots, h_m$ of appropriate arity in $\mathbb{M}$ to be the interpretations of the new function symbols in $Kr(\varphi)$ such that

$$\mathbb{M}, h_1, \ldots, h_m, s \models Kr(\varphi)$$

When φ is a sentence and s is the empty assignment we simply write $\mathbb{M} \models_{Sk}^{+} \varphi$ and $\mathbb{M} \models_{Sk}^{-} \varphi$.

Once again, we emphasize the rule for negation: it makes clear that to verify the sentence $\neg \psi$ amounts to falsify ψ and vice versa. It can be shown that

$$\mathbb{M} \models_{Sk}^{+} \neg \varphi \iff \mathbb{M} \models_{Sk}^{-} \varphi.$$

For illustration, consider the first-order sentence $\varphi = \exists x \forall y \exists z \forall w R(x, y, z, y)$ and an arbitrary model $\mathbb{M}$. By (Sk2), $Sk(\neg \varphi)$ is $Kr(\varphi)$, which is

$$\forall x \forall z \neg R(x, f_y(x), z, f_w(x, f_y(x), z))$$

Thus

$\mathbb{M} \models_{Sk}^{+} \neg \varphi$ if and only if $\mathbb{M} \models_{Sk}^{-} \varphi$ if and only if there are functions f and g such that

$$\mathbb{M}, f, g \models \forall x \forall z \neg R(x, f_y(x), z, f_w(x, f_y(x), z))$$

Thus the functions f and g which witness the truth of $\neg \varphi$ in $\mathbb{M}$ are the counterexamples to φ on $\mathbb{M}$.

7.2 Hintikka and Sandu: Independence-Friendly Logic (IF Logic)

Hintikka and Sandu (1989) introduce Independence-Friendly logic (IF logic) in order to express more patterns of dependencies (and independencies) of quantifiers than those allowed by ordinary first-order logic. The syntax of IF logic contains quantifiers of the form $(\exists x / W)$, and $(\forall x / W)$ where W is a finite set of variables. The idea is that the choice of a value for x is independent of the choices of the values for the variables in W. When $W = \varnothing$ we recover the standard quantifiers.

The usual example to introduce the idea of independence is the so-called epsilon-delta definition of a continuous function. A function f is continuous at a point x_0 if given any $\varepsilon > 0$ one can choose $\delta > 0$ so that for all y, when x_0 is within distance δ from y, then $f(x_0)$ is within distance ε from $f(y)$. Ignoring the restrictions on the quantifiers, this definition states the dependence of δ on both x_0 and ε. It is rendered in first-order logic by the following sentence, denoted by φ_c:

$$\forall x_0 \forall \varepsilon \exists \delta \forall y [|x_0 - y| < \delta \rightarrow |f(x_0) - f(y)| < \varepsilon].$$

Now it sometimes turns out that one can find a δ which works no matter what x_0 is. In this case the choice of δ depends only on ε but is independent of x_0. This phenomenon is known in mathematics as *uniform continuity*. It is expressed in IF logic by the following variant φ_{uc} of φ_c:

$$\forall x_0 \forall \varepsilon (\exists \delta / \{x_0\}) \forall y [|x_0 - y| < \delta \rightarrow |f(x_0) - f(y)| < \varepsilon].$$

The original interpretation of IF formulas and of independence is game-theoretical. Given a model $\mathbb{M}$, $G(\mathbb{M}, \varphi_{uc})$ is, unlike its relative $G(\mathbb{M}, \varphi_c)$, a game of imperfect information: the slash indicates that when choosing a value for δ from the universe of $\mathbb{M}$, Eloise does not "know" the value for x_0 chosen earlier by Abelard. The lack of knowledge is implemented at the level of strategies: we require that the Skolem function f_δ which corresponds to δ takes only ε as its argument instead of both x_0 and ε.

That is, $Sk(\varphi_{uc})$ is

$$\forall x_0 \forall \varepsilon \forall y [|x_0 - y| < f_\delta(\varepsilon) \rightarrow |f(x_0) - f(y)| < \varepsilon]$$

whereas $Sk(\varphi_c)$ is

$$\forall x_0 \forall \varepsilon \forall y [|x_0 - y| < g_\delta(x_0, \varepsilon) \rightarrow |f(x_0) - f(y)| < \varepsilon].$$

The definition of truth and falsity in terms of Skolem functions and counter-examples remains otherwise unchanged:

$\mathbb{M} \models_{Sk}^{+} \varphi_{uc}$ if and only if there exists a function $g : M \rightarrow M$ which is the interpretation of the function symbol g_δ such that

$$\mathbb{M}, g \models Sk(\varphi_{uc})$$

The falsity conditions are the same as those for φ_c.

Actually it turns out that φ_{uc} is equivalent to the ordinary first-order sentence φ^*:

$$\forall \varepsilon \exists \delta \forall x_0 \forall y [|x_0 - y| < \delta \rightarrow |f(x_0) - f(y)| < \varepsilon].$$

To see the equivalence, it is enough to notice that $Sk(\varphi^*)$ and $Sk(\varphi_{uc})$ are identical (modulo notational variance). The equivalence, however, should not induce us to think that the independence of an existential quantifier can always be expressed by a rearrangement of quantifiers resulting in an ordinary first-order sentence. It is well known that this is not the case. Here is an example known to logicians for more than half a century:

1. For every x and x', there exists a y depending only on x and a y' depending only on x' such that $Q(x, x', y, y')$ is true

In IF logic we express (1) by φ_H

$$\forall x \forall x' (\exists y / \{x'\})(\exists y' / \{x, y\}) Q(x, x', y, y').$$

It is easily checked that $Sk(\varphi_H)$ is

$$\forall x \forall x' Q(x, x', f(x), g(x'))$$

Let me mention right away that several formalisms have been invented to express (1). Henkin (1961), represents it by a *branching (Henkin) quantifier*

$$\begin{pmatrix} \forall x & \exists y \\ \forall x' & \exists y' \end{pmatrix} Q(x, x', y, y')$$

whose truth-conditions are given by:

$$\begin{pmatrix} \forall x & \exists y \\ \forall x' & \exists y' \end{pmatrix} Q(x, x', y, y') \Leftrightarrow \exists f \exists g \forall x \forall x' Q(x, x', f(x), g(x')).$$

The two formalisms are obviously equivalent in this particular case. Ehrenfeucht noticed that the "Ehrenfeucht sentence" φ_{eh}

$$\exists w \begin{pmatrix} \forall x & \exists y \\ \forall x' & \exists y' \end{pmatrix} R(x, x', y, y', w)$$

where $R(x, x', y, y', w)$ is the formula

$$w \neq y \wedge (x = x' \leftrightarrow y = y')$$

defines (Dedekind) infinity. Indeed, the truth-conditions of the branching quantifiers makes φ_{eh} equivalent with

$$\exists w \exists f \exists g \forall x \forall x' (w \neq f(x) \wedge x = x' \leftrightarrow f(x) = g(y)).$$

Now the formula $x = x' \rightarrow f(x) = g(x)$ states that $f = g$ whereas the other direction asserts that f is injective. All in all φ_{eh} asserts that there is an injective function f whose range is not the whole universe, i.e., the universe is (Dedekind) infinite. It is well known that infinity cannot be expressed in ordinary first-order logic.

To conclude this sketchy exposition of the Skolem semantics for IF logic, let me mention that the Skolemization and Kreiselization of an IF formula in the general case are completely identical to their first-order case, except for the clauses for the existential and universal quantifier, respectively:

- $Sk((\exists x / W)\psi) = Sub(Sk(\psi), x, f(y_1, \ldots, y_n))$, where $y_1, \ldots, y_n$ are all the free variables of $(\exists x / W)\psi$ minus the variables in W.
- $Kr((\forall x / W)\psi) = Sub(Kr(\psi), x, g(y_1, \ldots, y_m))$, where $y_1, \ldots, y_m$ are the free variables in $(\forall x / W)\psi$ minus the variables in W.

Sandu (1996) proves that truth can be defined in IF logic and Hintikka (1996) shows the significance of this result for the foundations of mathematics. Hodges (1997) gives a compositional semantics and Mann et al. (2011) collects the basic model-theoretical properties of IF logic.

7.3   Indeterminacy

Let us focus on the following IF sentences:

$$\varphi_{MP} = \forall x (\exists y / \{x\}) x = y$$

$$\varphi_{IMP} = \forall x (\exists y / \{x\}) x \neq y$$

and a finite structure $\mathbb{M}$ which contains two elements a and b. Otherwise the second player wins and the first looses. Both are variations of our earlier example $\forall x \exists y R(x, y)$ except that now the choice of a value for y is independent of the choice of a value for x. φ_{MP} defines the "Matching Pennies" game: two players turn a coin to Head or Tail without seeing each other. If their choices coincide, then the first player wins and the second looses. φ_{IMP} defines the Inverted Matching Pennies Game.

First we notice that $Sk(\varphi_{MP})$ is $\forall x x = c$, where c is a new individual constant; and $Kr(\varphi_{MP})$ is $\forall y d \neq y$ with d also a new constant.

Then we have both

- $\mathbb{M} \not\models^+_{Sk} \varphi_{MP}$, for it is impossible to find an individual a in M to be the interpretation of c so that it is identical with every individual in M;

and

- $\mathbb{M}, s \not\models^-_{Sk} \varphi_{MP}$, for it is impossible to find an individual b in M to be the interpretation of d such that b is distinct from any individual in M.

In game-theoretical terms: there is neither a winning strategy for Eloise nor one for Abelard in the underlying game. An analogical reasoning shows the indeterminacy of φ_{IMP} on every model with at least two elements. We can represent the two examples in the form of two matrices:

$$\varphi_{MP}: \begin{array}{c|c|c} & a & b \\ \hline a & (1,0) & (0,1) \\ \hline b & (0,1) & (1,0) \end{array} \qquad \varphi_{IMP}: \begin{array}{c|c|c} & a & b \\ \hline a & (0,1) & (1,0) \\ \hline b & (1,0) & (0,1) \end{array}$$

The rows represent the strategies (Skolem functions) of Eloise and the columns represent the strategies (counter-examples) of Abelard. These strategies are also known in game theory as *pure strategies*. The first matrix shows that if e.g. Eloise chooses to play strategy a and Abelard strategy b then, given that (a, b) does not satisfy $x = y$, Eloise gets payoff 0 and Abelard gets payoff 1.

The matrices represent the semantical games as strategic games $\Gamma = (S_\exists, S_\forall, u_\exists, u_\forall)$ where

- $S_\exists$ is the set of all possible (pure) strategies of Eloise (sequences of Skolem functions) in $G(\mathbb{M}, \varphi)$
- $S_\forall$ is the set of all possible (pure) strategies of Abelard (sequences of Kreisel counter-examples) in $G(\mathbb{M}, \varphi)$
- $u_\exists(\sigma, \tau)$ is the payoff of Eloise when she plays the strategy σ and Abelard plays the strategy τ. It is computed in the following way. The strategies σ and τ generate a sequence of individuals in M and an atomic or the negation of atomic formula which is a subformula ψ of φ. If the sequence satisfies ψ, then $u_\exists(\sigma, \tau)$ is 1; otherwise it is 0.
- $u_\forall(\sigma, \tau)$ is defined analogously.

7.4 Solution Concepts in Strategic Games

Strategic IF games are finite, win-lose two player games. We can analyze them by using solution concepts in classical game theory. The solution concepts are based on the notion of rationality of the players: each player prefers a greater payoff to a smaller one. The rationality of the players is common knowledge in the game.

One such concept is that of *dominant strategy*: if a strategy gives a player a better payoff than any other of her or his strategy with respect to every strategy of the opponent, then it is rational for the player to play it. There are no dominant strategies in our example.

Another concept is the *iterative elimination* of (weakly) dominated strategies. That is, a strategy may dominate some of the other strategies of the same player. If the player is rational, he or she will not play a dominated strategy. By iteratively eliminating the dominated strategies, we may hope to end up with a solution of the game. There are no dominated strategies to be eliminated in our example.

A notion which turns out to be useful in our case is that of *equilibrium*. Informally, a pair of strategies (σ, τ), where $\sigma \in S_\exists$ and $\tau \in S_\forall$ is an equilibrium if none of the players regrets his or her choice when these strategies are revealed. In our case there are no equilibria in either one of the games. For consider the pair of strategies (a, b) in the first game. After this pair is revealed, then Abelard would certainly regret his choice for he realizes that a better choice for him would have been a. And similarly for all the other pairs. Actually the indeterminacy of the sentences in our example matches the lack of an equilibrium in the underlying strategic games.

7.4.1 Equilibrium Semantics

A well known way to overcome the indeterminacy of certain games is to switch to mixed strategies, i.e., probability distributions over the players' pure strategies. Von Neumann's minimax theorem then shows that there is always an equilibrium in the game.

The idea to resolve the indeterminacy of IF sentences on finite models by appeal to von Neumann's theorem has been proposed for the first time (in the context of the branching quantifiers) by Blass and Gurevich (1986) following a suggestion by Ajtai. It has been studied in details for the first time in Sevenster (2006), developed in Sevenster and Sandu (2010) and Mann et al. (2011). Sandu (2012) gives an overview of the programme, and Sandu (2015, forthcoming) sketches some applications to Monty Hall and Lewis signalling problems. Here we shall keep the details to the minimum.

Let us fix a strategic IF game $\Gamma(\mathbb{M}, \varphi) = (S_\exists, S_\forall, u_\exists, u_\forall)$. A mixed strategy ν for player $p \in \{\exists, \forall\}$ is a probability distribution over S_p, that is, a function $\nu : S_p \to [0, 1]$ such that $\sum_{\tau \in S_i} \nu(\tau) = 1$. Given a mixed strategy μ for player $\exists$ and a mixed strategy ν for player $\forall$, the expected utility for player p is given by:

$$U_p(\mu, \nu) = \sum_{\sigma \in S_\exists} \sum_{\tau \in S_\forall} \mu(\sigma) \nu(\tau) u_p(\sigma, \tau).$$

Obviously a pure strategy is a mixed strategy which assign to it the probability 1 and assigns 0 to all the other pure strategies.

The pair (μ, ν), where μ is a mixed strategy for Eloise and ν is a mixed strategy for Abelard, is an equilibrium if the no regret condition above holds. More exactly:

1. For every mixed strategy μ' of Eloise, $U_\exists(\mu, \nu) \geq U_\exists(\mu', \nu)$
2. For every mixed strategy ν' of Abelard, $U_\forall(\mu, \nu) \geq U_\forall(\mu, \nu')$.

Von Neumann's Minimax Theorem (Von Neumann 1928) tells us that

- Every finite, constant sum, two-player game has an equilibrium in mixed strategies

A well known corollary of this theorem tells us that

- Every two such equilibria return the same expected utilities to the two players.

These two results ensure that we can talk about *the value of a strategic IF game* $\Gamma(\mathbb{M}, \varphi)$: it is the expected utility returned to player $\exists$ by any equilibrium in the strategic game $\Gamma(\mathbb{M}, \varphi)$. We can now define *the value of the IF sentence* φ in the finite model $\mathbb{M}$: it is the value of the strategic game $\Gamma(\mathbb{M}, \varphi)$. As a result each IF sentence receives a probabilistic value on every finite model.

The equilibrium semantics introduces a more fine grained structure on the space of indeterminate IF sentences. To see this, let us consider once again our earlier examples, φ_{MP} and φ_{IMP}. Let $\mathbb{M}$ be a set consisting of four elements, $\mathbb{M} = \{a_1, \ldots, a_4\}$ The strategic games may be displayed as

	a_1	a_2	a_3	a_4
a_1	$(1,0)$	$(0,1)$	$(0,1)$	$(0,1)$
a_2	$(0,1)$	$(1,0)$	$(0,1)$	$(0,1)$
a_3	$(0,1)$	$(1,0)$	$(1,0)$	$(0,1)$
a_4	$(0,1)$	$(0,1)$	$(0,1)$	$(1,0)$

	a_1	a_2	a_3	a_4
a_1	$(0,1)$	$(1,0)$	$(1,0)$	$(1,0)$
a_2	$(1,0)$	$(0,1)$	$(1,0)$	$(1,0)$
a_3	$(1,0)$	$(1,0)$	$(0,1)$	$(1,0)$
a_4	$(1,0)$	$(1,0)$	$(1,0)$	$(0,1)$

Let μ and ν be uniform probability distributions over $\{a_1, \ldots, a_4\}$. It may be shown that the pair (μ, ν) is an equilibrium in both games and that the value of φ_{MP} on $\mathbb{M}$ is $\frac{1}{4}$ and that of φ_{IMP} is $\frac{3}{4}$. Thus φ_{MP} gets value $\frac{1}{4}$ on $\mathbb{M}$ and φ_{IMP} gets value $\frac{3}{4}$. In the general case in which $\mathbb{M}$ has n elements, φ_{MP} gets value $\frac{1}{n}$ and φ_{IMP} gets value $\frac{n-1}{n}$.

7.5 Some Properties of the Equilibrium Semantics

Sevenster and Sandu (2010) show that the probabilistic interpretation of IF logic is a conservative extension of the old game-theoretical semantics in the following sense:

(E1) $\mathbb{M} \models^+_{Sk} \varphi$ iff the value of φ on $\mathbb{M}$ is 1
(E2) $\mathbb{M} \models^-_{SK} \psi$ iff the value of φ on $\mathbb{M}$ is 0.

From the way the equilibrium semantics has been defined, it is obvious that the maxim value is 1 and the minimum value is 0. What is less obvious, is that each rational number in the interval $(0, 1)$ is the value of an IF sentence on a certain model $\mathbb{M}$. There are various proofs of this result, e.g. Sevenster and Sandu (2010), Mann et al. (2011), and Barbero and Sandu (2014). The variant we prefer uses the IF sentence φ_{sig} :

$$\forall x \exists y (\exists z / \{x\})(S(x) \rightarrow \Sigma(y) \wedge R(z) \wedge z = x).$$

Theorem (Barbero and Sandu 2014) *For every integers m, n such that $0 \leq m < n$ there is a model $\mathbb{M} = (M, S^M, \Sigma^M, R^M)$ such that the value of φ_{sig} on $\mathbb{M}$ is $\frac{m}{n}$.*

For m and n as in the statement of the theorem, the model $\mathbb{M}$ has the form $\mathbb{M} = (M, S^M, \Sigma^M, R^M)$ where $M = \{s_1, \ldots, s_n, t_1, \ldots, t_m\}$, $R^M = S^M = \{s_1, \ldots, s_n\}$ and $\Sigma^M = \{t_1, \ldots, t_m\}$.

Intuitively, the sentence φ_{sig} models a "signaling" problem inspired from Lewis' work on conventions (Lewis, 1969):

– S represents a set of states
– Σ represents a set of signals (messages) that a Sender can send
– R represents a set of action that a Receiver can perform (after receiving a signal)

It can be shown that on every $\mathbb{M}$ which satisfies the conditions in the statement of the theorem we have: $\mathbb{M} \not\models^+ \varphi_{sig}$ and $\mathbb{M} \not\models^- \varphi_{sig}$.

Other variants of the above theorem may be found in Sevenster and Sandu (2010) and Sandu (2015, forthcoming).

Let $NE(\varphi, s, \mathbb{M})$ denote the value of the IF sentence φ on the finite model $\mathbb{M}$ with respect to the assignment s. Mann et al. (2011) produce a toolkit for computing $NE(\varphi, \mathbb{M})$. They prove, among other things, the following:

P1. $NE(\varphi \vee \psi, s, \mathbb{M}) = max(NE(\varphi, s, \mathbb{M}), NE(\psi, s, \mathbb{M}))$
P2. $NE(\varphi \wedge \psi, s, \mathbb{M}) = min(NE(\varphi, s, \mathbb{M}), NE(\psi, s, \mathbb{M}))$
P3. $NE(\neg \varphi, s, \mathbb{M}) = 1 - NE(\varphi, s, \mathbb{M})$.
P4. $NE(\exists x \varphi, s, \mathbb{M}) = max\{NE(\varphi, s(x/a), \mathbb{M}) : a \in M\}$
P5. $NE(\forall x \varphi, s, \mathbb{M}) = max\{NE(\varphi, s(x/a), \mathbb{M}) : a \in M\}$

We prefer an example borrowed from Mann et al. (2011) which illustrates how we compute the value of the IF sentence ψ

$$\exists u \forall w (u \neq w \vee \varphi_{MP})$$

on a finite model-set $\mathbb{M}$ with n elements. P4 and P5 tell us that

$$NE(\psi, \mathbb{M}) = max_a min_b \{NE(u \neq w \vee \varphi_{MP}, (u, a), (w, b), \mathbb{M}) : a, b \in M\}$$

By P1, we know that

$$NE(u \neq w \vee \varphi_{MP}, (u, a), (w, b), \mathbb{M}) \qquad =$$
$$max(NE(u \neq w, (u, a), (w, b), \mathbb{M}), NE(\varphi_{MP}, (u, a), (w, b), \mathbb{M}))$$

We know already that $NE(\varphi_{MP}, (u, a), (w, b), \mathbb{M})$ is $\frac{1}{n}$. By P1 we get

$$NE(u \neq w \vee \varphi_{MP}, (u, a), (w, b), \mathbb{M}) = max(NE(u \neq w, (u, a), (w, b), \mathbb{M}), \frac{1}{n})$$

From (E1) and (E2) we know that $NE(u \neq w, (u, a), (w, b), \mathbb{M})$ is 1 if $a \neq b$ and 0 otherwise. Thus for a fixed a, $min_b \{NE(u \neq w \vee \varphi_{MP}, (u, a), (w, b), M) : b \in M\}$ is reached when w is a and this minimum is $\frac{1}{n}$. We conclude that $max_a min_b$ is $\frac{1}{2}$.

In conclusion, each IF sentence which is indeterminate on a finite model receives a probabilistic value via von Neumann's MiniMax Theorem. It is now time to ask: What kind of probabilistic logic is probabilistic IF logic?

7.6 Probabilistic Semantics: Degrees of Rational Belief

There are various ways to combine logic and probabilities. One way is to leave the logical language intact but to add probabilistic features to the semantics. In this way we get a probabilistic semantics for the underlying language. We shortly review here a probabilistic semantics for the language of predicate logic (Leblanc 1994). It is presented as an alternative to the standard, model-theoretical semantics for the language of predicate logic. Other alternatives include substitutional semantics and truth-value semantics. All three interpret quantifiers substitutionally, that is, an existentially quantified statement is true if and only it has at least one true substitution instance, etc. As Leblanc emphasizes, the substitutional interpretation retains models, but only Henkin-models, that is, those models where each individual in the domain of discourse has a name in the language. The other two interpretations dispense with models altogether: truth-value semantics replaces models with truth-value assignments (functions) and probabilistic semantics replaces models with probability functions. Thus

> ...reference, central to standard semantics, is no concern at all of truth-value and probabilistic semantics; and truth, so central to standard semantics, is but a marginal concern of probabilistic semantics. (Leblanc 1994, p. 189).

Despite the fact that each semantics explicates logical entailment and logical truth in its own way, in all cases we have:

- A sentence is logically entailed by a set of sentences if and only if it is provable from that set
- A sentence is logically true if and only if it is provable.

In the remaining of this section, I will follow Leblanc (1994). Let L be a first-order language that in addition to the usual stock of predicate symbols contains also an infinite number of individual terms. The logical primitives of L are negation, conjunction and the universal quantifier; disjunction, implication and the existential quantifier are defined in the standard way. A term extension L^+ of L denotes any extension of L with countably many terms, that is, finitely many or $\aleph_0$ many terms in addition to those of L. (Note that by this definition L has 0 terms besides its own and it is thus its own term extension). We associate with L^+ a probability function P^+ that takes single sentences of L^+ as arguments (in the same way as truth-value functions do) and yield real numbers from the entire interval $[0, 1]$ as values. The classical truth values 1 and 0 are assimilated to the endpoints of the interval $[0, 1]$. A sentence's probability value is seen as a measure of its (un)certainty, with 0 representing maximal uncertainty and 1 maximal certainty.

Let L^+ be a term extension of L. A probability function for L^+ is any function P^+ from the sentences of L^+ to the reals such that the following constraints are satisfied:

C1. $P^+(\varphi) \geq 0$, for all $\varphi \in L$

C2. $P^+(\neg(\varphi \wedge \neg\varphi)) = 1$

C3. $P^+(A) = P^+(A \wedge B) + P^+(A \wedge \neg B)$

C4. $P^+(A) \leq P^+(A \wedge A)$

C5. $P^+(A \wedge B) \leq P^+(B \wedge A)$

C6. $P^+(A \wedge (B \wedge C)) \leq P^+((A \wedge B) \wedge C)$

C7. $P^+(A \wedge \forall x B) = Limit_{j \to \infty}(\varphi \wedge \Pi_{i=1}^{j}\psi(t_i^+/x))$

C1 and C2 are the counterparts of the axioms known as non-negativity and unit normalization in Kolmogorov (1933). C3–C6 are borrowed from Popper (1955). C7 is an adaptation by Bendall of an axiom in Gaifman (1964) which uses minima instead of limit (Cf. Leblanc 1994).

A sentence φ of L is defined to be *logically true* in the probabilistic sense if no matter what the term extension L^+ of L and probability function P^+ for L^+ we have $P^+(\varphi) = 1$. φ is *logically entailed* in the probabilistic sense by a set of sentences S of L if no matter the term extension L^+ of L and probability function P^+ for L^+ we have $P^+(\varphi) = 1$ if $P^+(\psi) = 1$ for every $\psi \in S$. (Equivalently but more simply, φ is logically true if φ takes the value 1 on every probability function for L).

In considering other alternative constraints, Leblanc discusses two additional principles mentioned by Kolmogorov as Additivity and Continuity. The sentence-theoretical counterpart of the latter is relevant for infinite disjunctions and thus does not apply here. And the statement-theoretical counterpart of the former is:

C8. If two statements φ and ψ of L^+ are logically incompatible in the standard sense, i.e., $\models \neg(\varphi \wedge \psi)$, then $P^+(\varphi \vee \psi) = P^+(\varphi) + P^+(\psi)$.

In addition, Leblanc also discusses the following principle of *Substitutivity*

C9. If φ and ψ are logically equivalent in the standard sense (i.e., $\models \varphi \leftrightarrow \psi$), then $P^+(\varphi) = P^+(\psi)$

that he finds quite natural in the context of logic. He shows (Theorem 4.40) that C1–C2 together with C8–C9 pick up the same probability function as C1–C6.

We are also told that Popper (1955) uses in place of C1–C2 the following three constraints:

C10. $P^+(\varphi \wedge \psi) \leq P^+(\varphi)$ (Monotonicity)
C11. There are at least two distinct statements φ and ψ of L^+ such that $P^+(\varphi) \neq P^+(\psi)$ (Existence).
C12. For each statement φ of L^+ there is a statement ψ of L^+ such that $P^+(\varphi) \leq P^+(\psi)$ and $P^+(\varphi \wedge \psi) = P^+(\varphi) \times P^+(\psi)$.

Drawing on earlier work, Leblanc (1994) mentions that C3–C6 and C10–C12 pick out the same probability function as C1–C6.

Among all these alternatives, we focus here on the system of constraints C1–C7. Here are few theorems in Leblanc (1994):

* The commutativity, associativity and idempotency (i.e., $P^+(\varphi) = P^+(\varphi \wedge \varphi)$) of conjunction,
* The probability of a conjunction is less or equal to the probability of each conjunct,
* Contradictions ($\varphi \wedge \neg\varphi$) get probability 0 and $\neg\varphi$ gets probability $1 - P^+(\varphi)$,
* $P^+(\varphi \rightarrow \psi) = 1$ iff $P^+(\varphi \wedge \neg\psi) = 0$ iff $P^+(\varphi) = P^+(\varphi \wedge \psi)$,
* The axioms of proposition logic get probability 1 and the rule of Modus Ponens holds, i.e., if $P^+(\varphi) = 1$ and $P^+(\varphi \rightarrow \psi) = 1$ then $P^+(\psi) = 1$
* All theorems (tautologies) of propositional logic gets probability 1,
* The principle of substitutivity (C9): if $\varphi \leftrightarrow \psi$ is a theorem of propositional logic, then $P^+(\varphi) = P^+(\psi)$,
* The axioms of predicate logic evaluate to 1,
* The soundness theorem of predicate logic for probabilistic semantics: If $S \vdash \varphi$ then no matter the term extension L^+ of L and probability function P^+ for L^+ we have that $P^+(\varphi) = 1$ if $P^+(\psi) = 1$ for each $\psi \in S$, that is, φ is logically entailed in the probabilistic sense.

7.7 Probabilistic Semantics: Statistical Knowledge (Relative Frequency)

The main ideas have been introduced in Bacchus (1990) and Halpern (1990). In the preceding case the syntax remained standard. In the present case, both the syntax of predicate logic is enriched with probabilistic features. The basic logical primitives are negation, disjunction and the existential quantifier. In addition to the standard syntactical clauses for first-order formulas, we have a new syntactical clause:

* When φ is a standard first-order formula and q is a rational number in the interval $[0, 1]$, then $Px(\varphi) \geq q$ is a formula with the intended interpretation: "the probability of selecting an x such that x satisfies φ is at least q". $Px(\varphi) = q$ is an abbreviation of

$$Px(\varphi) \geq q \wedge Px(\varphi) \leq q.$$

Here every free occurrence of x in φ is bound by the operator Px.

On the semantical side, we keep models (and the notion of reference), unlike in the previous framework, but we enrich them with a probability distribution over the individuals in the universe. That is, models are now triples $M = (D, I, P)$, where the domain of discourse D and the interpretation function I are standard, and P is a probability function such that $\Sigma_{d \in D} P(d) = 1$. Thus probabilities are associated in the first place with the individuals of the universe. As the result, each term of the form $Px(\varphi)$ receives a probabilistic value in a way to be sketched below.

First, ordinary first-order formulas are interpreted in the model $M = (D, I, P)$ relative to an assignment g in the standard way.

Second, we have one additional clause:

$$M, g \models Px(\varphi) \geq q \text{ iff } \Sigma_{d \in D: M, g[x \to d] \models \varphi} \geq q.$$

That is, the probability of selecting an x that satisfies φ is the sum of the probabilities of the elements in the universe that satisfy φ. To get a better grasp on what is going on, here is an example. Consider a model whose universe contains nine marbles: five are black and four are white. Let us assume that P assigns a probability of $1/9$ to each marble, which captures the idea that one is equally likely to pick any marble. Suppose the language contains a unary predicate B whose interpretation is the set of black marbles. The sentence $PxB(x) = 5/9$ is true in this model regardless of the assignment.

It is easy to see that the following principles hold (are true in every probabilistic model):

Ax1 $Px(\varphi) \geq 0$
Ax2 $Px(\varphi) + Px(\neg\varphi) = 1$
Ax3 $Px(\varphi) + Px(\psi) \geq Px(\varphi \vee \psi)$
Ax4 $Px(\varphi \wedge \psi) = 0 \to Px(\varphi) + Px(\psi) = P(\varphi \vee \psi).$

Ax1 follows straight from the definitions. Ax2 follows from the fact that negation is classical, complementary negation and the fact that the probabilities of the individuals in the universe sum up to 1. Ax2. Ax3 and Ax4 are also straightforward: when the set of individuals satisfying φ is disjoint from the set of individuals satisfying ψ, then the probability of each individual is taken into account, and we get Ax4. If they are not disjoint, then the probability of the individuals which are in both sets are counted only once, and we get Ax3. Notice also that by Ax2 and Ax4, $Px(\varphi \vee \neg\varphi)$ is 1. Similar argument shows that $Px(\varphi \wedge \neg\varphi) = 0$.

The axioms (Ax1)–(Ax4) are known as *Kolmogorov axioms of probabilities.*

We notice that if $\varphi(x)$ and $\psi(x)$ are logically equivalent, then for every model $\mathbb{M}$, the set of individuals satisfying φ in $\mathbb{M}$ is the same as the set of individuals satisfying ψ. Thereby $Px\varphi$ and $Px\psi$ must be identical in $\mathbb{M}$. In other words, the counterpart of the Principle of Substitutivity (C9) holds for $Px\varphi$. Given that $Px\varphi$ satisfies Ax1 and Ax2 which are the counterparts of (C1) and (C2), and it satisfies

Ax4 which is the counterpart of (C8), we conclude on the basis of Theorem 4.40 in (Leblanc 1994), that $Px\varphi$ satisfies the counterparts of constraints (C1)–(C6).

As a final observation, we should notice also that when φ is a sentence of the underling first-order language, then Px does not bind any free variable in φ. In that case, either φ is true in the underlying model $\mathbb{M}$ or it is false. In the former case, $\{d \in D : M, g\,[x \mapsto d] \models \varphi\} = D$ and thus $Px(\varphi) = 1$. In the latter case, a similar argument shows that $Px(\varphi) = 0$. In other words, for sentences the only values available are 0 or 1. (Cf. Bacchus (1990), and Halpern (1990))

At this moment let us get back to the difference between the two probabilistic interpretations introduced so far. Interpretationally the difference between $Px\varphi$ and $P\varphi$ (we dropped the subscript) is illustrated by the pair

1. The probability that a randomly chosen brick is black is $\frac{5}{9}$
   and
2. The probability that b (a particular brick) is black is $\frac{5}{9}$.

(1) as we have seen, is rendered by $PxB(x) = 5/9$. But (2) does not have a free variable to be bound by Px and cannot, for this reason, by modelled in the present framework. We could of course prefix it with Px but our remarks above show that the only values that the resulting sentence could take are 0 or 1. This is not what (2) says. I guess that this is one of the main reasons why (2) is taken to illustrate the degree of belief conception (Halpern, 1990, see also our conclusion).

Bacchus (1990) and Halpern (1990) give extensions where the probability operator P quantifies over more than one variable.

7.8 IF Probabilities: Some Comparisons

It is perhaps easier to say what IF probabilities do not express. They do not express the statistical probability of an individual x to have a property φ, for the simple reason that probabilistic IF logic does not have the syntactical resources to do so: unlike$Px\varphi$, $NE(\varphi, \mathbb{M})$ is a concept in the metalanguage. We could lift $NE(\varphi, \mathbb{M})$ to the object language by enriching the syntax with clauses of the form $NE(\varphi) = r$. The corresponding semantical clause would be:

- $\mathbb{M} \models NE(\varphi) = r$ if and only if $NE(\varphi, \mathbb{M}) = r$.

Even in such an extension, $NE(\varphi)$ behaves like a sentential and not like a variable binding operator. Syntactically $NE(\varphi)$ resembles more the degree of belief operator $P(\varphi)$.

On the other side, the interpretation of $NE(\varphi)$, presupposes, like the interpretation of $Px\varphi$, models and model-theoretical semantics. Both $NE(\varphi)$ and$Px\varphi$ are relativized to models. There is, however, one basic difference between the two, apart from the expressive power of the underlying language: probability distributions in IF logic are not over the individuals of an underlying universe but over the

Skolem functions and Kreisel counter-examples in that universe. Thus we can say that IF probabilistic logic combines features from both the degree of belief and the statistical frequency approaches. Let us look at some of the details.

First we look at the counterparts of (Ax1)–(Ax4) obtained by replacing $Px\varphi$ by $NE(\varphi)$.

Using (P1)–(P3) it is easy to show that all of them are valid, that is, hold in every model. For instance, for the counterpart of (Ax4), suppose that $NE(\varphi \wedge \psi) = 0$ is true in a model $\mathbb{M}$. Then by (E2), $\mathbb{M} \models^-_{Sk} \varphi \wedge \psi$. From the semantics of IF logic, it follows that either $\mathbb{M} \models^-_{Sk} \varphi$ or $\mathbb{M} \models^-_{Sk} \psi$. If the former, then we infer using (E2) and (P1) that

$$NE(\varphi) = 0 = max\{NE(\varphi), NE(\psi)\} = NE(\varphi \vee \psi)$$

holds in $\mathbb{M}$. The case $\mathbb{M} \models^-_{Sk} \psi$ is similar.

Thus *probabilistic IF logic obeys Kolmogorov's axioms.*

We now look at the counterparts in IF logic of the constraints (C1)–(C6) keeping in mind Leblanc's observation to the effect that Kolmogorov's axiom (C8), the Principle of Substitutivity (C9) and the constraints (C1)–(C2) pick up the same probability function as (C1)–(C6).

When φ is an ordinary (slash-free) first-order sentence, it follows from (E1) and (E2) that the only probabilistic values taken by φ are 0 and 1. In this case the game-theoretical negation becomes classical, ordinary negation and all the counterparts in IF logic of the constraints (C1)–(C6) are easily shown to be satisfied.

It is the phenomenon of independence which leads to indeterminacy in IF logic and to the violation of the counterpart of Principle of Substitutivity (C9). The counterpart of this principle in IF logic says that if two IF sentences are logically equivalent, then their probabilistic values are the same in every model. In symbols:

Substitutivity   If $\models \varphi \equiv \psi$ then for every model $\mathbb{M}$:

$$M \models NE(\varphi) = NE(\psi)$$

where $\models \varphi \equiv \psi$ is defined as

$$\mathbb{M} \models^+ \varphi \text{ iff } \mathbb{M} \models^+ \psi$$

for every model $\mathbb{M}$. (Recall we exclude 1-element models).

Now take $\varphi = \forall x (\exists y / \{y\}) x = y$ and $\psi = \exists y \forall x x = y$. We recall that φ fails to be true in any model and so does ψ. Let $\mathbb{M}$ be any model set with two elements. Then, as pointed out earlier:

$$\mathbb{M} \models NE(\varphi) = \tfrac{1}{2} \text{ and } \mathbb{M} \models NE(\psi) = 0.$$

(Here we have used (E2).)

One could object to the counter-example on the basis of the fact that our notion of logical equivalence does not cover one element models. It can be shown, however, that even with this broader notion of logical equivalence, substitutivity fails. To see this, consider the IF sentences φ_{MP} and $(\varphi_{IMP} \vee \forall x \forall y x = y)$. In a 1-element model both φ_{MP} and $\forall x \forall y x = y$ are true. Hence $(\varphi_{IMP} \vee \forall x \forall y x = y)$ is also true. On the other side, on any model with strictly more than one element, φ_{IMP} is indeterminate and $\forall x \forall y x = y$ is false, which makes $(\varphi_{IMP} \vee \forall x \forall y x = y)$ indeterminate. We already know that in such a model φ_{MP} is indeterminate too. So the antecedent of the Substitutivity principle is satisfied for the broader notion of logical equivalence. (I owe this insight to Fausto Barbero) Now let $\mathbb{M} = \{a_1, \ldots, a_4\}$. By P1:

$$\mathbb{M} \models NE(\varphi_{IMP} \vee \forall x \forall y x = y) = max\,\{NE(\varphi_{IMP}),$$

$$NE(\forall x \forall y x = y)\} = max\left(\frac{3}{4}, 0\right) = \frac{3}{4}.$$

On the other side, we have seen that

$$\mathbb{M} \models NE(\varphi_{MP}) = \frac{1}{4}.$$

Substitutivity also fails for a stronger notion of logical equivalence, i.e., agreement on both truth and falsity on all models $\mathbb{M}$:

$$\mathbb{M} \models^+ \varphi \text{ iff } \mathbb{M} \models^+ \psi$$
$$\mathbb{M} \models^- \varphi \text{ iff } \mathbb{M} \models^- \psi.$$

φ_{MP} and φ_{IMP} will serve as a counterexample also in this case. Recall that both of them are indeterminate on all models (with at least 2-elements). Thus they are trivially strongly logical equivalent. Let $\mathbb{M} = \{a_1, \ldots, a_4\}$. On the other side, we know from our observations in Sect. 7.4 that the value of φ_{MP} and φ_{IMP} in $\mathbb{M}$ are $\frac{1}{4}$ and $\frac{3}{4}$ respectively.

It is easy to see that any indeterminate IF sentence φ is a counter-example to both $NE(\varphi \vee \neg\varphi) = 1$ and $NE(\varphi \wedge \neg\varphi) = 0$. For let $\mathbb{M}$ be a model set with two elements. Then

$$\mathbb{M} \models NE(\varphi_{MP} \vee \neg\varphi_{MP}) = max\,\{NE(\varphi_{MP}), NE(\neg\varphi_{MP})\} = max\left\{\frac{1}{2}, \frac{1}{2}\right\} = \frac{1}{2}$$

and

$$\mathbb{M} \models NE(\varphi_{MP} \wedge \neg\varphi_{MP}) = min\,\{NE(\varphi_{MP}), NE(\neg\varphi_{MP})\} = \frac{1}{2}.$$

But then we get immediately a counter-example to (C2). Let $\mathbb{M}$ a model set with two elements. Then

$$\mathbb{M} \models NE(\neg(\varphi_{MP} \wedge \neg\varphi_{MP})) = 1 - NE(\varphi_{MP} \wedge \neg\varphi_{MP}) = \frac{1}{2} \neq 1.$$

It is easy to see that φ_{MP} is also a counter-example to the validity of the counterpart of (C3):

$$NE(\varphi) = NE(\varphi \wedge \psi) + NE(\varphi \wedge \neg\psi).$$

Again, on any model $\mathbb{M}$ with two elements we have:

$$\mathbb{M} \models NE(\varphi_{MP} \wedge \varphi_{MP}) = min\,\{NE(\varphi_{MP}), NE(\varphi_{MP})\} = \frac{1}{2}.$$

We also know that

$$\mathbb{M} \models NE(\varphi_{MP} \wedge \neg\varphi_{MP}) = \frac{1}{2}.$$

Hence

$$\mathbb{M} \models NE(\varphi_{MP} \wedge \varphi_{MP}) + NE(\varphi_{MP} \wedge \neg\varphi_{MP}) = 1 \neq NE(\varphi_{MP}).$$

Let me point out that in probabilistic semantics the constraint $P(\neg\varphi) = 1 - P(\varphi)$ is obtained as a theorem using (C2) and (C3). Although the counterparts of (C2) and (C3) fail in IF logic, the counterpart of the principle $P(\neg\varphi) = 1 - P(\varphi)$ still holds, as witnessed by (P3).

Let me finally say something about implication in probabilistic IF logic, where $\varphi \rightarrow \psi$ is defined as $\varphi\neg \vee \psi$.

Both the degrees of belief and the statistical frequency systems satisfy the following principles:

1. $P(\varphi \rightarrow \psi) = 1$ iff $P(\varphi \wedge \neg\psi) = 0$
2. $P(\varphi \rightarrow \psi) = 1$ iff $P(\varphi) = P(\varphi \wedge \psi)$
3. $P(\varphi \rightarrow \varphi \wedge \varphi) = 1$
4. $P(\varphi \wedge \psi \rightarrow \varphi) = 1$
5. $P(\varphi \wedge \psi \rightarrow \varphi) = 1.$

For a proof of these statements on the basis of (C1–C6), see Leblanc (1994). The proof of all of them relies essentially on (C3). For instance, for the proof of (2) notice that by (C3), $P(\varphi) = P(\varphi \wedge \psi) + P(\varphi \wedge \neg\psi)$. Hence if $P(\varphi) = P(\varphi \wedge \psi)$, then $P(\varphi \wedge \neg\psi) = 0$ and from (1) we get the desired result, $P(\varphi \rightarrow \psi) = 1$.

Although the counterpart of (C3) fails in IF logic, the counterpart of (1) still holds in virtue of (E1) and (E2). But the counterparts of (2)–(5) are not valid, as may be easily seen taking $\varphi = \varphi_{MP}$ and $\psi = \varphi_{IMP}$.

7.9 Conclusions

All the three probabilistic systems introduced in this paper satisfy Kolmogorov's axioms for probabilities. If we consider only the first-order subfragment of IF logic, then $P\varphi, Px\varphi$ and $NE(\varphi)$ satisfy also the Substitutivity principle (C9) and thereby all the axioms (C6)–(C6). But the full language of IF violates the constraints (C9), and in particular the constraints (C2) and (C3).

The three systems are intended to model different phenomena.

We have already pointed out that $Px\varphi$ is intended to model statistical knowledge as illustrated by the sentence

1. The probability that a randomly chosen brick is black is $\frac{5}{9}$
   whereas $P\varphi$ is intended to express the degree of belief conception illustrated by
2. The probability that b (a particular brick) is black is $\frac{5}{9}$.

It is not quite clear how (2) could be analyzed in Leblanc's probabilistic semantics, apart from simply assigning it the primitive probabilistic value $\frac{5}{9}$. Bacchus (1990) and Halpern (1990) propose a more interesting framework in terms of possible worlds and a probability distribution over them. In this new framework, the value of $P\varphi$ is the sum of the probabilities of the possible worlds in which φ is true.

$NE(\varphi)$ arises in the context of the interaction of standard quantifiers in sentence like $\forall x(\exists y/\{x\})x = y$ interpreted in a model set $\mathbb{M}$. We ask the question: what are the odds that when you draw a ball from a set $\mathbb{M}$ and, after you put it back, I will do the same, without seeing the ball you drew, the choices coincide? There is no uncertainity about the world, as in the other two frameworks, but uncertainty about what the other will do. The "odds" in this case are determined by certain rationality assumptions about the behaviour of the players.

References

Bacchus F (1990) LP, A Logic for representing and reasoning with statistical knowledge. Comput Intell 6:209–231

Barbero F, Sandu G (2014) Signalling in independence-friendly logic. J IGPL 22:638–664

Blass A, Gurevich Y (1986) Henkin quantiers and complete problems. Ann Pure Appl Logic 32:1–16

Gaifman H (1964) Concerning measures of first-order calculi. Isr J Math 2:1–18

Halpern J (1990) An analysis of first-order logics of probability. Artif Intell 46:311–350

Henkin L (1961) Some remarks on infinitely long formulas. In: Bernays P (ed) Infinitistic methods. Pergamon Press, Oxford/London/New York/Paris, pp 167–183

Hintikka J (1974) Quantifiers vs. quantification theory. Linguist Inq 5:153–177

Hintikka J (1996) The principles of mathematics revisited. Cambridge University Press, Cambridge, UK

Hintikka J, Sandu G (1989) Informational independence as a semantical phenomenon. In: Fenstad JE et al (eds) Logic, methodology and philosophy of science VIII. Elsevier, Amsterdam

Hodges W (1997) Compositional semantics for a language of imperfect information. J IGPL 5:539–563

Kolmogorov AN (1933) Grundbegriffe der Wahrscheinlichkeitsrechnung. Julius Springer, Berlin

Leblanc H (1982) Popper's 1955 axiomatization of absolute probability. Pac Philos Q 63:133–145

Leblanc H (1994) Alternatives to first-order semantics. In: Gabbay D, Guenther F (eds) Handbook of philosophical logic, vol I. Kluwer Academic Publishers, Dordrecht, pp 189–274

Lewis DK (1969) Convention: a philosophical study. Harvard University Press, Cambridge

Mann AL, Sandu G, Sevenster M (2011) Independence-friendly logic: a game-theoretic approach. London Mathematical Society, Lecture Note Series 396. Cambridge University Press, Cambridge

Popper KR (1955) Two autonomous axiom systems for the calculus of probabilities. Br J Philos Sci 6:51–57, 176, 351

Sandu G (1996) IF first-order logic, Kripke, and 3-valued logic. Appendix in Hintikka, pp 254–270

Sandu G (2011) Game-theoretical semantics. In: Horsten L, Pettigrew R (eds) The continuum companion to philosophical logic. Continuum International Publishing Group, pp 216–270

Sandu G (2012) Independence-friendly logic: dependence and independence of quantifiers in logic. Philos Compass 7(10):691–711

Sandu G (2015, forthcoming) Languages with imperfect information. In: Gosh S, Verbrugge R, van Benthem J (eds) Models of strategic reasoning: logics, games and communities. FoLLI-LNAI State-of-the-Art Survey, LNCS 8972. Springer, Heidelberg

Sevenster M (2006) Branches of imperfect information: logic, games, and computation. PhD thesis, University of Amsterdam, Amsterdam

Sevenster M, Sandu G (2010) Equilibrium semantics of languages of imperfect information. Ann Pure Appl Logic 161(5):618–631

Von Neumann J (1928) Zur Theorie der Gesellschaftsspiele. Math Ann 100:295–320

Chapter 8
A Remark on a Relational Version
of Robinson's Arithmetic Q

Mihai Ganea

8.1 Why Study Relational Arithmetical Languages and Theories?

The standard first-order arithmetical language L_A has the primitive symbols $\{\overline{0}, s, +, \cdot\}$, where $\overline{0}$ is an individual constant and $s, +, \cdot$ are symbols for successor, addition and multiplication (see Mendelson (2001), pp. 154–5). The less-than relation (symbolized by $\leq$) is sometimes taken as primitive and sometimes is introduced by the convention $t \leq u =_{df} \exists z\,(t + z = u)$, where z is the first variable not occurring in the terms t, u. Other minor variations are possible, such as the use of the individual constant $\overline{1}$ instead of the symbol for successor (this occurs when the natural numbers are identified as the non-negative part of a discretely ordered ring). This language and the axiomatic theories expressed in it have been subjected to intensive study (see Hájek and Pudlák 1991; Kossak and Schmerl 2006) for surveys of some of the vast body of knowledge accumulated).

A relational arithmetical language is obtained from L_A by substituting some of its functional symbols with symbols for the graphs of those functions or for other arithmetical relations. I will write $L_A[f_i/R_i]$ for the language obtained by such a substitution, where R_i are the relation symbols replacing the function symbols f_i. Broadly speaking, a relational arithmetical theory T is a first-order theory expressed in a relational language L_T such that T does not prove that the relation symbols in L_T express total functions.

There are at least two reasons why relational languages and theories constitute interesting objects of study. The first is to identify formal theories that somehow escape Gödel's second incompleteness theorem. This is the goal of a series

M. Ganea (✉)
Department of Philosophy, University of Toronto, Toronto, Canada
e-mail: mihai.ganea@utoronto.ca

© Springer International Publishing Switzerland 2015
I. Pârvu et al. (eds.), *Romanian Studies in Philosophy of Science*, Boston Studies
in the Philosophy and History of Science 313, DOI 10.1007/978-3-319-16655-1_8

of papers by Willard, who managed to show that there exist consistent axiom systems expressed in $L_A[\cdot/M]$ that prove their own semantic tableaux consistency (see Willard 2005, 2006). Strengthening these results to consistency relative to Hilbert-style deduction encounters a difficulty pointed out by Solovay in a private communication (according to Willard 2006, p. 474). Solovay's result is the prelude to the theorem proved in Sect. 8.2: specifically he showed that any theory that can interpret the relational version of the theory Q described below cannot prove its Hilbert-style consistency.

A second reason for studying relational arithmetics is the search for natural theories of intermediate r.e. degree, recently re-emphasized in (Hart 2010):

> The r.e. degrees represent all possible theories, whether produced by people, or by Kantian rational beings very different from us, or never produced by any sentient beings. But all the deductive theories actually produced by people have been either decidable, or else have been of the same degree as elementary number theory and the first-order predicate calculus. The latter is the Turing highest of all the r.e. degrees; if we could decide theoremhood in any theory of that degree, then we could decide theoremhood in any axiomatic theory. Why should human reason always devise axiomatic theories only of the lowest or of the highest possible degree but never of any of the wealth of intermediate degrees? (p. 274)

Hart notes that the existence of theories of intermediate degrees is known since (Feferman 1957), but also that Feferman's examples are 'pretty artificial' (p. 234, fn. 25). In fact it is known that there exist *finitely* axiomatized theories of any r.e. degree (see Peretyat'kin 1997, Theorem 0.7.5, p. 13), but of those known none are 'natural'. Since the first-order theory of addition (Presburger arithmetic) is decidable and finitely axiomatizable (by the axioms of Q which describe addition recursively) and Q itself is of maximum r.e. degree, it seems that the search for Hart's desired theories should take place among the sub-theories of Q that extend its addition axioms. One way we could hope to achieve the required difficulty degree for such a theory would be to choose axioms that describe not functions, but graphs of possibly non-total functions. As we shall see below, this cannot be accomplished by replacing the multiplication function and its defining axioms by axioms characterizing its graph.

8.2   An Alternative Proof of a Result by Visser

Let Q^H be the first-order theory in the language $L_A[+/A, \cdot/M]$ with primitive symbols $\{\overline{0}, s, A, M\}$, where A and M are ternary relations (for the graphs of addition and multiplication), and the following axioms:

1) $A(x, y, u) \wedge A(x, y, v) \supset u = v$,
2) $M(x, y, u) \wedge M(x, y, v) \supset u = v$,
3) $s(x) \neq \overline{0}$,
4) $s(x) = s(y) \supset x = y$,
5) $x \neq \overline{0} \supset \exists y\, (x = s(y))$,
6) $A(x, \overline{0}, x)$,

7) $\exists u\,(A\,(x,y,u) \wedge z = s(u)) \equiv A\,(x,s(y),z)$,

8) $M\,(x,\bar{0},\bar{0})$,

9) $\exists u\,(M\,(x,y,u) \wedge A\,(u,x,z)) \equiv M\,(x,s(y),z)$.

This relational version of Robinson's arithmetic Q is due to Hájek (or rather it is his interpretation of a suggestion by Grzegorczyk[1]). In what follows I will detail an observation attributed to Albert Visser in (Švejdar 2009a, p. 94), namely that there exists an interpretation of Q in Q^H that avoids cut shortening, unlike the interpretation of Q in Q^- given in (Švejdar 2007). Together with the fact that the proof of Theorem 4 in (Ganea 2009) actually yields an interpretation of Q^H in TC^+, this result gives an answer to the 'historical puzzle' stated in Sect. 4 of (Ganea 2009): what proof of the interpretability of Q in F (announced by Tarski in 1953) was possible before Solovay's invention of cut shortening in 1976?

The idea of the interpretation described below is different than that attributed to Visser and consists in shifting the arithmetical structure described by Q^H to the positive numbers, thus freeing 0 to play the role of the 'undefined' value for the addition and multiplication functions.

Definition

a) $\bar{0}^* = x \equiv_{df} s\,(\bar{0}) = x$,

b) $s^*(x) = y \equiv_{df} \left(x = \bar{0} \wedge y = \bar{0}\right) \vee \left(x \neq \bar{0} \wedge y = s(x)\right)$,

c) $p(x) = y \equiv_{df} \left(x = \bar{0} \wedge y = \bar{0}\right) \vee \left(x \neq \bar{0} \wedge x = s(y)\right)$,

d) $x + y = z \equiv_{df} \left[\left(x = \bar{0} \vee y = \bar{0}\right) \wedge z = \bar{0}\right] \vee$
$$\left\{\left(x \neq \bar{0} \wedge y \neq \bar{0}\right) \wedge \left[\exists u\,(A\,(p(x),p(y),u) \wedge z = s(u)\right) \vee$$
$$\left(\neg\exists u A\,(p(x),p(y),u) \wedge z = \bar{0}\right)\right]\right\},$$

e) $x \cdot y = z \equiv_{df} \left[\left(y = \bar{0}^* \wedge z = \bar{0}^*\right)\right.$
$$\left.\vee \left(\left(\left(x = \bar{0} \vee y = \bar{0}\right) \wedge y \neq \bar{0}^*\right) \wedge z = \bar{0}\right)\right] \vee$$
$$\left\{\left[\left(x \neq \bar{0} \wedge y \neq \bar{0}\right) \wedge y \neq \bar{0}^*\right] \wedge \left[\exists u\,(M\,(p(x),p(y),u) \wedge z = s(u)\right)\right.$$
$$\left.\vee \left(\neg\exists u M\,(p(x),p(y),\ u) \wedge z = \bar{0}\right)\right]\right\}.$$

In other words, $\bar{0}$ is the dominant argument for addition (if either argument is undetermined, then so is the sum), whereas for multiplication $\bar{0}^*$ is dominant in the second position (a product is $\bar{0}^*$ if the second argument is $\bar{0}^*$, regardless of whether the first argument is undetermined or not).

---

[1] I neglect the last axiom of the 'official' version of Q^H, which is simply a notational convention regarding the relational symbol '$\leq$'. See (Švejdar 2007) for an account of how the study of relational versions of Q was initiated and (Ganea 2009; Švejdar 2009a) for the links between these theories and the theories of concatenation TC and F. It is interesting to note that the events described by Švejdar apparently had no direct connection with the Willard-Solovay correspondence regarding the link between relational theories and the second incompleteness theorem.

Theorem Q^H verifies the axioms of Q for the language $\left\{\overline{0}^*, s^*, +, \cdot\right\}$.

Proof The domain of the interpretation will be universal, i.e. given by the formula $x = x$. It is obvious that p and s^* are provably well-defined as functions and that one can prove in Q^H the formulas $s^*(x) \neq \overline{0}^*, s^*(x) = s^*(y) \supset x = y$ and $x \neq \overline{0}^* \supset \exists y \, (x = s^*(y))$, i.e. the axioms of Q for successor.

It is also clear that the relation $x + y = z$ is provably functional: a z satisfying it exists for every choice of x, y and it is also unique. If $x = \overline{0}$ or $y = \overline{0}$, then $x + y = z$ if and only if $z = \overline{0}$ by (d). If $x \neq \overline{0}$ and $y \neq \overline{0}$, then either $\exists u A \, (p(x), p(y), u)$ or $\neg \exists u A \, (p(x), p(y), u)$. In the first case, u is unique by (1) and the functionality of p, so take $z = s(u)$, also unique by the functionality of s. In the second case, $z = \overline{0}$.

Q^H also proves that $x + \overline{0}^* = x$. If $x = \overline{0}$, then $\overline{0} + \overline{0}^* = \overline{0}$. If $x \neq \overline{0}$, then $x = s(u)$ for some u by (5), $A\left(u, p\left(\overline{0}^*\right), u\right)$ by (6) and therefore $x + \overline{0}^* = x$ by (d).

If $x = \overline{0}$ then $x + s^*(y) = s^*(x + y) = \overline{0}$ by (b) and (d). If $y = \overline{0}$, then $s^*(y) = \overline{0}$ by (b) and $x + s^*(y) = \overline{0} = x + y = s^*(x + y)$ by (d) and (b). Suppose then $x \neq \overline{0}, y \neq \overline{0}$ and $A(p(x), p(y), z)$ for some z. By (7), this means that we have $A(p(x), y, s(z))$, and therefore by (b) and (d) we have $s^*(x + y) = s(s(z)) = x + s(y) = x + s^*(y)$. If $x \neq \overline{0}, y \neq \overline{0}$, and there is no z such that $A(p(x), p(y), z)$, then $x + y = \overline{0} = s^*(x + y)$. It also follows that there is no u such that $A\left(p(x), p\left(s^*(y)\right), u\right)$, since otherwise we would have $A(p(x), p(s(y)), u)$ by (b), $A(p(x), y, u)$ by (c), $A(p(x), s(p(y)), u)$ since $y \neq \overline{0}$, and $A(p(x), p(y), p(u))$ by (7) and (c). Hence $x + s^*(y) = \overline{0} = s^*(x + y)$, and the second addition axiom of Q is verified.

Q^H verifies that (e) defines a proper function through a reasoning similar to the case of the addition function: definition (e) distinguishes exclusive and exhaustive combinations of arguments and ensures that in each case there exists a unique value for the function. In particular, if $y = \overline{0}^*$, then $x \cdot y = \overline{0}^*$, and so the first multiplication axiom of Q is verified.

Only the last multiplication axiom remains:

(i) If $y \neq \overline{0}^*$, then either

    (ia) $x = \overline{0}$ or

    (ib) $x \neq \overline{0}$. In the first sub-case, $x \cdot s^*(y) = \overline{0}$ by (e), given that $s^*\left(\overline{0}^*\right) = s\left(\overline{0}^*\right) \neq \overline{0}^*$ and also $x \cdot y + x = \overline{0}^* + \overline{0} = \overline{0}$ by (d) and (e). In the second sub-case, let $x = s(z), z = p(x)$. Suppose that $\exists u \left(M \left(p(x), p\left(s^*\left(\overline{0}^*\right)\right), u\right)\right)$. Then by (e) we have $x \cdot s^*(y) = s(u)$. But $M \left(p(x), p\left(s^*\left(\overline{0}^*\right)\right), u\right)$ entails $M\left(z, s\left(\overline{0}\right), u\right)$ by (a) and (b), and $\exists v \left(M \left(z, \overline{0}, v\right) \wedge A \left(v, z, u\right)\right)$ by (9). Since $v = \overline{0}$ by (8), it follows that $A\left(\overline{0}, z, u\right)$ and $x \cdot \overline{0}^* + x = s(u) = x \cdot s^*(y)$ by (d). If $\neg\exists u\left(M\left(z, p\left(s\left(\overline{0}\right)\right), u\right)\right)$, then $\neg\exists v A\left(\overline{0}, z, v\right)$ by (9) and so $x \cdot s^*\left(\overline{0}^*\right) = \overline{0} = x \cdot \overline{0}^* + x$ by (d) and (e).

(ii)   If $y = \overline{0}$, then $s^*(y) = \overline{0}$ by (b) and $x \cdot s^*(y) = \overline{0}$ by (e). But since $x \cdot y = \overline{0}$ as well, we have $x \cdot y + x = \overline{0} = x \cdot s^*(y)$. Suppose then

(iii) $y \neq \overline{0}^*, y \neq \overline{0}$. We have $s^*(y) = s(y)$ by (b) and $y = s(z)$ for some z by (5), $z \neq \overline{0}$.

   (iiia)   If $x = \overline{0}$ then have $x \cdot y + x = \overline{0} = x \cdot s^*(y)$ by (d) and (e)
   (iiib)   Suppose then that $x \neq \overline{0}, x = s(r), r = p(x)$. We have $M(p(x), p(s(y)), v)$ if and only if $M(r, s(z), v)$. By (2) and (9) such a unique v exists if and only if there exists a unique w such that $M(r, z, w)$ and $A(w, r, v)$. If v exists then $x \cdot y + x = x \cdot s^*(y) = s(v)$ since the products involved do not have any argument with a dominant value. If such a v does not exist, then $x \cdot s^*(y) = \overline{0}$ and either $x \cdot y = \overline{0}$ or $x \cdot y + x = \overline{0}$. Either way, we have $x \cdot s^*(y) = x \cdot y + x$ and thus the theorem is proved.

Inspection of Theorem 4 in (Ganea 2009) shows that what is in fact proved is that TC^+ interprets Q^H rather than the weaker variant Q^-. The only non-trivial point is the converse of the bi-conditional (9), but this is covered by the argument given in point (iv) of the proof. We thus obtain the desired interpretation of Q in F that does not rely on Solovay's technique. Furthermore, Solovay's strengthening of Pudlák's second incompleteness theorem described in (Willard 2006, p. 474) follows immediately: a theory interprets Q^H if and only if it interprets Q itself. Unfortunately, the theorem also indicates that no extension of Presburger arithmetic using a ternary relation M whose properties are compatible with those of the graph of multiplication can have intermediate r.e. degree.

References

Feferman S (1957) Degrees of unsolvability associated with classes of formalized theories. J Symb Log 22:161–175

Ganea M (2009) Arithmetic on semigroups. J Symb Log 74(1):265–278

Hájek P, Pudlák P (1991) Metamathematics of first-order arithmetic, Perspectives in mathematical logic. Springer, Berlin

Hart WD (2010) The evolution of logic. Cambridge University Press, Cambridge, UK

Kossak R, Schmerl J (2006) The structure of models of Peano arithmetic, vol 50, Oxford logic guides. Clarendon, Oxford

Mendelson E (2001) Introduction to mathematical logic, 4th edn. CRC Press, Boca Raton

Peretyat'kin M (1997) Finitely axiomatizable theories. Plenum, New York

Švejdar V (2007) An interpretation of Robinson arithmetic in its Grzegorczyk's weaker variant. Fundamenta Informaticae 81:347–354

Švejdar V (2009a) Relatives of Robinson arithmetic. In: Peliš M (ed) The Logica yearbook 2008. College Publications, London, pp 253–263

Švejdar V (2009b) On interpretability in the theory of concatenation. Notre Dame J Form Log 50(1):87–95

Willard E (2005) An exploration of the partial respects in which an axiom system recognizing only addition as a total function can verify its own consistency. J Symb Log 70(4):1171–1209

Willard E (2006) A generalization of the second incompleteness theorem and some exceptions to it. Ann Pure Appl Log 141:472–496

Chapter 9
The Simple Majority Rule in a Three-Valued Logic Framework

Adrian Miroiu

9.1 Introduction

In the past decade a growing number of papers on different issues in social choice theory appealed to formal techniques originating in various branches of modern formal logic.[1] First-order (Grandi and Endriss 2013) and higher-order logic (Nipkow 2009), modal logic (Ågotnes et al. 2009; Pauly 2007), many-valued and fuzzy logics (Barrett and Salles 2011) helped provide more rigorous formalizations of social choice problems and, as a result, gain a deeper understanding of the field (Endriss 2011).

The study of judgment aggregation, initiated by (List and Pettit 2002) is an exemplar case. It focuses on the way in which sets of judgments held by the members of a group can be aggregated to form a collective set of judgments. A judgment set is a subset of a given "agenda". The agenda is a set of propositions upon which a collective judgment is sought. An individual's judgment set contains exactly those propositions in the agenda that the individual believes to be true. The large literature on judgment aggregation usually assumes that the evaluations of the propositions allow for a proposition to be either true of false. However, an increasing

---

[1] Of course, logic had a much older role in modern social choice theory, going back to the early 1940s. In an interview published in Arrow et al. (2011). K.J. Arrow remembers an episode from his student years. In his last term he took a course on logic with the Polish logician A. Tarski. A testimony of the influence of Tarski on the young Arrow is to be found in the 1940 author preface to the English edition of *An Introduction to Logic* (Tarski 1959), where Tarski thanks K. J. Arrow for his help in reading proofs.

A. Miroiu (✉)
National School of Political and Administrative Studies, SNSPA, Bucharest, Romania
e-mail: ad_miroiu@yahoo.com

© Springer International Publishing Switzerland 2015
I. Pârvu et al. (eds.), *Romanian Studies in Philosophy of Science*, Boston Studies in the Philosophy and History of Science 313, DOI 10.1007/978-3-319-16655-1_9

number of papers focus on multi-valued evaluations: the truth values of a proposition range over a larger set, allowing for intermediate degrees of truth in between true and false. Dokow and Holzman (2010) work with non-binary aggregators that have multi-valued judgment aggregation as one natural interpretation. Pauly and van Hees (2006) and van Hees (2007) studied judgment aggregation in the framework of many-valued logic, drawing in particular on Post's many-valued systems. Duddy and Piggins (2013) used a Łukasiewicz-type multi-valued logic.

In this paper I shall follow a different approach to the use of many-valued logic in social choice. It was initiated nearly a half of century ago by Murakami (1966, 1968), who first considered logical mechanisms originating in many-valued logic in his pioneering study of representative decision-making.[2] On this approach, propositions do not describe issues, but attitudes of the individuals toward issues, like for example how to choose between two alternatives; thus, the atomic proposition p_1 is taken to describe the attitude of the individual A_1 with respect to the alternatives x and y, and p_2 to describe the attitude of the individual A_2 with respect to the same alternatives, etc. Attitudes can have two values: for example, A_1 either prefers x to y or prefers y to x; or can have multiple values, when intermediate cases are allowed (for example A_1 can be indifferent between x and y.)

The intuitive idea is that logical operators are similar to aggregation rules. An aggregation rule gives a social preference for each distribution of preferences of the members of a certain group. Similarly, logical operators like disjunction and conjunction give an aggregate truth-value for each distribution of truth-values of the propositions they connect. In general, if π is a binary logical operator, then $\pi(pq)$ gives the aggregate attitude of the group formed of the individuals A_1 and A_2 with respect to the two alternatives x and y.

Naturally, the question that immediately comes into one's mind is if logical operators corresponding to well-known aggregation rules like for example the simple majority rule or the absolute majority rule can be identified in a logical framework. Following this path, I rely on a Łukasiewiczian three-valued logic to define logical operators that can be easily compared with such aggregation rules like for example the simple and the absolute majority rules, the Jury rule or the extended Pareto rule.

On this account, compound propositions receive a quite different interpretation. Consider for example the proposition $\pi(\pi(p_1p_2)\pi(p_3p_4))$. Its meaning is this: by appealing to the aggregation rule π a collective decision of the group formed of the individuals A_1 and A_2 is reached. Similarly, a collective decision of the group formed of the individuals A_3 and A_4 is reached. Then these decisions are aggregated in a higher-order group formed of the two groups. It looks natural to try to interpret iterated applications of the decision rules in terms of representative systems or democracy. Murakami defined a representative system 'as a hierarchy of voting procedures, each of which may be called a council.Every individual casts a ballot

---

[2]Fine (1972) and Fishburn (1971) developed his approach.

or ballots in one council or councils. A decision in each council is represented in a higher council whose decision is, in turn, represented in a still higher council and so on' (Murakami 1966).

The paper is organized as follows. Section 9.2 presents the framework. The standard operators (negation, implication, disjunction and conjunction) in a Łukasiewicz-type three-valued logic are introduced. In Sect. 9.3 a new binary operator μ is also introduced; its intended meaning is that of the simple majority rule. Majority rule is usually studied in the general case, when the group of people who are to make a decision is large. The two-member groups are viewed as degenerate cases that do not deserve a special attention. Here I take the reverse path: I start with the binary rule and only then move to the general, n-ary case. The binary majority operator μ^2 has a very clear interpretation as a binary logical operator, just like the standard logical operators conjunction, disjunction and implication. In analogy with May's (1954) famous characterization of the majority rule, here μ^2 is characterized in terms of four properties: commutativity, self-duality, monotonicity and responsiveness. If the framework is slightly extended to allow for individual attitudes toward three alternatives then we can easily obtain a toy counterpart of Arrow's impossibility theorem. It shows that no binary operator can be both unanimous and transitive. The main argument in Sect. 9.4 is to prove that the binary operator μ^2 can be extended to the n-ary case, in the sense that all applications of the majority rule μ to a sequence consisting in n members can be defined in terms of the binary majority rule μ^2. Far from being a degenerate case, the binary majority rule operator is able to account for all n-ary applications. It is argued, however, that other binary operators corresponding to other voting rules cannot be so extended. Examples include the absolute majority rule operator and the Jury rule operator. Section 9.5 concludes.

9.2 The Framework

As usual (Urquhart et al. 2001), we start with a countable (infinite) set of propositions $\Sigma = \{p, q, r, \ldots p_1. \ldots p_n, \ldots\}$, called a propositional signature. Θ is a finite set of operator names, sometimes also called (propositional) functions. A function ρ attaches to each operator in Θ a non-negative integer called its arity. A propositional language is a pair $\mathbf{L} = (\Theta, \rho)$. The set $\mathbf{L}_\Sigma$ of formulas of the language $\mathbf{L}$ given a signature Σ is defined inductively as follows:

(1) Members of Σ are $\mathbf{L}_\Sigma$ formulas:
(2) If $\theta \ \varepsilon \ \Theta$, $\rho(\theta) = m > 0$ and $\psi_1, \ldots \psi_m$ are $\mathbf{L}_\Sigma$ formulas, then $\theta(\psi_1, \ldots \psi_m)$ is also a $\mathbf{L}_\Sigma$ formula.

Say negation to the unary operator $\sim$; binary operators are, e.g. well-known logical operators like implication $\rightarrow$, conjunction $\wedge$, disjunction $\vee$, equivalence $\equiv$.

For a given signature, a propositional letter can take one of the following three values: $1, 0, -1$. In Łukasiewicz's three-valued logic,[3] they refer to truth, possible, and false. However, the interpretation I shall constantly have in mind is different. Suppose that $p_1, \ldots p_n$ express the attitudes of n individuals $A_1, \ldots A_n$ with respect to two alternatives x and y. Value 1 means that the individual prefers alternative x to y; value -1 means that the individual prefers alternative y to x, and value 0 carries the meaning that the individual is indifferent between the two alternatives. A n-ary operator π^n represents an aggregation of the attitudes of the n individuals. A matrix $\mathbf{M}_\theta$ for an n-ary operator $\theta \, \varepsilon \, \Theta$ is function which attaches to each n-tuple $(a_1, \ldots a_n)$, where $a_i \varepsilon \{1, 0, -1\}$, a member of $\{1, 0, -1\}$. A matrix $\mathbf{M}_\Theta$ is the collection of all $\mathbf{M}_\theta$, for $\theta \, \varepsilon \, \Theta$. Łukasiewicz's three-valued logic is a pair $\mathbf{L}3 = (\mathbf{L}, \mathbf{M})$, where $\mathbf{L}$ is a propositional language and $\mathbf{M}$ is a matrix for Θ. For the most important logical operators, $\mathbf{M}_\Theta$ is given by:

p	p	Fp	Vp
1	-1	-1	1
0	0	-1	1
-1	1	-1	1

$\rightarrow$	1	0	-1
1	1	0	-1
0	1	1	0
-1	1	1	1

$\wedge$	1	0	-1
1	1	0	-1
0	0	0	-1
-1	-1	-1	-1

$\vee$	1	0	-1
1	1	1	1
0	1	0	0
-1	1	0	-1

With the above interpretation of the meanings of p and q in mind, the expression $p \vee q$ expresses the attitude of the group formed of the individuals A and B relative to the alternatives x and y. The group's attitude follows the most favorable attitude to x of its members: if A or at least B prefers x, then the group also prefers x; if the most favorable attitude to x of the members of the group is indifference, then the group is indifferent between x and y, but if both A and B prefer y, then the group {A, B} also prefers y. Analogously, in the case of the expression $p \wedge q$ the attitude of the group {A, B} follows the least favorable attitude of its members toward x.

---

[3] I write 1, 0 and -1 as usually done is social choice theory instead of the usual values: 1, ½ and 0 in order to emphasize that a social choice interpretation is here intended.

Now let $\Sigma = \{p_1, p_2 \ldots p_n, \ldots\}$ be a signature. A value assignment is a sequence $\mathbf{a}_\Sigma = a_1 a_2 \ldots a_{n-1} a_n \ldots$, with $a_i \varepsilon \{1, 0, -1\}$. In what follows, I shall also appeal to an initial segment $\mathbf{a}_{\Sigma n} = a_1 a_2 \ldots a_{n-1} a_n$ of a value assignment and call it the value assignment for the sequence $\mathbf{p} = p_1 p_2 \ldots p_{n-1} p_n$ of propositional variables.

In this paper the following notation is used: when focusing on an n-ary operator π^n ($n \geq 1$), I shall write $\pi^n (p_1 p_2 \ldots p_{n-1} p_n)$ and appeal to parentheses to clearly distinguish the expressions it applies to. When the arity n of an operator π^n is not the main focus, I shall simply write π instead of π^n. This convention applies to logical operators too. But the usual notations will be also appealed to when the operators $\wedge$, $\vee$, $\rightarrow$, and $\equiv$ will be used in the definitions of the properties of the operators. The expression $p \Rightarrow q$ is short for: $(p \rightarrow q) \wedge \sim (q \rightarrow p)$.

9.3 The Binary Majority Rule

In this section I shall discuss one binary operator μ^2 called the (simple) majority rule. It is defined as follows[4]: $\mu^2(pq) = \text{sgn}(p + q)$. The sgn function is given by: (i) if $n > 0$, then $\text{sgn}(n) = 1$; (ii) if $n < 0$, then $\text{sgn}(n) = -1$; and (iii) if $n = 0$, then $\text{sgn}(n) = 0$.

μ^2	1	0	-1
1	1	1	0
0	1	0	-1
-1	0	-1	-1

The operator μ^2 has the following intuitive interpretation. Suppose again that p and q express the attitudes of two voters A and B with respect to the two alternatives x and y. So, intuitively μ^2 aggregates the preferences of the two voters as follows: if both prefer an alternative, then they prefer it collectively; if both are indifferent, then their joint preference is also indifference; if they have opposite preferences, then they jointly do not prefer any alternative. Finally, if only one voter is indifferent, then their common preference is the preference of the other voter.

Below I shall prove that μ^2 is the only binary operator that satisfies some attractive properties. For the beginning, let me introduce a battery of such properties.

- A binary operator is independent if

$$(q \equiv r) \rightarrow (\pi(pq) \equiv \pi(pr)); \text{ and}$$
$$(p \equiv r) \rightarrow (\pi(pq) \equiv \pi(rq))$$

---

[4]This is the restriction of the general definition of the simple majority rule to the two member groups case; see the definition of μ in the next section.

Independence is built in the very definition of logical operators: when they aggregate propositions only truth-value of the compound propositions is relevant. Therefore in the remainder of this paper we shall take independence as satisfied by default.

- A binary operator π^2 is commutative if $\pi^2(pq) \equiv \pi^2(qp)$.

The operators $\wedge$, $\vee$, $\equiv$ are commutative; e.g., $(p \equiv q) \equiv (q \equiv p)$ states that $\equiv$ is commutative. Commutativity requires that the individuals A and B who hold attitudes toward the two alternatives x and y must be equally treated: it does not matter if we first consider A's attitude as expressed by p and then B's attitude as expressed by q, or the other way round.[5]

- A binary operator is self-dual if $\pi^2(pq) \equiv\sim \pi^2 (\sim p \sim q)$.

Self-duality entails that the alternatives x and y must be equally treated.[6] Dictatorial and anti-dictatorial operators are self-dual. A dictatorial operator always gives the value of one of its arguments; and anti-dictatorial operator always gives the opposite value of one of its arguments. Example of a dictatorial binary operator: $\pi^2(pq) \equiv p$; example of an anti-dictatorial binary operator: $\pi^2(pq) \equiv\sim p$. Operators $\wedge$ and $\vee$ are not self-dual.

Some binary operators are both commutative and self-dual, such as:

α^2	1	0	-1
1	1	0	0
0	0	0	0
-1	0	0	-1

The operator α^2 is the binary absolute majority rule. We have $\alpha^2(pq) = 1$ if both $p = 1$ and $q = 1$; $\alpha(pq) = -1$ if both $p = -1$ and $q = -1$, and $\alpha^2(pq) = 0$ in all the other cases. I shall return to the absolute majority operator in the next section.

- A binary operator is monotonic if:

$$(p \to r) \to \left(\pi^2(pq) \to \pi^2(rq)\right); \text{ and}$$
$$(q \to r) \to \left(\pi^2(pq) \to \pi^2(pr)\right)$$

Examples: α^2 is monotonic; $\wedge$ and $\vee$ are also monotonic; but $\to$ is not monotonic.

- A binary operator is responsive if $\left((\pi^2(pq) \equiv\sim \pi^2(pq)) \wedge (p \Rightarrow r)\right) \to \pi^2(rq)$

By responsiveness, if π^2 gives indifference, and one of its arguments is replaced with another argument having a higher value (0 or 1 instead of -1, or 1 instead of 0), then the value of π^2 moves to 1. The operators μ^2 and

---

[5]A commutative logical operator corresponds then to an anonymous rule.

[6]Clearly, a self-dual logical operator corresponds to a neutral rule.

$\eta^2 = \max(p + q - 1, -1)$ are responsive; conjunction $\wedge$, disjunction $\vee$ and implication $\rightarrow$ are not responsive. The matrix for η^2 is given below[7]:

η^2	1	0	−1
1	1	0	−1
0	0	−1	−1
−1	−1	−1	−1

Note: if π^2 is self-dual and monotonic, then responsiveness entails:

$$((\pi^2(pq) \equiv \sim \pi^2(pq)) \wedge (r \Rightarrow p)) \rightarrow \sim \pi^2(rq)$$

(and similarly for q). So, by responsiveness if $\pi^2(pq) = 0$ and one of its arguments is replaced by a proposition with a higher value, then the value of the aggregate $\pi^2(pq)$ moves to 1 (while if an argument is replaced by a proposition with a lower value then the value of $\pi^2(pq)$ goes down to −1).

Another important property is unanimity. Suppose that both arguments p and q of π^2 have the same value a; then $\pi^2(pq)$ has the same value a:

- A binary operator is unanimous if $(p \equiv q) \rightarrow (\pi^2(pq) \equiv p)$

Conjunction $\wedge$, disjunction $\vee$, α^2 and μ^2 are examples of unanimous operators. Implication and η^2 are not unanimous.

The first result to be presented is Theorem 1. It states that the binary operator μ^2 can be characterized in terms of the properties of commutativity, self-duality, monotonicity and responsiveness: μ^2 satisfies each of them, and no other logical operator satisfies them all.

Theorem 1 π^2 is a commutative, self-dual, monotonic and responsive binary operator if and only if $\pi^2 = \mu^2$.

Proof One direction of the proof is straightforward: μ^2 satisfies the four properties. For the converse direction, suppose that the binary operator π^2 is commutative, self-dual, monotonic and responsive. We want to prove that $\pi^2(pq) = \text{sgn}(p + q)$. We have nine cases: (11), (10), (1−1), (01), (00), (0−1), (−11), (−10), and (−1−1). Since π^2 is commutative, cases (10) and (01); (1−1) and (−11); (0−1) and (−10) are similar. Since π^2 is self-dual, cases (10) and (−10); (01) and (0−1); and (11) and (−1−1) are also similar. So we need to analyze only cases (11), (10), (1−1), and (00).

Case 1: (00). Since π^2 is self-dual, we have $\pi^2(00) \equiv \sim \pi^2(\sim 0 \sim 0) \equiv \sim \pi^2(00)$. But $\pi^2(00) \equiv \sim \pi^2(00)$ only if $\pi^2(00) = 0 = \text{sgn}(0 + 0)$.

Case 2: (10). We have $\pi^2(00) = 0$; responsiveness entails that $\pi^2(10) = 1 = \text{sgn}(1 + 0)$.

---

[7]Duddy and Piggins (2013) call η^2 the Łukasiewicz triangular norm T_L and appeal to it to characterize the deductively closed and free from veto power collective judgments. An important property of η^n is that it is associative.

Case 3: (11). As shown in case 2, $\pi^2(10) = 1$. We also have that $0 \to 1$, and so by the monotonicity of π^2 we get $\pi^2(11) = 1 = \text{sgn}(1 + 1)$.

Case 4: (1–1). By self-duality, we have $\pi^2(1 - 1) = \sim \pi^2(-11)$. By commutativity $\pi^2(-11) = \pi^2(1 - 1)$ and so $\pi^2(1 - 1) = \sim \pi^2(1 - 1)$, which can only hold when $\pi^2(1 - 1) = 0 = \text{sgn}(1 + (-1))$.

Theorem 1 is the counterpart in this formalism of the well-known axiomatization of the simple majority rule given in (May 1954). I appealed to a weaker version of his responsiveness axiom, which required the addition of a special monotonicity axiom.

One may attempt to characterize other logical operators in an analogous way. Here are two examples: conjunction $\wedge^2$ and disjunction $\vee^2$. I first introduce two new properties, top and bottom group preferences:

Top Group preference (TGP). If $p \Rightarrow q$, then $\pi^2(pq) \equiv q$.
Bottom Group Preference (BGP). If $p \Rightarrow q$, then $\pi^2(pq) \equiv p$.

By **TGP** the group always prefers the option most favorable to the alternative x, while by **BGP** the group always prefers the option most favorable to the alternative y.

Notice that no operator can be commutative, self-dual and also satisfy one of the axioms **BGP** and **TGP**. To see this, take for example **BGP** and consider the case (01). We have $0 \Rightarrow 1$ and so $\pi^2(01) = 0$ by **BGP**. Now self-duality requires that $\sim \pi^2(0 - 1) = 0$ and so $\pi^2(0 - 1) = 0$. But if π is commutative then $\pi^2(-10) = 0$. Since $-1 \Rightarrow 0$ by applying again **BGP** we get that $\pi^2(0 - 1) = -1$ – contradiction. As observed above, self-duality entails that the alternatives x and y must be treated equally; but both **TGP** and **BGP** entail that one of the two alternatives enjoys a special status.

The following theorem characterizes the two logical binary operators conjunction and disjunction:

Theorem 2

(a) π^2 is commutative, unanimous and satisfies **BGP** if and only if $\pi^2 = \wedge^2$.
(b) π^2 is commutative, unanimous and satisfies **TGP** if and only if $\pi^2 = \vee^2$.

Proof I give the proof of part (a); the proof for part (b) is similar. Clearly $\wedge^2$ satisfies the three properties. For the converse direction of the proof let us assume that π^2 has the three properties. We consider all possible cases:

Cases (11), (00) and (–1–1): since π^2 is unanimous, we have $\pi^2(aa) = a = \wedge^2(aa)$.

Cases (0–1) and (–10): since $-1 \Rightarrow 0$, by **BGP** we get that $\pi^2(-10) = -1 = \wedge^2(-10)$; by commutativity we have $\pi^2(-10) = \pi^2(-10) = -1 = \wedge^2(-10)$.

Cases (01) and (10): $0 \Rightarrow 1$ entails by **BGP** that $\pi^2(01) = 0 = \wedge^2(10)$.
Commutativity gives again $\pi^2(10) = \pi^2(01) = 0 = \wedge^2(10)$.

Cases $(1-1)$ and (-11): in the same way as above we have $\pi^2(-11) = 1 = \wedge^2(-11)$ because $-1 \Rightarrow 1$. Since π^2 is commutative it follows that $\pi^2(1-1) = \pi^2(-11) = 1 = \wedge^2(1-1)$.

As mentioned above, in the present framework the analysis is restricted to only two alternatives x and y. But let for a moment try to broaden it. Suppose that p^1, q^1 etc. express the attitudes of the individuals A, B etc. on the relation between the alternatives x and y; p^2, q^2 etc. express the attitudes of the individuals A, B etc. on the relation between the alternatives y and z; and p^3, q^3 etc. express the attitudes of the individuals A, B etc. on the relation between the alternatives x and z. I assume that all the individuals A, B etc. have consistent attitudes, and so the following must hold:

$$((p^1 \rightarrow p^2) \wedge (p^2 \rightarrow p^3)) \rightarrow (p^1 \rightarrow p^3) \tag{9.1}$$

$$((q^1 \rightarrow q^2) \wedge (q^2 \rightarrow q^3)) \rightarrow (q^1 \rightarrow q^3) \tag{9.2}$$

Say that a binary operator π is transitive if:

$$(\pi(p^1q^1) \wedge \pi(p^2q^2)) \rightarrow \pi(p^3q^3)$$

The following result is the counterpart of Arrow's theorem in this framework:

Theorem 3 If π is unanimous and transitive, then it is dictatorial.

Proof Assume that $p^1 = 1$, $p^2 = 0$, and also $q^1 = 0$, $q^2 = 0$. Then by unanimity we get $\pi(p^1q^1) = \pi(11) = 1$ and also $\pi(p^2q^2) = \pi(00) = 1$. (Here the fact that π satisfies Independence was assumed.) By the transitivity of π if follows that $\pi(p^3q^3) = 1$. Since π is unanimous, $\pi(p^3q^3) = 1$ must hold whenever $p^3 \equiv q^3$. We have three cases. First, if $p^3 = 1$ and $q^3 = 1$, then $\pi(11) = 1$. Secondly, if $p^3 = 0$ and $q^3 = 0$, then $\pi(00) = 1$. Observe that in both cases the expressions (9.1) and (9.2) have value 1: the individuals A and B have consistent preferences. Third, put $p^3 = -1$ and $q^3 = -1$. We get $\pi(-1-1) = 1$. But in this case we face a contradiction, because as we can easily check the expressions (9.1) and (9.2) do not hold: both have the value 0, i.e. both individuals A and B fail to hold transitive attitudes toward the three alternatives x, y and z. However, the possibility to construct the case when $p^1 = 1$, $p^2 = 0$, $p^3 = -1$ and $q^1 = 1$, $q^2 = 0$, $q^3 = -1$ and so to produce a contradiction is ruled out if π is dictatorial. For if π is dictatorial, we must have for example $\pi(p^1q^1) = p^1$, $\pi(p^2q^2) = p^2$ and $\pi(p^3q^3) = p^3$. Given (9.1), this does not allow for $\pi(p^1q^1) = 1$, $\pi(p^2q^2) = 0$ and $\pi(p^3q^3) = -1$; and similarly if π takes always the value of q.

9.4 *n*-ary Operators

In this section I introduce some *n*-ary operators originating in social choice theory and study their relation to the corresponding binary operators: the simple majority rule μ^n, the absolute majority rule α^n, the extended Pareto rule ε^n and the jury rule λ^n. Their definitions are as follows. Let $\mathbf{p} = p_1 p_2 \ldots p_{n-1} p_n$ be a sequence of propositional variables (the propositional letters in the sequence need not be different). As suggested above, intuitively p_i expresses the attitude of some voter i with respect to two alternatives x and y, with value $p_i = 1$ meaning that i prefers alternative x to y, $p_i = -1$ meaning that i prefers alternative y to x, and $p_i = 0$ carrying the meaning that i is indifferent between the two alternatives. Then $\mu^n(\mathbf{p}) = 1$ if more members of the society prefer x to y; $\mu^n(\mathbf{p}) = -1$ if more voters prefer y to x; and $\mu^n(\mathbf{p}) = 0$ if the votes are equally distributed between x and y. By the absolute majority rule α^n, an alternative is preferred by the society if it is preferred by more than half of the total number of members of the society; and the society is indifferent between two alternatives if none is preferred by more than half of the total members of the society.[8] By the extended Pareto rule ε^n an alternative is preferred by a society if all its members prefer it, and is indifferent in all the other cases. By the jury rule λ^n an alternative is preferred by a society if none of its members opposes it and at least some person prefers it, and is indifferent in all the other cases.

More formally, we have:

- The *n*-ary simple majority rule μ^n is defined by: $\mu^n(\mathbf{p}) = \text{sgn}\left(\sum_{i=1}^{n} p_i\right)$.
- The *n*-ary absolute majority α^n is given by: $\alpha^n(p_1 p_2 \ldots p_{n-1} p_n) = 1$ if more than $n/2$ of the p_i's have value 1; $\alpha^n(p_1 p_2 \ldots p_{n-1} p_n) = -1$ if more than $n/2$ of the p_i's have value -1; and $\alpha^n(p_1 p_2 \ldots p_{n-1} p_n) = 0$ in all the other cases.
- The *n*-ary extended Pareto rule ε^n is defined by: $\varepsilon^n(\mathbf{p}) = 1$ if $p_i = 1$ for all i; $\varepsilon^n(\mathbf{p}) = -1$ if $p_i = -1$ for all i and $\varepsilon^n(\mathbf{p}) = 0$ in all the other cases.
- The *n*-ary jury rule λ^n is defined by: $\lambda^n(\mathbf{p}) = 1$ if $p_i \geq 0$ for all i, and $p_i = 1$ for some i; $\lambda^n(\mathbf{p}) = -1$ if $p_i \leq 0$ for all i and $p_i = -1$ for some i; and $\lambda^n(\mathbf{p}) = 0$ in all the other cases.

The binary operators are special cases of the *n*-ary ones. Specifically, note that $\mu^2 = \lambda^2$ and $\alpha^2 = \varepsilon^2$. However, it is interesting to study the converse relation: is it possible to extend binary operators to the *n*-ary case? For some binary operators this operation can be done in a quite straightforward manner by appealing to the property of associativity. For example, it is usual to extend conjunction $\wedge$ and disjunction $\vee$ to *n* arguments as follows:

---

[8]The binary case is awkward: since for a society formed of exactly two members more than half equals two, by α^2 an alternative is collectively preferred if it is preferred by both individuals, while indifference occurs in all the other cases.

$$\wedge\,(p_1 p_2 \ldots p_{n-1} p_n) = \min\,(p_1, p_2 \ldots p_{n-1}, p_n)\Big) = \wedge\,(p_1 \wedge (p_2 \ldots p_{n-1} p_n));$$
$$\vee\,(p_1 p_2 \ldots p_{n-1} p_n) = \max\,(p_1, p_2 \ldots p_{n-1}, p_n)\Big) = \vee\,(p_1 \vee (p_2 \ldots p_{n-1} p_n)).$$

When $n = 3$, we have in the case of conjunction: $\wedge\,(p_1 p_2 p_3) = (p_1 \wedge (p_2 \wedge p_3))$. Other operators also have this property. One example is the operator η introduced in the above section; the social choice operator ε (the extended Pareto rule) can also be extended in the same way:

$$\varepsilon\,(p_1 p_2 \ldots p_{n-1} p_n) = \varepsilon\,(p_1 \varepsilon\,(p_2 \ldots p_{n-1} p_n))$$

But operators like the simple majority rule μ are not associative. Consider three propositions, p_1, p_2 and p_3, taken to express the attitudes of three persons A, B and C concerning the alternatives x and y. Suppose that the individual A votes for y while B and C vote for x. Then $p_1 = -1$, $p_2 = p_3 = 1$. By definition, $\mu\,(p_1 p_2 p_3) = \mathrm{sgn}\,(p_1 + p_2 + p_3)$. Then we must get $\mu\,(p_1 p_2 p_3) = \mathrm{sgn}\,(-1 + 1 + 1) = 1$. However, $\mu\,(p_1 \mu\,(p_2 p_3)) = \mu\,(-1\mu(11)) = \mu\,(-11) = 0$, which shows that the attempt to extend the binary majority rule operator to the ternary case fails if we want to appeal to the standard method based on the property of associativity.

In this section I describe an alternative method to extend μ to the n-ary case. Then I prove that α^2 and λ^2 behave quite differently: they resist all attempts to be extended. This means that, e.g. in the case of the absolute majority rule α^2 we cannot construct any logical expression σ_α with the property that it contains only occurrences of the binary operator α^2 and $\alpha^n\,(p_1 p_2 \ldots p_{n-1} p_n) \equiv \sigma_\alpha$ is true for all value-assignments.

Let $\mathbf{p} = p_1 p_2 \ldots p_{n-1} p_n$ be a sequence of propositional variables. By definition we must have $\mu^n\,(\mathbf{p}) = \mu^n\,(p_1 p_2 \ldots p_{n-1} p_n) = \mathrm{sgn}\left(\sum_{i=1}^{n} p_i\right)$. I shall denote by $\mathbf{p}^{-i}$ the sequence resulting from $\mathbf{p}$ by deleting p_i from it. We need to construct a logical expression σ_μ with the property that contains only occurrences of the binary operator μ^2 and $\mu^n\,(p_1 p_2 \ldots p_{n-1} p_n) \equiv \sigma_\mu$ is true for all value-assignments. I first give two helpful lemmas.

Lemma 1

(a) If $\mu^n\,(\mathbf{p}) = 1$ and $p_i < 1$ for some $i = 1, \ldots n$ then $\mu^{n-1}\,(\mathbf{p}^{-i}) = 1$.
(b) If $\mu^n\,(\mathbf{p}) = 1$, then $\mu^{n-1}\,(\mathbf{p}^{-i}) \geq 1$ for each $i = 1, \ldots n$ and $\mu^{n-1}\,(\mathbf{p}^{-i}) = 1$ for some i.

Proof For (a), the proof is immediate once we appeal to the definition of μ^n. Observe that $\mu\,(\mathbf{p}) = 1$ entails that $\sum_{i=1}^{n-1} p_i \geq 1 - p_n$. Then clearly $\sum_{i=1}^{n-1} p_i \geq 1$ if $p_n < 1$ and so $\mathrm{sgn}\left(\sum_{i=1}^{n-1} p_i\right) = 1 = \mu\,(\mathbf{p}^{-n})$. For (b) suppose, without loss of generality, that $i = n$. By definition, $\mu^n (\mathbf{p}) = \text{sgn} \left(\sum_{i=1}^{n-1} p_i + p_n \right)$. If $\mu^n (\mathbf{p}) = 1$, it

follows that $\sum_{i=1}^{n-1} p_i + p_n \geq 1$. Now we can only have $\mu (\mathbf{p}^{-n}) < 0$ if $\text{sgn} \left(\sum_{i=1}^{n-1} p_i \right) =$

-1, and thus $\left(\sum_{i=1}^{n-1} p_i \right) < 0$; but in this case $\sum_{i=1}^{n-1} p_i + p_n < 1$ for each value of p_n –

contradiction. But we cannot have $\mu^{n-1} (\mathbf{p}^{ik}) = 0$ for each i. Consider the following cases.

Case 1: $p_i = 1$ for all i. Then by unanimity $\mu^{n-1} (\mathbf{p}^{-i}) = 1$ for each i.
Case 2: there is some i such that $p_i < 1$. But then by (a) we have $\mu^{n-1} (\mathbf{p}^{-i}) = 1$.

The following dual propositions can be proved in a similar way:

(a) if $\mu (\mathbf{p}) = -1$ and $p_i > -1$ for some $i = 1, \ldots n$, then $\mu^{n-1} (\mathbf{p}^{-i}) = -1$.
(b) if $\mu (\mathbf{p}) = -1$, then $\mu (\mathbf{p}^{-i}) \leq 0$ for each $i = 1, \ldots n$ and $\mu (\mathbf{p}^{-i}) = -1$ for some i.

Lemma 2 $\mu^n (\mathbf{p}) = \mu^n (\mu^{n-1} (\mathbf{p}^{-1}), \ldots \mu (\mathbf{p}^{-n}))$

Proof The proof of the lemma is by induction on the number of members of the sequence $\mathbf{p}$. First, let $n = 2$, and so $\mathbf{p} = (p_1 p_2)$. But $\mu^2 (p_1 p_2) = \mu^2 (\mu (\mathbf{p}^{-1}) \mu^2 (\mathbf{p}^{-2}))$. Since as noted above the operator μ satisfies the property of unanimity we get $\mu (\mathbf{p}^{-1}) = \mu (p_2) = p_2$ and $\mu (\mathbf{p}^{-2}) = \mu (p_1) = p_1$ and so $\mu^2 (p_1 p_2) = \mu^2 (p_2 p_1)$, which is true by commutativity.

Now let $n > 2$. Suppose that s of the members of $\mathbf{p}$ have value 1, z of its members have value 0, and m of its members have value -1, and $n = s + z + m$. Since μ is commutative, we can write $\mathbf{p}$ as follows: $p_1, \ldots p_s, p_{s+1} \cdots p_{s+z}, p_{s+z+1} \cdots p_{s+z+m} = p_n$. We have three cases.

Case 1: $\mu^n (\mathbf{p}) = 1$. Then by Lemma 1 all $\mathbf{p}^{-i}$'s are such that $\mu^{n-1} (\mathbf{p}^{-i}) = 1$ or $\mu^{n-1} (\mathbf{p}^{-i}) = 0$, and there is some i such that $\mu^{n-1} (\mathbf{p}^{-i}) = 1$. Then clearly

$$\mu^n \left(\mu^{n-1} (\mathbf{p}^{-1}), \ldots \mu^{n-1} (\mathbf{p}^{-n}) \right) = 1 \text{ because } \sum_{i=1}^{n} \mathbf{p}^{-i} \geq 1.$$

Case 2: $\mu^n (\mathbf{p}) = -1$. The proof is just like in case 1.
Case 3: $\mu^n (\mathbf{p}) = 0$. We have two subcases:
   Subcase 3a: $s = m = 0$, and so $z = n$. Observe also that $\mu^n (\mathbf{p}) = \mu^{n-1} (\mathbf{p}^{-i})$ $(i = 1, \ldots n)$, because deleting a p_i with value 0 does not change the value of μ. Therefore by unanimity the result is proved. Subcase 3b: $s = m \neq 0$. We have in z cases $\mu^{n-1} (\mathbf{p}^{-i}) = 0$, in s cases $\mu^{n-1} (\mathbf{p}^{-i}) = -1$ and in m cases $\mu^{n-1} (\mathbf{p}^{-i}) = 1$. Since $s = m$, we get $\text{sgn} \left(\sum_{i=1}^{n} \mathbf{p}^{-i} \right) =$

$$\text{sgn}\left(\sum_{i=1}^{s}\mathbf{p}^{-i} + \sum_{i=s+1}^{s+z}\mathbf{p}^{-i} + \sum_{i=s+z+1}^{s+z+m=n}\mathbf{p}^{-i}\right) = \text{sgn}\,(s-m+0) = 0 \text{ and so}$$

$$\mu^{n}\left(\mu^{n-1}\left(\mathbf{p}^{-1}\right),\dots\mu^{n-1}\left(\mathbf{p}^{-n}\right)\right) = \text{sgn}(0) = 0.$$

Theorem 4 below states that all applications of the majority rule μ to a sequence consisting in n members can be defined in terms of iteratively applying μ to sequences containing only two members, i.e. in terms of the binary majority rule μ^2.

Theorem 4 Let $\mathbf{p} = p_1 \dots p_n$. Then $\mu(\mathbf{p})$ is equivalent with an expression σ_{μ} which contains only the binary majority rule.

The proof is by induction on the number n of members of sequence $\mathbf{p}$. First, let $n = 3$. Then $\mathbf{p} = p_1 p_2 p_3$. All sequences $\mathbf{p}^{-i}$ have exactly two members. We can easily check that $\mu^3\,(\mathbf{p}) \equiv \mu^2\left(\mu^2\left(\mu^2\left(\mu^2\left(\mathbf{p}^{-1}\right)\mu^2\left(\mathbf{p}^{-2}\right)\right)\right.\right.$ $\mu^2\left(\mu^2\left(\mathbf{p}^{-1}\right)\mu^2\left(\mathbf{p}^{-3}\right)\right)\right)\mu^2\left(\mu^2\left(\mathbf{p}^{-2}\right)\mu^2\left(\mathbf{p}^{-3}\right)\right))$. The expression in the right part of the equivalence states that we first apply μ^2 to each of the three subsets $\{p_2, p_3\}$, $\{p_1, p_3\}$ and $\{p_1, p_2\}$ of $\{p_1, p_2, p_3\}$: $\mu^2(p_2 p_3)$; $\mu^2(p_1 p_3)$; $\mu^2(p_1 p_2)$. Then we apply μ^2 to each pair of these expressions and get: $\mu^2(\mu^2(p_2 p_3)\mu^2(p_1 p_3))$; $\mu^2(\mu^2(p_2 p_3)\mu^2(p_1 p_2))$; $\mu^2(\mu^2(p_1 p_3)\mu^2(p_1 p_2))$. Third, apply μ^2 to the first two expression thus obtained: $\mu^2(\mu^2(\mu^2(p_2 p_3)\mu^2(p_1 p_3))\mu^2(\mu^2(p_2 p_3)\mu^2(p_1 p_2)))$. Finally, apply μ^2 to this expression and the remaining $\mu^2(\mu^2(p_1 p_3)\mu^2(p_1 p_2))$.

Now let $n > 3$. By induction the proposition holds for all the sequences with at most $n-1$ members. Since each of the sequences $\mathbf{p}^{-i}$ has $n-1$ members, $\mu^{n-1}\left(\mathbf{p}^{-i}\right)$ is equivalent by induction with an expression containing only μ^2. Write $\sigma\left(\mathbf{p}^{-i}\right)$ for each of these n expressions. Now form all n sets of $n-1$ such expressions. For example, $\{\sigma\left(\mathbf{p}^{-1}\right),\dots\sigma\left(\mathbf{p}^{-(n-1)}\right)\}$ contains all but $\sigma\left(\mathbf{p}^{-n}\right)$. By induction, $\mu^{n-1}\left(\sigma\left(\mathbf{p}^{-1}\right)\dots\sigma\left(\mathbf{p}^{-(n-1)}\right)\right)$ is an expression equivalent with some expression containing only μ^2. I write $\Sigma(\mathbf{p})^{-n}$ for it. Next, define $\mu^{n-1}\left(\Sigma(\mathbf{p})^{-2}\dots\Sigma(\mathbf{p})^{n}\right)$; again by induction, it is equivalent with an expression containing only μ^2. Finally, put $\sigma_{\mu} = \mu^2\left(\mu^{n-1}\left(\Sigma(\mathbf{p})^{-2}\dots\Sigma(\mathbf{p})^{n}\right),\ \Sigma(\mathbf{p})^{-1}\right)$. I prove that $\mu^n\,(\mathbf{p}) \equiv \sigma_{\mu}$.[9]

Case 1: $\mu^n\,(\mathbf{p}) = 1$. By Lemma 2 we also have $\mu^n\left(\mu^{n-1}\left(\mathbf{p}^{-1}\right),\dots\mu^{n-1}\left(\mathbf{p}^{-n}\right)\right) = 1$ and so $\mu^n\left(\sigma\left(\mathbf{p}^{-1}\right),\dots\sigma\left(\mathbf{p}^{-n}\right)\right) = 1$. But by Lemma 1 all $\sigma(\mathbf{p}^{-i})$'s are such that $\sigma\left(\mathbf{p}^{-i}\right) = 1$ or $\sigma\left(\mathbf{p}^{-i}\right) = 0$, and there is some $\sigma\left(\mathbf{p}^{-i}\right)$ such that $\sigma\left(\mathbf{p}^{-i}\right) = 1$. Therefore for each $\Sigma(\mathbf{p})^{-i}$ we have that $\Sigma(\mathbf{p})^{-i} \geq 0$ and $\Sigma(\mathbf{p})^{-i} = 1$ for some i. So for $\sigma_{\mu} = \mu^2\left(\mu^{n-1}\left(\Sigma(\mathbf{p})^{-2}\dots\Sigma(\mathbf{p})^{n}\right), \Sigma(\mathbf{p})^{-1}\right)$ we have two possibilities: a) $\Sigma(\mathbf{p})^{-1} = 0$. Then we must have $\Sigma(\mathbf{p})^{-i} = 1$ for some $i \geq 2$, which entails that $\mu^{n-1}\left(\Sigma(\mathbf{p})^{-2}\dots\Sigma(\mathbf{p})^{n}\right) = 1$ and thus $\sigma_{\mu} = \mu^2\,(1,0) = 1$; b) $\Sigma(\mathbf{p})^{-1} = 1$.

---

[9]Note that for $n = 3$, σ_{μ} is exactly the expression used in the first step of this proof.

Then $\mu^{n-1}\left(\Sigma(\mathbf{p})^{-2}\dots\Sigma(\mathbf{p})^{n}\right) = a \geq 0$, which gives again $\sigma_{\mu} = \mu^{2}(a,1) = 1$.

Case 2: $\mu^{n}(\mathbf{p}) = -1$. The proof is just like in case 1.

Case 3: $\mu^{n}(\mathbf{p}) = 0$. Then by Lemma 2 we have $\mu^{n}\left(\sigma\left(\mathbf{p}^{-1}\right), \dots \sigma\left(\mathbf{p}^{-n}\right)\right) = 0$. The definition of μ^{n} entails that the number of $\sigma\left(\mathbf{p}^{-i}\right)$'s with the property that $\sigma\left(\mathbf{p}^{-i}\right) = 1$ is equal with the number of $\sigma\left(\mathbf{p}^{-i}\right)$'s with the property that $\sigma\left(\mathbf{p}^{-i}\right) = -1$. A similar argument entails that the number of $\Sigma(\mathbf{p})^{-i}$'s with the property that $\Sigma(\mathbf{p})^{-i} = 1$ is equal with the number of $\Sigma(\mathbf{p})^{-i}$'s with the property that $\Sigma(\mathbf{p})^{-i} = -1$. We have three subcases:

(a) $\Sigma(\mathbf{p})^{-i} = 1$. Then in the sequence $\Sigma(\mathbf{p})^{-2}, \dots \Sigma(\mathbf{p})^{n}$ formed of $n-1$ members the number of $\Sigma(\mathbf{p})^{-i}$'s such that $\Sigma(\mathbf{p})^{-i} = -1$ is larger than the number of $\Sigma(\mathbf{p})^{-i}$'s such that $\Sigma(\mathbf{p})^{-i} = 1$ and so $\mu^{n-1}\left(\Sigma(\mathbf{p})^{-2}\dots\Sigma(\mathbf{p})^{n}\right) = -1$. Then clearly $\sigma_{\mu} = \mu^{2}\left(\mu^{n-1}\left(\Sigma(\mathbf{p})^{-2}\dots\Sigma(\mathbf{p})^{n}\right), \Sigma(\mathbf{p})^{-1})\right) = \mu^{2}(-1,1) = 0 = \mu^{n}(\mathbf{p})$.

(b) $\Sigma(\mathbf{p})^{-1} = -1$. By an analogous argument we conclude that $\sigma_{\mu} = \mu^{2}(1,-1) = 0 = \mu^{n}(\mathbf{p})$.

(c) $\Sigma(\mathbf{p})^{-1} = 0$. Then in the sequence $\Sigma(\mathbf{p})^{-2}, \dots \Sigma(\mathbf{p})^{n}$ formed of $n-1$ members the number of $\Sigma(\mathbf{p})^{-i}$'s such that $\Sigma(\mathbf{p})^{-i} = -1$ is equal with the number of $\Sigma(\mathbf{p})^{-i}$'s such that $\Sigma(\mathbf{p})^{-i} = 1$ and so $\mu^{n-1}\left(\Sigma(\mathbf{p})^{-2}\dots\Sigma(\mathbf{p})^{n}\right) = 0$. Then $\sigma_{\mu} = \mu^{2}\left(\mu^{n-1}\left(\Sigma(\mathbf{p})^{-2}\dots\Sigma(\mathbf{p})^{n}\right), \Sigma(\mathbf{p})^{-1})\right) = \mu^{2}(0,0) = 0$.

Remark We can easily see that if $\mathbf{p}$ consists in just one member p, we can put $\mu(\mathbf{p}) = \mu(pp)$, and by unanimity we get $\mu(\mathbf{p}) \equiv p$. Therefore the unary case is also covered.

Theorem 5 λ^{2} cannot be extended to the n-ary case.

Proof The proof consists in showing that there is no expression σ_{λ} which contains only the binary function λ^{2} and $\lambda^{n}(\mathbf{p}) = \sigma_{\lambda}(\mathbf{p})$ for all $\mathbf{p} = p_{1}\dots p_{n}$. We only need to consider the simplest case when we have three propositional variables p_{1}, p_{2} and p_{3} and show that $\lambda^{3}(p_{1}p_{2}p_{3})$ is not definable in terms of λ^{2}. This means that there is no expression σ_{λ} with the property that $\lambda^{3}(p_{1}p_{2}p_{3}) = \sigma_{\lambda}(p_{1}p_{2}p_{3})$ for all propositional variables p_{1}, p_{2} and p_{3} and σ_{λ} contains only occurrences of λ^{2}. Since λ^{2} and μ^{2} are identical, we can replace in σ_{λ} all occurrences of λ^{2} with occurrences of μ^{2}. The proof has three steps. In the first step I show that σ_{λ} satisfies self-duality; in the second step I show that it satisfies monotonicity. Finally, I prove that if $\lambda^{3}(p_{1}p_{2}p_{3}) = \sigma_{\lambda}(p_{1}p_{2}p_{3})$ for all propositional variables p_{1}, p_{2} and p_{3} then we get a contradiction.

First, I show by induction on the complexity of σ_{λ} that self-duality is satisfied. Suppose first that $\sigma_{\lambda} = \mu^{3}(p_{1}p_{2}p_{3})$. Then clearly self-duality is preserved, for μ^{3} is self-dual. Let $\sigma_{\lambda} = \mu^{3}\left(\sigma_{\lambda}^{1}(p_{1}p_{2}p_{3}), \sigma_{\lambda}^{2}(p_{1}p_{2}p_{3}), \sigma_{\lambda}^{3}(p_{1}p_{2}p_{3})\right)$.

By induction $\sigma_\lambda^i (\sim p_1 \sim p_2 \sim p_3) = \sim \sigma_\lambda^i (p_1 p_2 p_3)$ $(i = 1, 2, 3)$. Then $\sim \sigma_\lambda (\sim p_1 \sim p_2 \sim p_3) = \sim \mu^3 \big(\sigma_\lambda^1 (\sim p_1 \sim p_2 \sim p_3), \sigma_\lambda^2 (\sim p_1 \sim p_2 \sim p_3),$ $\sigma_\lambda^3 (\sim p_1 \sim p_2 \sim p_3)\big) = \sim \mu^3 \big(\sim \sigma_\lambda^1 (p_1 p_2 p_3), \sim \sigma_\lambda^2 (p_1 p_2 p_3)$ $, \sim \sigma_\lambda^3 (p_1 p_2 p_3)\big) = \mu^3 \big(\sigma_\lambda^1 (p_1 p_2 p_3), \sigma_\lambda^2 (p_1 p_2 p_3), \sigma_\lambda^3 (p_1 p_2 p_3) = \sigma_\lambda$. In the second step we can proceed in a similar way to show that monotonicity is also holds.

Now let us move to the final step of the proof.[10] First, notice the following property of λ^3.

Let $p_1^1 = 1$, $p_2^1 = 1$ and $p_3^1 = -1$. By definition, $\lambda^3 = \big(p_1^1 p_2^1 p_3^1\big) = 0$. Let $p_2^2 = -1$. Then $\lambda^3 \big(p_1^1 p_2^2 p_3^1\big) = 0$. Similarly, if $p_3^3 = 1$, then $\lambda^3 \big(p_1^1 p_2^1 p_3^3\big) = 0$; and if $p_1^4 = -1$, then $\lambda^3 \big(p_1^4 p_2^2 p_3^3\big) = 0$. But suppose that λ^3 is the result of extending λ^2 to the ternary case, i.e. there is some σ_λ such that $\lambda^3 (p_1 p_2 p_3) = \sigma_\lambda (p_1 p_2 p_3)$ for all propositional variables p_1, p_2 and p_3. Then λ^3 must be monotonic and self-dual. Moreover, note that $p_1^1 = \sim p_1^4$, $p_2^1 = \sim p_2^2$, and $p_3^1 = \sim p_3^3$; so $\lambda^3 \big(p_1^1 p_2^1 p_3^1\big) = \sim \lambda^3 \big(\sim p_1^4 \sim p_2^2 \sim p_3^3\big)$ by self-duality. I shall prove that if $\sigma_\lambda \big(p_1^1 p_2^1 p_3^1\big) = 0$, then $\sigma_\lambda \big(p_1^1 p_2^2 p_3^1\big) \neq 0$ or $\sigma_\lambda \big(p_1^1 p_2^1 p_3^3\big) \neq 0$ or $\sigma_\lambda \big(p_1^4 p_2^2 p_3^3\big) \neq 0$. The proof is on induction on the complexity of σ_λ.

Case 1. $\sigma_\lambda (p_1 p_2 p_3) = \mu^3 (p_1 p_2 p_3)$. Since $\sigma_\lambda \big(p_1^1 p_2^1 p_3^1\big) = 0$, we have $\mu^3 \big(p_1^1 p_2^1 p_3^1\big) = 0$. But μ^3 is responsive, and we have $(p_2^2 \rightarrow p_2^1) \wedge \sim (p_2^1 \rightarrow p_2^2)$; so $\mu^3 \big(p_1^1 p_2^2 p_3^1\big) = -1$, which gives $\sigma_\lambda \big(p_1^1 p_2^2 p_3^1\big) \neq 0$.

Case 2. $\sigma_\lambda (p_1 p_2 p_3) = \mu^3 \big(\sigma_\lambda^1 (p_1 p_2 p_3), \sigma_\lambda^2 (p_1 p_2 p_3), \sigma_\lambda^3 (p_1 p_2 p_3)\big)$. By induction, the property holds for all σ_λ^i $(i = 1, 2, 3)$. Since all σ_λ^i are monotonic, given the values of p_j^k's we have:

(1) $\sigma_\lambda^i \big(p_1^1 p_2^2 p_3^1\big) \rightarrow \sigma_\lambda^i \big(p_1^1 p_2^1 p_3^1\big)$

(2) $\sigma_\lambda^i \big(p_1^1 p_2^1 p_3^1\big) \rightarrow \sigma_\lambda^i \big(p_1^1 p_2^1 p_3^3\big)$

(3) $\sigma_\lambda^i \big(p_1^4 p_2^2 p_3^3\big) \rightarrow \sigma_\lambda^i \big(p_1^1 p_2^2 p_3^3\big)$

But by the definition of λ^3,

(4) $\mu^3 \big(\sigma_\lambda^1 \big(p_1^1 p_2^1 p_3^1\big), \sigma_\lambda^2 \big(p_1^1 p_2^1 p_3^1\big), \sigma_\lambda^3 \big(p_1^1 p_2^1 p_3^1\big)\big)$

$\qquad = \mu^3 \big(\sigma_\lambda^1 \big(p_1^1 p_2^2 p_3^1\big), \sigma_\lambda^2 \big(p_1^1 p_2^2 p_3^1\big), \sigma_\lambda^3 \big(p_1^1 p_2^2 p_3^1\big)\big),$

and similarly for the other two cases. So

(5) $\sigma_\lambda^i \big(p_1^1 p_2^2 p_3^1\big) = \sigma_\lambda^i \big(p_1^1 p_2^1 p_3^1\big),$

because otherwise μ^3's being responsive would contradict the above equivalence. A similar argument applies for the other two cases. However, by the inductive hypothesis we must have

(6) $\sigma_\lambda^i \big(p_1^1 p_2^1 p_3^1\big) \neq 0$ or $\sigma_\lambda^i \big(p_1^1 p_2^2 p_3^1\big) \neq 0$ or $\sigma_\lambda^i \big(p_1^1 p_2^2 p_3^3\big) \neq 0$ or $\sigma_\lambda^i \big(p_1^4 p_2^2 p_3^3\big) \neq 0$.

---

[10]It is inspired by the necessity part of Fine's proof of his Theorem 3 in (Fine 1972); I appeal to a very simplified version of his property of zigzaggedness.

From (5) and (6) we get:

(7) $\sigma_\lambda^i \left(p_1^1 p_2^2 p_3^1 \right) = \sigma_\lambda^i \left(p_1^4 p_2^2 p_3^3 \right) \neq 0$

But notice that $p_1^1 =\sim p_1^4$, $p_2^1 =\sim p_2^2$, and $p_3^1 =\sim p_3^3$; so we must also have $\sigma_\lambda^i \left(p_1^1 p_2^1 p_3^1 \right) =\sim \sigma_\lambda^i \left(\sim p_1^4 \sim p_2^2 \sim p_3^3 \right)$ by the self-duality of σ_λ^i – which contradicts (7).

Theorem 6 α^2 cannot be extended to the n-ary case.

Proof Again, it suffices to show that α^2 cannot be extended to α^3. Suppose that there is some expression σ_α which contains only the binary absolute majority rule α^2 and $\alpha^n (\mathbf{p}) = \sigma_\alpha (\mathbf{p})$ for all $\mathbf{p} = p_1 p_2 p_3$. Suppose that at the profile $\mathbf{p}$ we have $p_1 = 0, p_2 = p_3 = 1$. If σ_α contains at least one occurrence of p_1, then α^2 gives value 0, and the definition of α^2 entails that all other subsequent applications must result in the same value, so $\sigma_\alpha (\mathbf{p}) = 0$. But clearly we must have $\alpha^3 (\mathbf{p}) = 1$ – contradiction. Therefore, we must construct σ_α such that it contains no occurrence of p_1. If on the other hand we take into account a value assignment $\mathbf{p}'$ such that $p_2 = 0, p_1 = p_3 = 1$ (in this case we also have $\alpha^3 (\mathbf{p}') = 1$) we must conclude that σ_α is such that it contains no occurrence of p_2; a similar argument shows that σ_α does not satisfy the required property if it includes an occurrence of p_3. Therefore no σ_α satisfies the property that $\sigma_\alpha (\mathbf{p}) = \alpha (\mathbf{p})$ for all assignments $\mathbf{p}$.

9.5 Conclusion

Supposing that the language we use consists in propositions that express the attitudes of individuals on an issue (how to choose between the alternatives x and y), a Łukasiewiczian three-valued logic framework can be shown to be rich enough to allow for the reconstruction of many aggregation rules, the simple and the absolute majority rules among them. I argued that, on this account, in its primary use simple majority rule applies to groups of people consisting in only two members and so it can be modeled as a binary logical operator. I characterized it by means of some simple properties, in analogy with the famous result presented by May (1954) and also showed that an impossibility theorem can be obtained in this framework.

One of the missing steps in the appeal to simple majority rule, as applied to groups of people formed of an arbitrary number n ($n > 2$) of members, is its relation to the binary rule. I proved that the n-ary logical operator corresponding to simple majority rule can be obtained by extending the binary operator.[11]

---

[11]The three-valued logic is functionally complete in the following sense (Słupecki 1972): let $\mathbf{F}$ be a set of logical operators that contains all the unary operators and at least one essential operator. A binary operator is essential if it takes on all the values from $\{1, 0, -1\}$. Then each binary operator is definable in terms of the logical operators in $\mathbf{F}$. In fact, we need not consider all the unary operators as given. Słupecki (Słupecki 1967) showed that all the binary logical operators can be defined by adding to the classical $\rightarrow$ (implication) and (negation) operators a new unary operator

The paper relies on a logical framework in which the propositions of the language used are interpreted as expressing the attitudes of the individuals toward certain issues. The aggregation of attitudes is then expressed by means of logical operators. Compound propositions are taken to express the attitudes of complex groups: some are formed of individuals, while others are higher-order and have also groups as their members. The aggregation of attitudes is an attempt to describe such complex situations of group decisions.

References

Ågotnes T, van der Hoek W, Wooldridge M (2009) On the logic of preference and judgment aggregation. Auton Agent Multi Agent 22:4–30

Arrow KJ, Sen A, Suzumura K (eds) (2011) Handbook of social choice and welfare, vol 2. North-Holland, Amsterdam

Barrett R, Salles M (2011) Social choice with fuzzy preferences. In: Arrow KJ, Sen A, Suzumura K (eds) Handbook of social choice and welfare, vol 2. North-Holland, Amsterdam

Dokow E, Holzman R (2010) Aggregation of non-binary evaluations. Adv Appl Math 45:487–504

Duddy C, Piggins A (2013) Many-valued judgment aggregation: characterizing the possibility/impossibility boundary. J Econ Theory 148:793–805

Endriss U (2011) Logic and social choice theory. In: Gupta A, van Benthem J (eds) Logic and philosophy today, vol 2. College Publications, London

Fine K (1972) Some necessary and sufficient conditions for representative decision on two alternatives. Econometrica 40:1083–1090

Fishburn PC (1971) The theory of representative majority decision. Econometrica 39:273–284

Grandi U, Endriss U (2013) First-order logic formalisation of impossibility theorems in preference aggregation. J Philos Log 42:595–618

Hähnle R (2001) Advanced many-valued logic. In: Gabbay DM, Guenthner F (eds) Handbook of philosophical logic, vol 2, 2nd edn. Kluwer, Dordrecht

List C, Pettit P (2002) Aggregating sets of judgments: an impossibility result. Econ Philos 18:89–110

May KO (1954) A set of independent necessary and sufficient conditions for simple majority decisions. Econometrica 20:680–684

Murakami Y (1966) Formal structure of majority decision. Econometrica 34:709–718

Murakami Y (1968) Logic and social choice. Dover, New York

Nipkow T (2009) Social choice theory in HOL: arrow and Gibbard-Satterthwaite. J Autom Reason 43:289–304

Pauly M (2007) Axiomatizing collective judgment sets in a minimal logical language. Synthese 158:233–250

Pauly M, van Hees M (2006) Logical constraints on judgment aggregation. J Philos Log 35:569–585

T. It has the property that $Tp = 0$ for all values of p. Murakami (1968) proved a similar result: any binary voting rule can be defined in terms of iterate applications of μ, a negation operator and a constant rule: $Vp = 1$ for all values of p. Clearly, it is sufficient to show that the operators T and $\rightarrow$ are definable. First, implication is defined by: $p \rightarrow q = \mu (\sim pqVp)$; second, Slupecki's unary operator T is simply: $Tp = \mu (p \sim p)$. However, note that the ternary μ^3 is used. It is also necessary to show that the result extends to the n-ary case.

Słupecki J (1967) The full three-valued propositional calculus (1936). In: McCall S (ed) Polish logic 1920–1939. Oxford University Press, Oxford, pp 335–337

Słupecki J (1972) Completeness criterion for systems of many-valued propositional calculus (in Polish) (1939); English translation in Stud. Logica 30:153–157.

Tarski A (1959) An introduction to logic. Oxford University Press, London

Urquhart A (2001) Basic many-valued logic. In: Gabbay DM, Guenthner F (eds) Handbook of philosophical logic, vol 2, 2nd edn. Kluwer, Dordrecht

van Hees M (2007) The limits of epistemic democracy. Soc Choice Welf 28:649–666

Chapter 10
A Free Logic for Fictionalism

Mircea Dumitru

10.1 Statement of Purpose

This work is part of a larger project in which I deal with issues in metaphysics and philosophy of language concerning fictional objects. The main point that I am going to make in this paper is this: in order to articulate the logical principles which govern the discourse on fictional objects what we need is a sort of free logic.

Non-existing objects, arbitrary objects, and fictional objects have received a considerable amount of philosophical attention lately. My interest here will be with fictional objects only. If we look at the recent literature we can see that fictionalism is a lively issue nowadays in metaphysics, philosophical logic, philosophy of language, and aesthetics.

Important works are examining diverse strategies which are meant to deal with those kinds of Meinongian objects. Let me mention only two approaches I find important in the field: Terence Parson's theory of non-existing objects,[1] and Kit Fine's theory of arbitrary objects.[2]

Odd as it may seem *prima facie*, there is a strong connection between fiction and fictional objects and characters, on the one hand, and representation on the other hand.[3] Works of fiction (stories, tales, novels, and even paintings and sculptures) are meant by their creators to represent. The audience, to whom those works are

---

[1]The theory is put forward and developed in (Parsons 1980).

[2]In connection with Parson's theory and with my own line of argumentation in this paper see especially (Fine 1984).

[3]Cf. (Sainsbury 2010).

M. Dumitru (✉)
Department of Theoretical Philosophy, University of Bucharest, Bucharest, Romania
e-mail: mircea.dumitru@unibuc.ro

© Springer International Publishing Switzerland 2015
I. Pârvu et al. (eds.), *Romanian Studies in Philosophy of Science*, Boston Studies
in the Philosophy and History of Science 313, DOI 10.1007/978-3-319-16655-1_10

addressed, interprets the message of the works using this convention according to which the authors of the works intend to represent something through those works. One important issue which emerges here will be what is a correct account and form of representation which could make room and justice not only to real objects which exist out there in the real world but also to these fictional objects and characters which are the referents of our terms in fictional discourse?

It is very likely that this kind of fictional representation has not the pursuit and attainment of truth as its epistemic value. Fictions play other important roles in our cognitive, aesthetic, and emotional life. They are not meant to be truth-apt. As Sainsbury puts it in his recent book devoted to fictionalism: "One thing we need to say is that attaining truth is not an essential aim of fictional representation. We value fictions for other properties. This connects fiction with the philosophical notion of fictionalism."[4]

Fictionalism has a genuine explanatory virtue which can be grasped if we think of the following dialectics brought about by philosophical arguments in various quarters of contemporary philosophy. Thus, for many of us, when in philosophical mode, it is extremely hard to buy into the full ontological existence of, say, unobservable things, or abstract things, or nonfactual (and merely possible) things, or even moral values. But the very same people, when again in their philosophical mode, have a hard time in rejecting the forms of discourse which are about those sui-generis objects. People have to be fair and do justice to the basic fact of the intuitive meaningfulness of the languages of mathematics, physics, modality, or morality. Fictionalism will help us in this regard: the things on which we think in those fields of inquiry are important, and they have to be accepted by us, even if they do not qualify ontologically, semantically, or epistemologically as being truth-apt or as truth-makers or truth-bearers.

Important fictionalist philosophical accounts of scientific theories and mathematics have been expounded and notoriously defended by, among some contemporary philosophers, Bas van Fraassen,[5] and Hartry Field,[6] respectively. The fictionalism that emerges from these works invites us not to believe a given ontologically disputed form of discourse as being true, but to accept that form of discourse for being "empirically adequate", i.e. capable of making correct predictions of observable facts, or for its usefulness in showing us a correct way to facts which do not essentially involve numbers, even if we describe them in numerical and mathematical terms.

A weaker version of fictionalism will endorse the contrast between the existence of facts, related to what is going on in a story, and the unreality and unfacticity of the story as such. This is pretty much related to understanding fictionalism as a view which involves (implicitly) prefixing a fiction operator to the sentences of some discourse whose ontology is dubious or contentious. The effect of such an operator

---

[4](Sainsbury 2010), p. 2.

[5]Cf. (Bas van Fraassen 1980).

[6]Cf. (Field 1980).

is to make a sentence true out of something which is not truth-apt. Truth is signaling here what is of a certain importance for us in the corresponding discourse. The strategy has been used with good results in the domain of modality. According to (Rosen 1990), statements which prima facie commit to merely possible objects and merely possible worlds should be taken to mean whatever those statements convey when prefixed by a fiction operator which reads "According to the possible worlds story ...".

Fictional objects and fictionalism provide a very interesting and substantial material for metaphysics (ontology), philosophy of language and philosophical logic. My paper here addresses some of those intricate issues disentangling certain logical principles that govern the meaningful discourse on fictional objects.

10.2 An Account of Fictional Objects

For present purposes, my concern is with objects such as *Holmes*, *Dracula*, and the like. If they are the kind of objects that can be native to a story or other such context we may call them fictional.

Fictional objects are to be carefully distinguished from other non-fictional non-existent objects. Not every non-existing object is a fictional object. In order to make a clear distinction between fictional objects and some other non-existing objects which are not fictional objects the following proposal is helpful.[7]

Main ontological thesis concerning fictional objects: fictional objects are essentially objects of reference, i.e. objects created through a story or a narrative and introduced via a cluster of descriptions.

The following items substantiate the thesis:

(i)   The fictional objects ontologically depend upon the descriptions which are used in order to introduce those fictional objects. We mean this in the sense that there can be no fictional objects without a mark or a symbol by which they are introduced.

(ii) Fictional objects are essentially tied to those marks. There can be no fictional object without the particular mark by which it is introduced.

(iii) Fictional objects are created through their marks being created.

(iv) We may allow that fictional objects with the same internal content be introduced by different marks.

(v)   The position that I want to defend is an anti-realist position: features of fictional objects are ultimately to be explained in terms of features of their marks.

However, the view on fictional objects I have in mind is also supportive for descriptivism. Fictional discourse may very well be a proper place for descriptivism.

---

[7]Cf. (Fine 1984).

Definite descriptions play an essential role in introducing fictional objects. Fictional terms seem to have a major irreducible descriptive content.

This being noticed, the next legitimate question is what kind of description theories do we need to articulate the principles which govern the discourse about fictional objects? For reasons that I will go over briefly a classical descriptivist account such as Russell's theory of definite descriptions won't do here, and if I am right the kind of theory that helps here is a kind of free description theory.

10.3   What Sort of Description Theory Do We Need for Making Good Sense of the Principles Involved in the Fictional Discourse?

A major motivation for developing free logic has always been to provide a basis for theories of definite descriptions.[8] The best known theory in classical (non-free) logic is obviously Russell's. Russell's answer to the problems posed by definite descriptions was to deny them the status of singular terms, and to regard an expression of the form 'the so and so' as needing to be eliminated in context, where the most important principles governing this elimination are

(R1) The so and so exists iff exactly one thing is a so and so

   and

(R2) The so and so is such and such iff there is exactly one so and so and it is such and such.
   Formally,

$$\left(\text{R1}'\right)\ E!\,(\iota v)\,A \leftrightarrow (\exists v)\ (A \& (\forall w)\ (A\,(w/v) \rightarrow w = v))$$

   and

$$\left(\text{R2}'\right)\ B\,((\iota v)\,A) \leftrightarrow (\exists v)\,(A \& (\forall w)\ (A\,(w/v) \rightarrow w = v)\,\& Bv)\ (v \text{ is free in } A.)$$

The theory tells us how to treat descriptions whose scopes are uniquely fulfilled. It also tells us that there are two ways in which 'E!(ιv)A' can fail to be true: (i) if no object at all fulfils the scope of A; and (ii) if more than one object does.

After reigning philosophical logic and philosophy of language for over 40 years, Russell's theory has been targeted by some objections that made people see more clearly the status of his theory and its diverse implications. However I wouldn't

---

[8]Cf. (Lambert 2003).

say that they put the theory entirely to rest. Thus, objections coming mainly from Strawson[9] include things of the following sorts:

(i)   A sentence which contains an improper description is not false, as declared by Russell's theory. Rather, it is a sentence through which the speaker fails to refer to anything, and so fails to make a complete statement.

(ii) According again to Strawson, Russell's theory supports the view that a sentence which contains an empty denoting phrase *implies* a sentence which asserts the entity to which the denoting phrase purports to refer, whereas for Strawson himself such a sentence in which an empty definite description occurs *presupposes* a sentence which asserts the existence of that entity.

(iii) Strawson emphasizes the idea that many descriptions are context-bound. However, in Russell's theory one can hardly make room for such pragmatic features.

(iv) Some other authors, notably Donnellan,[10] have pointed out that there are cases in which definite descriptions are not used descriptively, but rather in a way in which they are like names of individuals. On the other hand, though, the Russellian approach does not capture what a speaker says when he/she utters a sentence in which a definite description is used referentially as opposed to attributively.

(v)   If we allowed descriptions of the form $(\iota v)A$ as substituends for the singular term t in the following identity sentence, the logical principle of identity $t = t$ would be violated, in case the identity $(\iota v)A = (\iota v)A$ is false, if the scope is not uniquely fulfilled, i.e. if the description is improper. However, it is worth emphasizing that Russell gets round this by regarding definite descriptions as not being genuine singular terms at all but 'improper symbols' which may look like singular terms but in fact are not, so they cannot be substituted for the singular term t in the principle above.

(vi) Russell fails to treat what looks like singular terms and behaves like singular terms as singular terms. He makes a distinction between the real logical form of a sentence and its apparent or grammatical form. The grammatical form of a sentence may mislead us as to its true logical form. Now, it's debatable whether or not this is a problem for Russell's theory of definite descriptions. So, if we can find a theory of descriptions in which they are treated as genuine singular terms then we can overcome this drawback of Russell's theory. Free logic with definite descriptions provides such a logical framework, and it seems to be preferable for that reason.

(vii) Last but not least, if we explain away via Russell's approach sentences in which fictional terms occur, then the existential sentences we come up with will give us wrong results, since the existential quantifier in classical logic will have its usual objectual and ontological committing reading. Thus,

---

[9] See (Strawson 1950).

[10] See (Donnellan 1966).

'Othello killed Desdemona' will be paraphrased as 'The noble Moor in the service of the Venetian state who does such and such killed Desdemona'. (Of course, 'Desdemona' can be explained away likewise). And now the definite description 'The noble Moor' will be eliminated in context by an existential sentence to the effect that 'There is an exactly one noble Moor who does such and such, and whoever is a noble Moor who does such and such killed Desdemona'. But of course, if you read the existential quantifier the same standard objectual way I read it, and if your ontology is like mine on this particular aspect then you will reject the existential sentence. However, my intuition is to say that 'Othello killed Desdemona' is a true sentence, at least in Shakespeare's story. Hence, something went wrong with the existentially committing Russellian analysis.

What are we supposed to do then? The answer that I explore is that positive free logic with free descriptions is a serious solution to the problem which is worth exploring. I'll take a look at it against the background of a family of free logic systems.

10.4   A Crash Course in Free Logic[11]

Logic free of existential presuppositions is a branch of philosophical logic which has been developed in the last forty years. Existential presuppositions are linked with singular and general terms. Accordingly, the concept of a free logic was understood as 'logic free of existential presuppositions with respect to its singular and general terms'. Standard first order logic with '=' (FOL=) is almost fully free with respect to its general terms or predicates. There is only one exception, though, namely universal terms or predicates like '$Px \vee \sim Px$' or '$x = x$'. In (FOL=) '$(\exists x)(Px \vee \sim Px)$' and '$(\exists x)(x = x)$' are valid. We can read the latter as 'something exists', and this seems to express a truth of ontology rather than a truth of logic.

The main concern of free logic has been existential presuppositions with respect to singular terms. For in standard FOL = we have for every singular term t and variable v the valid formula: $\models (\exists v)(v = t)$. And due to the ontological commitment of singular terms, in FOL = we have rules for quantifiers, such as existential introduction ($\exists$I) and universal elimination ($\forall$E), which are not sound if the terms do not refer to actual existing things.

---

[11]A rich resource for the topic is (Morscher and Hieke 2001), which my exposition in this section is based upon.

Against this motivational background, an adequate definition of free logic has to include three components.[12] Thus, a logical system L_F is a free logic if and only if (iff)

(1) L_F is free of existential presuppositions with respect to the singular terms of the language of L_F;

(2) L_F is free of existential presuppositions with respect to the general terms of the language of L_F; and

(3) the quantifiers of L_F have existential import.

It is appropriate to speak about a family of systems of free logic. The distinctive feature of those systems is the fact that singular terms which are empty or non-denoting, provided they do not refer to existing things, have a legitimate place in this family of logic systems. Moreover, the theorems of a free logic system are valid even if the singular terms which occur in them are empty.

There are three types of free logic systems. The criterion according to which we can distinguish between those types is whether or not elementary sentences containing empty singular terms are true or false or else lack any truth-value at all.

(FL−) A logical system L_{F-} is a *negative free logic* iff L_{F-} is a free logic and every atomic sentence of L_{F-} containing at least one empty singular term is false.

(FL+) A logical system L_{F+} is a *positive free logic* iff L_{F+} is a free logic and there is at least one true atomic sentence of L_{F+} containing at least one empty singular term.

(FLn) A logical system L_{Fn} is a *neutral free logic* iff L_{Fn} is a free logic and every atomic sentence of L_{Fn} containing at least one empty singular term has no truth-value at all.

Alongside with those three types of free logic systems, the following three semantic approaches that have been developed for free logic systems are now well-entrenched:

(S1) Semantics with a *partial interpretation function* and a *total valuation function*.

(S2) Semantics with an *inner* and an *outer domain*: this uses a *total interpretation function* and a *total valuation function*.

(S3) *Supervaluation semantics*: this type of semantics uses a *partial* and a *total interpretation function* and a *total* and *two partial valuation functions*.

---

[12]In giving this compact presentation of free logic systems I draw on the excellent systematic and historic synopsis one can get in (Morscher and Simons 2001).

10.4.1 Semantic Systems of Free Logic

	Semantics with a partial interpretation function and a total valuation function	Inner and outer domain	Supervaluation
Interpretation function	Partial	Total	Partial & total
Valuation function	Total	Total	Total & two partial valuation functions

Each type of semantic system specifies its own type of models M. As usual, a model M consists in a domain D and in an interpretation function I, which is associated with a valuation function V. I is always defined on the set of descriptive symbols, i.e. non-logical predicates and individual constants of the language of that free logic system. What is distinctive for the semantics for free logic is that I is not supposed to assign an *existing* object to each individual constant. I therefore assigns to some individual constant t of L_F either *a non-existing object* or *no object at all*; in the second case $I(t)$ remains undefined, and I therefore is a partial function. The valuation functions V which are based on the interpretation functions I are always defined on the set of closed formulae of L_F. They can be either total or partial.

(S1) Semantics with a partial interpretation function and a total valuation function

An M^{pitv}-model is an ordered pair. It comprises a possibly empty domain D and a partial function I^{pitv}, i.e. $M^{pitv} = (D, I^{pitv})$, such that

(1) for every individual constant t of the language of L_F: either I^{pitv} does not assign anything at all to t and $I^{pitv}(t)$ thereby remains undefined or $I^{pitv}(t) \in D$;
(2) for every n-place predicate P^n of L_F : $I^{pitv}(P^n) \subseteq D^n$;
(3) for every object $d \in D$ there is an individual constant t of the language of L_F such that $I^{pitv}(t) = d$. [The interpretation function I^{pitv} of an M^{pitv}-model provides a 'full' (or complete) interpretation of the associated domain D.]

Next we define truth and falsehood in a model M^{pitv} for every closed formula A of the language of L_F. We do this by defining a total valuation function V^{pitv} from the set of closed formulae of L_F into the set {T,F} of truth-values as follows:

(1) $V^{pitv}(P^n t_1, t_2, \ldots, t_n) = $ T iff for every t_i $(1 \le i \le n)$: $I^{pitv}(t_i)$ is defined and $< I^{pitv}(t_1), I^{pitv}(t_2), \ldots, I^{pitv}(t_n) > \in I^{pitv}(P^n)$;
(2) $V^{pitv}(t_1 = t_2) = $ T iff $I^{pitv}(t_1)$ is defined and $I^{pitv}(t_2)$ is defined and $I^{pitv}(t_1) = I^{pitv}(t_2)$.
(3) $V^{pitv}(E!t) = $ T iff $I^{pitv}(t)$ is defined.
(4) $V^{pitv}(\sim A) = $ T iff $V^{pitv}(A) \ne $ T;
(5) $V^{pitv}(A \rightarrow B) = $ T iff $V^{pitv}(A) \ne $ T or $V^{pitv}(B) = $ T or both;
(6) $V^{pitv}(\forall v A) = $ T iff for every individual constant t: if $I^{pitv}(t)$ is defined then $V^{pitv}(A(t/v)) = $ T.
(7) $V^{pitv}(A) = $ F iff $V^{pitv}(A) \ne $ T.

It is worth noticing that the interpretation of quantifiers as shown in clause (6) above is substitutional. Hence, it is obligatory that the interpretation functions of the models provide a complete interpretation. The semantic concepts of validity, logical consequence, and satisfiability are defined in the usual way.

The system L_{F-} of *negative free logic* is adequate, i.e. sound and complete, with respect to the semantics with a partial interpretation function and a total valuation function. However, by a change of clauses (1) and (2) above in the definition of the valuation function V^{pitv} we can adapt M^{pitv}-models in such a way that they can be used for proving the adequacy of systems of positive free logic (in the way done by Hughes Leblanc and Robert K. Meyer).[13]

(S2) Semantics with an inner and an outer domain

We define an M^{iod}-model as a triple: $M^{iod} = (D_o, D_i, I^{iod})$. D_o and D_i are two disjoint and possibly empty sets of objects. D_o is called the *outer domain*, and D_i is called the *inner domain*, whose union is non-empty:

(i) $D_o \cap D_i = \varnothing$
(ii) $D_o \cup D_i \neq \varnothing$.

We define D as the union: $D = D_o \cup D_i$.
The *interpretation function* I^{iod} is a total function which is defined thus:

(1) for every individual constant t of L_F, $I^{iod}(t) \in D$;
(2) for every n-place predicate P^n of L_F, $I^{iod}(P^n) \subseteq D^n$;
(3) for every object $d \in D_i$ there is an individual constant t of L_F such that $I^{iod}(t) = d$.

The valuation function V^{iod} is also total and it assigns a truth-value, i.e. T or F, to each closed formula of L_F relative to an M^{iod}-model. V^{iod} is defined recursively as follows:

(4) $V^{iod}(P^n t_1, t_2, \ldots, t_n) = $ T iff $< I^{iod}(t_1), I^{iod}(t_2), \ldots, I^{iod}(t_n) > \in I^{iod}(P^n)$;
(5) $V^{iod}(t_1 = t_2) = $ T iff $I^{iod}(t_1) = I^{iod}(t_2)$.
(6) $V^{iod}(E!t) = $ T iff $I^{iod}(t) \in D_i$.
(7) $V^{iod}(\sim A) = $ T iff $V^{iod}(A) \neq $ T;
(8) $V^{iod}(A \rightarrow B) = $ T iff $V^{iod}(A) \neq $ T or $V^{iod}(B) = $ T or both;
(9) $V^{iod}(\forall v A) = $ T iff for every individual constant t: if $I^{iod}(t) \in D_i$ then $V^{iod}(A(t/v)) = $ T.
(10) $V^{iod}(A) = $ F iff $V^{iod}(A) \neq $ T.

M^{iod}-models are used mainly for *positive free logic*. L_{F+} is adequate with respect to M^{iod}-models (as it was proved by Hughes Leblanc and Richmond Thomason).[14]

---

[13] See, for instance (Leblanc and Meyer 1970).

[14] See, for instance (Leblanc and Thomason 1968).

(S3) Supervaluation semantics

Meinongianism is not appealing for everybody, which makes the inner/outer domain semantics not a favorite for all. Then the question arises how to develop a semantics appropriate for a positive free logic without using the inner/outer domain semantics? *Supervaluation semantics* comes as a compelling solution to this. It starts with models of the same type as in the first approach; however, atomic sentences containing empty singular terms are allowed to be truth-valueless. But this would result in a rejection of classical laws of logic, and in order to avoid this effect, the models are 'completed'. This way, the truth-value gaps which result in the first part of the valuation process are removed.

This way we build a new type of models: $M^{sv} = (D, I^{sv})$. D is again a possibly empty set of objects and I^{sv} is a partial interpretation function like I^{pitv}. As in the case of M^{pitv}-models, the conditions we impose here on I^{sv} are the same as the conditions we imposed earlier on I^{pitv}. Thus:

(1) for every individual constant t of the language of L_F: either I^{sv} does not assign anything at all to t and $I^{sv}(t)$ thereby remains undefined or $I^{sv}(t) \in D$;
(2) for every n-place predicate P^n of L_F: $I^{sv}(P^n) \subseteq D^n$;
(3) for every object $d \in D$ there is an individual constant t of the language of L_F such that $I^{sv}(t) = d$.

However, unlike V^{pitv}, the valuation function V^{sv} associated with M^{sv}-models is also a partial function (like I^{sv}) and *its domain is restricted to atomic formulae* of L_F. Consequently, V^{sv} is a partial function from closed atomic formulae of L_F into the set $\{T, F\}$ of truth-values; it is defined as follows:

(1a) If for any t_i ($1 \le i \le n$), $I^{sv}(t_i)$ is defined, then $V^{sv}(P^n t_1, t_2, \ldots, t_n) = T$ iff $< I^{sv}(t_1), I^{sv}(t_2), \ldots, I^{sv}(t_n) > \in I^{sv}(P^n)$;
(1b) If for at least one t_i ($1 \le i \le n$), $I^{sv}(t_i)$ is undefined, then $V^{sv}(P^n t_1, t_2, \ldots, t_n)$ is undefined.
(2a) If both $I^{sv}(t_1)$ and $I^{sv}(t_2)$ are defined, then $V^{sv}(t_1 = t_2) = T$ iff $I^{sv}(t_1) = I^{sv}(t_2)$.
(2b) If either $I^{sv}(t_1)$ or $I^{sv}(t_2)$ is undefined but the other is defined then $V^{sv}(t_1 = t_2) = F$.
(2c) If neither $I^{sv}(t_1)$ nor $I^{sv}(t_2)$ is defined, then $V^{sv}(t_1 = t_2)$ is undefined.
(3) $V^{sv}(E!t) = T$ iff $I^{sv}(t)$ is defined, and $V^{sv}(E!t) = F$ iff $I^{sv}(t)$ is undefined.

We define now the concept of a *completion* (i.e. a complete supermodel) of an M^{sv}-model:

$M^{csv} = (D', I^{csv})$ is a *completion* of $M^{sv} = (D, I^{sv})$ iff

(1) $D' \ne \varnothing$;
(2) $D \subseteq D'$;
(3) for every n-place predicate P^n: $I^{sv}(P^n) \subseteq I^{csv}(P^n)$;
(4) for every individual constant t: if $I^{sv}(t)$ is defined then $I^{csv}(t) = I^{sv}(t)$;
(5) for every individual constant t: $I^{csv}(t) \in D'$.

(1) through (4) say that M^{csv} is a supermodel of M^{sv}, and clause (5) says that I^{csv} is a total function and M^{csv} is, accordingly, 'complete'.

Now, from the point of view of an M^{sv}-model of which M^{csv} is a completion, the valuation function V^{csv} of an M^{csv}-model is a total function from all the closed formulae of L_F into the set $\{T,F\}$ of truth-values. V^{csv} therefore depends on V^{sv}. It is defined as follows:

(1) If A is a closed atomic formula of L_F and $V^{sv}(A)$ is defined, then $V^{csv}(A) = V^{sv}(A)$.

(2) If A is a closed atomic formula of L_F and $V^{sv}(A)$ is undefined, then $V^{csv}(A)$ is determined independently of V^{sv} in the usual way for complete models as follows:

(2a) If A is a closed atomic formula of the form $(P^n t_1, t_2, \ldots, t_n)$, then $V^{csv}(P^n t_1, t_2, \ldots, t_n) = T$ if $<I^{csv}(t_1), I^{csv}(t_2), \ldots, I^{csv}(t_n) > \in I^{csv}(P^n)$, and $V^{csv}(P^n t_1, t_2, \ldots, t_n) = F$ if $< I^{csv}(t_1), I^{csv}(t_2), \ldots, I^{csv}(t_n) > \notin I^{csv}(P^n)$.

(2b) If A is a closed atomic formula of the form $t_1 = t_2$, then $V^{csv}(t_1 = t_2) = T$ iff $I^{csv}(t_1) = I^{csv}(t_2)$.

(2c) If A is a closed atomic formula of the form $E!t$, then $V^{sv}(E!t)$ is always defined. Hence, clause (1) will do the job, i.e. for each individual constant $t : V^{csv}(E!t) = V^{sv}(E!t)$.

(3) $V^{csv}(\sim A) = T$ iff $V^{csv}(A) = F$;

(4) $V^{csv}(A \rightarrow B) = T$ iff $V^{csv}(A) = F$ or $V^{csv}(B) = T$ or both;

(5) $V^{csv}(\forall v A) = T$ iff for every individual constant t: if $I^{csv}(E!t) = T$ then $V^{csv}(A(t/v)) = T$.

Next, we define the *supervaluation* $S(M^{sv})$ as a partial function from closed formulae of L_F into the set $\{T,F\}$ of truth-values as follows:

(1) $S(M^{sv})(A) = T$ iff $V^{csv}(A) = T$ for very completion M^{csv} of M^{sv}.

(2) $S(M^{sv})(A) = F$ iff $V^{csv}(A) = F$ for very completion M^{csv} of M^{sv}.

(3) $S(M^{sv})(A)$ is undefined otherwise, i.e. iff $V^{csv}(A) = T$ for at least one completion M^{csv} of M^{sv} and $V^{csv'}(A) = F$ for at least one completion $M^{csv'}$ of M^{sv}.

Lastly, we define logical consequence in terms of supervaluations in the following way: a closed wff of L_F is *logically supertrue* iff for all M^{sv}-models $M^{sv}:S(M^{sv})(A) = T$.

A closed formula B of L_F is a *logical consequence* of a class C of closed formulae of L_F iff for all M^{sv}-models M^{sv}: if $S(M^{sv})(A) = T$ for each $A \in C$, then $S(M^{sv})(B) = T$.

A set C of closed formulae of L_F is *supersatisfiable* iff there is at least one model M^{sv} such that $S(M^{sv})(B) = T$ for each $B \in C$.

Bas van Fraassen has been used supervaluation semantics in order to prove soundness and completeness of *positive free logic with=*.[15]

---

[15]See (van Fraassen 1966).

10.5   What Is a Free Description and How Does It Help?

In order to articulate the principles which govern the fictional discourse a promising approach is positive free logic; for we want at least some sentences containing fictional terms be true in intended interpretations ('in the story', 'in the novel', and the like). However, since the ontological account of fictional objects I sketched before makes those object essentially dependent upon features of their mark through which the objects are introduced into discourse, i.e. makes them objects of reference, fictional singular terms which refer to those objects are essentially tied to descriptions. And as I already pointed out a Russellian construal of descriptions will make sentences containing names of fictional objects literally false. So what we need are descriptive paraphrases of sentences in which fictional singular terms occur, such that improper singular descriptions be allowed as genuine constituents of those paraphrases. In a couple of words, what are needed here are free descriptions.

Indeed, what free logic does is to free us from ontologically committing ourselves to the existential presupposition of classic description theories. Not all 'real' singular terms have to refer. Consequently, we are allowed to put forward description theories which legitimate the view that improper descriptions are genuine singular terms lacking reference.

This is a way out from Russell's theory of definite descriptions. And not only because we collide head on with Russell's view that after all definite descriptions are incomplete symbols not to be identified with a sub-class of genuine proper names. But also because if we look at Russell's theory from a free logical point of view then his views turn out to be more akin to negative free logic. To the contrary, Frege's theory of definite descriptions is closer in spirit to free logic and it is more akin to positive free logic. The fundamentals of Frege's theory as laid down in (Frege 1893) will give ground to this claim. Indeed, Frege took descriptions to be genuine singular terms, and he considered descriptions and simple names as instantiating the general category of proper names. In an ideal scientific language Frege does not find any proper place for empty singular terms. They occur, however, in natural languages in at least two ways: (i) there are proper names which do not refer to anything which really exists (such as 'Holmes', 'Zeus' a.s.o.); (ii) there are improper descriptions which can occur in perfectly meaningful sentences: 'the planet closer to the Sun than Mercury' (the description is improper for it is contingently empty) or 'the greatest prime number' (the description is improper for it is necessarily empty). Frege's way out from this drawback is to secure a referent for descriptions that would otherwise look suspicious due to their non-referring status. Basically, to do this Frege stipulates a solution whose artificiality is obvious. But Frege was not misled by his own move, and he did not attempt anything like a linguistic analysis of the actual usage of improper descriptions. What he meant was a scientific revision of improper use of the language via scientifically better substitutes for problematic phrases in the vernacular. There are two suggested moves that Frege makes in order to circumvent problems of the sort created by descriptions which lack reference. He stipulates that all improper descriptions designate an arbitrarily chosen object, such as the empty set or the number 0 or the truth-value F. Frege construes thus the identity sentence $(\iota v)\, A = (\iota v)\, B$ as true when nothing responds to the condition A and nothing responds to B, and likewise if more than one thing is A and more than one thing is B. Or else, he incorporates some set theory or something similar to it (viz. his theory of value-ranges) into his logical theory. Accordingly, if the predicate A is uniquely satisfied then $(\iota v)A$ denotes the unique object denoted by t such that $A(t/v)$, and if A is not uniquely satisfied then $(\iota v)A$ will denote the set $\{v/A\}$ of things that satisfy A.

If we go beyond the conceptual and technical differences that really separate Frege from Russell with regard to the proper logical analysis of singular descriptions, we will find this common classic existential presupposition which is shared by both that in order to be considered real or genuine the singular terms have to refer to something which is either a real existing thing or it is artificially constructed or stipulated.

Free logic liberates us from this assumption. Nowadays, there is a great variety of free descriptions theories. All of them incorporate the following principle which is commonly known as

$$\textit{Lambert's Law} : (\forall v)\,(v = (\iota w)\, A \leftrightarrow (\forall w)\ (A \rightarrow w = v)).$$

What the law says is that a description is proper when its scope is uniquely fulfilled.

As such *Lambert's Law* does not hold in standard first order logic. However, if we assume

$$\textit{Hintikka's Law} : \ \mathrm{E}!t \leftrightarrow (\exists v)\,(v = t)$$

whose import is the equivalence of singular existence with the existence of an individual one can derive Russell's theory, and furthermore the negative free logic implicature of Russell's theory, provided we add the principle $(\iota v)A = (\iota v)A$.

On the other hand, starting from a positive free logic with self-identity, one can add the same minimal assumption which is added by every free theory of definite descriptions, viz. *Lambert's Law*, and get a Fregean positive free description theory. As a matter of fact, a whole hierarchy of definite description theories can be generated starting from the theory containing only *Lambert's Law* as the minimal theory. What is the nature of this hierarchy is not something very well understood.[16] In any rate, free logic is an excellent place to formulate and assess different competing definite description theories, and having in view the essential ties between fictional names and definite descriptions, that makes free logic an ideal background against which one can advance and evaluate metaphysical theses about fictional objects.

---

[16]Cf. (Lambert 2001).

10.6 Conclusions

The *main ontological thesis concerning fictional objects* adopted in this paper has been that fictional objects are essentially objects of reference, i.e. objects created through a story or a narrative and introduced via a cluster of descriptions.

This creates an essential connection between terms that introduce into discourse fictional objects and (singular) definite singular descriptions. It seems to me obvious that people have created in various contexts pieces of meaningful discourse about those alleged entities that fictional objects are. How could one account for *that* if the corresponding descriptions actually do not refer to anything? The descriptions are just literally empty.

The proposal which I am making in my paper is that a free logic interpretation of the descriptions is appropriate to solve the puzzle. More specifically, a brand of positive free logic will help us to make sense of our semantic intuitions that according to the background discourse sentences in which fictional terms occur may be very well true, even if the supposed existential assumption on which descriptions are built is not satisfied. Free logic seems to me a good option to deal with this very intricate issue in philosophy of language and philosophical logic.[17]

Bibliography

Donnellan K (1966) Reference and definite descriptions. In: Martinich AP (ed) Philosophy of language. Oxford University Press, Oxford, pp 235–247

Field H (1980) Science without numbers: a defense of nominalism. Basil Blackwell, Oxford

Fine K (1984) Review of Parsons *Non-existent objects*. Philos Stud 45:95–142

Frege G (1893) Grundgesetze der Arithmetik

Lambert K (2001) Free logics. In: Goble L (ed) The Blackwell guide to philosophical logic. Blackwell, London, pp 258–279

Lambert K (2003) Free logic. Selected essays. Cambridge University Press, Cambridge

Leblanc H, Meyer RK (1970) On prefacing $(\forall X)A \supset A(Y/X)$ with $(\forall Y)$. A free quantification theory without identity. Zeitschrift für matematische Logik und Grundlagen der Mathematik 16:447–462, Reprinted in Leblanc H (1982) Existence, truth, and provability. State University of New York Press, Albany, pp 58–75

Leblanc H, Thomason RH (1968) Completeness theorems for some presupposition-free logics. Fundam Math 62:125–164, Reprinted in Leblanc H (1982) Existence, truth, and provability. State University of New York Press, Albany, pp 22–57

Morscher E, Hieke A (eds) (2001) New essays in free logic. In honour of Karel Lambert. Kluwer Academic Press, Dordrecht

Morscher E, Simons P (2001) Free logic: a fifty-year past and an open future. In: Morscher E, Hieke A (eds) New essays in free logic. In honour of Karel Lambert. Kluwer Academic Press, Dordrecht, pp 1–34

Parsons T (1980) Nonexistent objects. Yale University Press, New Haven

---

[17]The proposal developed here was initially presented in a paper published in *Logica Yearbook* 2004, 95–107.

Rosen G (1990) Modal fictionalism. Mind 99(395):327–54

Sainsbury RM (2010) Fiction and fictionalism. Routledge, London

Strawson P (1950) On referring. In: Martinich AP (ed) Philosophy of language. Oxford University Press, Oxford, pp 219–234

van Fraassen BC (1966) The completeness of free logic. Zeitschrift für matematische Logik Grundlagen der Mathematik 12:219–234

van Fraassen BC (1980) The scientific image. Clarendon, Oxford

Part IV
Quantum Phenomena, Scientific Realism, and Emergence

Chapter 11
Quantum Mechanics: Knocking at the Gates of Mathematical Foundations

Radu Ionicioiu

11.1 What Is the Problem? (Is There a Problem?)

From its very inception quantum mechanics generated a fierce debate regarding the meaning of the mathematical formalism and the world view it provides (Bohr, 1935, 1984; Einstein et al., 1935; Wheeler and Zurek, 1984). The new quantum *Weltanschauung* is characterized, on the one hand, by novel concepts like wave-particle duality, complementarity, superposition and entanglement, and on the other by the rejection of classical ideas such as realism, locality, causality and non-contextuality. For instance, quantum correlations with no causal order (Oreshkov et al., 2012) challenge Reichenbach's principle of common cause (Cavalcanti and Lal, 2014).

The disquieting feeling one has at the contact with quantum theory was echoed by several physicists, including the founding fathers: *"Anyone who is not shocked by quantum mechanics has not understood it"* (Bohr); *"I think I can safely say that nobody understands quantum mechanics"* (Feynman). Consequently, there is an unsolved tension between what we predict and what we understand (Adler, 2014; Laloë, 2012). Although the predictions of quantum mechanics are by far and away unmatched (in terms of precision) by any other theory, understanding "what-all-this-means" is lacking. Briefly, we would like to have a story behind the data, to understand the meaning of the formalism.

There is a wide spectrum of positions concerning the problem of quantum foundations (Echenique-Robba, 2013), with attitudes ranging from "there is no problem, don't waste my time" to "we do have a problem and we don't know how to solve it":

R. Ionicioiu (✉)
Department of Theoretical Physics, National Institute of Physics and Nuclear Engineering, 077125 Bucharest–Măgurele, Romania
e-mail: r.ionicioiu@theory.nipne.ro

© Springer International Publishing Switzerland 2015
I. Pârvu et al. (eds.), *Romanian Studies in Philosophy of Science*, Boston Studies in the Philosophy and History of Science 313, DOI 10.1007/978-3-319-16655-1_11

Actually quantum mechanics provides a complete and adequate description of the observed physical phenomena on the atomic scale. What else can one wish? (van Kampen, 2008) Quantum theory is based on a clear mathematical apparatus, has enormous significance for the natural sciences, enjoys phenomenal predictive success, and plays a critical role in modern technological developments. Yet, nearly 90 years after the theory's development, there is still no consensus in the scientific community regarding the interpretation of the theory's foundational building blocks. (Schlosshauer et al., 2013)

The confusion around the meaning of the formalism and the lack of an adequate solution to the measurement problem (among others) resulted in a plethora of interpretations. Apart from the (once) dominant Copenhagen interpretation – informally known as "shut-up-and-calculate" – there are numerous others: pilot wave (de Broglie-Bohm), many-worlds (Everett, 1957), consistent histories (Griffiths, 2014), transactional (Kastner, 2013), relational (Rovelli, 1996), Ithaca, quantum Bayesianism (Fuchs et al., 2014) etc.

Historically, the tone of the discussion was set by the Bohr-Einstein debate on the foundations of quantum theory (Bohr, 1935, 1984; Einstein et al., 1935). Einstein lost the conceptual battle due to his persistence to understand quantum phenomena in classical terms like realism, locality and causality. In this respect Einstein was wrong, but his mistake was fertile, as often happens, since the Bohr-Einstein debate, and the ensuing EPR argument (Bohr, 1935; Einstein et al., 1935), paved the way for the seminal results of Bell-CHSH (Bell, 1964; Clauser et al., 1969) and Kochen-Specker (Bell, 1966; Kochen et al., 1967). Although Bohr was correct – one cannot understand QM using classical notions – he was right in an unfruitful way, as his view dominated for decades and inhibited any rational discussion on the foundations of QM; in the words of Gell-Mann: *"Bohr brainwashed a whole generation of physicists into thinking that the problem of quantum mechanics has been solved"* (by the Copenhagen interpretation) (Gell-Mann, 1979).

Nonetheless, the discussions on quantum foundations never disappeared completely (although they were considered disreputable for a long time) and at present the field enjoys a renewed interest. This revival is due, first, to new experimental methods enabling to actually perform several *Gedanken* experiments (classic as well as new ones) (Ionicioiu and Terno, 2011; Jacques et al., 2007; Tang et al., 2012; Wheeler, 1984). And second, due to the recently emerged field of quantum information and its focus on the role played by information in physical systems. It is somehow ironic that philosophical discussions – started almost a century ago by Bohr, Einstein and other founding fathers – turn out to be essential nowadays in establishing the security (via Bell's inequality violation) of real life quantum cryptosystems.

The structure of the article is the following. I briefly review the relationship between physics and mathematics, the origin of mathematical concepts (like number, geometry etc.) and their evolution in time. Our main conjecture states that a new quantum ontology, based on non-classical mathematical concepts, is necessary for solving the existing quantum paradoxes and achieving a better understanding of quantum foundations. Finally, I discuss two research directions which could achieve this goal.

11.2 Physics and Mathematics

As discussed before, the theorems of Bell-CHSH (Bell, 1964; Clauser et al., 1969), Kochen-Specker (Bell, 1966; Kochen et al., 1967) and Leggett-Garg (Leggett and Garg, 1985) are major results in quantum foundations since they reframe the problem from philosophical arguments into precise observables which can be measured in the lab. These tests have been performed numerous times and with different physical systems. The conclusion is clear: quantum experiments cannot be explained in terms of classical desiderata such as *local realism* (Bell-CHSH), *non-contextuality* (Kochen-Specker) and *macroscopic realism* (Leggett-Garg).

Given this profound conflict between quantum experiments and our intuition, it is useful to take a step back and have a look at the structure of physical theories. One of the things we take for granted is the fundamental role of mathematics. Indeed, all theories have a mathematical framework in which the observed phenomena are translated. Galileo was the first to emphasize the crucial role played by mathematics in the description of natural phenomena:

> Philosophy is written in that great book which ever lies before our eyes – I mean the universe – but we cannot understand it if we do not first learn the language and grasp the symbols, in which it is written. This book is written in the mathematical language, and the symbols are triangles, circles and other geometrical figures, without whose help it is impossible to comprehend a single word of it; without which one wanders in vain through a dark labyrinth.

This brings us to the question: what is the origin of mathematics? How do we form, how do we arrive at our mathematical concepts? For example, constructs like number, geometry or vector space did not exist, historically, 10,000 years ago. There are several possible answers. For a Platonist, mathematical objects exist in an ideal, platonic world and we discover them by (allegedly) having access to these pre-existing forms.

There is a second position – which we support here – regarding mathematical concepts. According to this ansatz, mathematical concepts are distilled, or abstracted, from our interaction with the external world (Mac Lane, 1981).[1] This view has two consequences. First, it naturally answers Wigner's dilemma about the unreasonable effectiveness of mathematics in describing physical phenomena (Wigner, 1960). And second, it provides an insight towards solving the conflict between our classical intuition and quantum experiments. Our main mathematical concepts are distilled from a fundamentally classical world. As such, these constructs have a definite classical flavour and, consequently, cannot capture irreducible quantum aspects – after all, they were not designed to deal with these phenomena in the first place.

---

[1] I'm grateful to Prof. I. Pârvu for bringing to my attention Mac Lane's article (Mac Lane, 1981) which shares a similar view of mathematical concepts.

Therefore we replace the question *why quantum mechanics is paradoxical and defies our classical intuition?* by asking instead *what type of mathematical concepts can we distill from quantum experiments?* This perspective shift helps us to break away from preconceived notions inherited from classical physics. Instead of imposing classical prejudices on the description of quantum phenomena, one aims to find the natural logico-mathematical concepts emerging from those experiments.

This insight has its roots in the seminal article of Birkhoff and von Neumann on quantum logic (Birkhoff and von Neumann, 1936) in which they showed that propositions about quantum particles obey a different type of logic. More exactly, the distributive law (valid in classical logic) fails in quantum logic, $p \wedge (q \vee r) \neq (p \wedge q) \vee (p \wedge r)$. Putnam masterfully captured this perspective shift in the title of his famous article *Is logic empirical?* (Putnam, 1968).

The recent topos program (Doering and Isham, 2011) and the quasi-set program (French and Krause, 2010) follow similar lines of thought. The common idea behind these programs is that quantum phenomena require new mathematical structures. The topos approach generalizes the concept of set by using the topos category instead. In the quasi-sets program the indistinguishability of quantum particles is build-in right from the start in the concept of quasi-set. Other approaches include paraconsistent logic (da Costa and de Ronde, 2013), many-valued logic (Pykacz, 2014) and sheaves (Abramsky and Brandenburger, 2011).

11.3   Mathematics: Evolving Concepts

> The 'paradox' is only a conflict between reality and your feeling of what reality 'ought to be'.
>
> Feynman

The previous ansatz and the intrusion of the empirical into logic and set theory – the very foundations of mathematics – is unsettling for many people. Mathematics, and even more so logic, has still an aura of absolute truth, uncontaminated by the contingency of real world phenomena. It is illuminating to see in perspective how our mathematical concepts evolved historically and how they were also shaped by our prejudices.

A textbook example are the rational numbers. For Pythagoras the number was, perforce, a rational number; other types of numbers were inconceivable. Greek mathematics was plunged into a profound crisis with the discovery that the diagonal of the unit square is irrational (according to the legend, Hippasus was drowned for this discovery). For modern mathematics rational numbers are just one possible type of numbers, among others. So from our perspective it is difficult to understand the depth of the crisis.

Two key ideas of Pythagorean philosophy will clarify this difficulty. First, for Pythagoras the universe is a Kosmos, therefore ordered, harmonious (in opposition to Chaos). And second, the number is the measure of everything. As a consequence, any two numbers should be commensurable, hence their ratio has to be a rational number. In a Kosmos everything *should be* expressed as a rational number, since only commensurable quantities can exist.

This reveals the dynamics of the paradox. A preconceived idea (the universe is ordered, a Kosmos) has a logical consequence that all quantities (e.g., lengths) should be commensurable. A Kosmos, by definition, cannot have incommensurable lengths.

Interestingly, we can find echoes of this conceptual crisis in everyday language. From a purely technical term with a precise meaning – a number which is not a ratio of two integers – *irrational* became a cognitive attribute. The link between the two meanings (mathematical and cognitive) is straightforward. For Pythagoreans, an irrational number is an inconceivable concept, which cannot be thought of logically (i.e., rationally). Ironically, today one can logically prove theorems about irrational numbers (a clearly rational activity).

A similar evolution happened repeatedly in the history of mathematics. Before the discovery of complex numbers it was impossible to imagine a number whose square is negative (hence imaginary number). Likewise, Euclidean geometry reigned supreme for more than two millennia. Euclidean geometry was *the* geometry, the only (logically) possible one. One can hardly underestimate this prejudice – even Gauss (known as *Princeps mathematicorum*) hesitated to publish his ideas about non-Euclidean geometry. Today, Euclidean geometry is just one possible geometry, among others.

The message thus becomes apparent: whenever we stumble upon a paradox, we need to critically examine our prejudices and be ready to extend our concepts.

11.4 Realism: From Classical to Quantum

When it comes to atoms, language can be used only as in poetry.
The poet, too, is not nearly so concerned with describing facts
as with creating images and establishing mental connections.

Bohr

Let us consider now the problem of realism. Classical systems are endowed with well-defined properties and a measurement only reveals them, ideally without changing them. This is, in essence, classical realism: systems have intrinsic, pre-existing attributes which are independent of the measuring apparatus. Thus one can talk about the objects *possessing* the properties.

This intuition is challenged by quantum experiments, via Bell-CHSH, Kochen-Specker and Leggett-Garg theorems. There are several statements in the literature along these lines: quantum mechanics forces us to abandon realism, quantum systems do not have pre-existing properties, the measurement creates the properties. These statements are in a sense true, but also somehow vague, resulting in several misleading claims. In the following we aim to clarify these aspects.

First, quantum mechanics does not compel us to abandon realism understood in a very broad sense: there is a world out-there. There is no need to fall into solipsism or subjectivism. Notwithstanding certain (unwarranted) claims like "consciousness collapses the wave-function", quantum mechanics does not deny the existence of an external reality. Without this minimal notion of reality one cannot do science, and even less predict or talk about what the Universe looked like before Earth and human observers came into existence. We adopt the following definition of realism:

There exists an external world independent of our consciousness, but not (necessarily) of our actions.

A second, rough-and-ready notion of realism is: *something is real if I can kick it and it kicks back*. Thus I can talk about an electron being real, since I can kick it (change its state with an external field) and it kicks back (it emits bremsstrahlung radiation).

True, quantum reality is a very different beast from the classical world – certain aspects are affected by our actions and thus one can roughly talk about the non-existence of pre-defined properties. In this sense a measurement "brings into existence" these attributes.

Second, quantum objects – photons, electrons, protons – do have certain intrinsic properties (thus invariant upon measurement). Electric charge, total spin, rest mass, leptonic and baryonic numbers are such examples. The very fact that one can talk about a photon versus an electron shows that these two entities are different and can be differentiated based on certain characteristics. An electron has always an electric charge e and spin-$\frac{1}{2}$, whereas a photon is always chargeless and has spin-1. Without the existence of certain intrinsic properties words like "photon", "electron" or "proton" would be meaningless.

Nevertheless, it is also true that quantum systems do not have, prior to measurement, other type of properties, like the spin component on a given axis (e.g., S_z). For an electron, measuring the spin component along a direction **n** will randomly produce either $+1$ or -1 (in units of $\frac{\hbar}{2}$). Thus the spin component is not predefined, does not exist prior to the measurement; equivalently, the measurement does not reveal a pre-existing attribute.

In the classical world the measurement is passive, like reading a book (pre-existing text). In the quantum realm the measurement is active, it elicits an answer. We probe the system and we obtain an answer according to the question we ask – there is no answer before asking the question. This is the meaning of Peres dictum "unperformed experiments have no results" (Peres, 1978).

There is an interesting twist to the previous statement, namely quantum-controlled experiments (Ionicioiu and Terno, 2011). In this case one can have the answer before we know the question, but the answer has to be consistent with the subsequently revealed question. As a result, we need to *interpret* what we measure. In other words, the answer (the measurement result) is meaningless without the context (the question asked).

The difference between classical and quantum view of reality can be summarized as follows:

| **classical**: | reality = pre-existing $\Rightarrow$ | measurement *reveals* |
| **quantum**: | reality $\neq$ pre-existing $\Rightarrow$ | measurement *creates* |

Recently, (Kochen, 2013) attempted to clarify this confusing state of affairs by drawing a distinction between two types of properties: *intrinsic* and *extrinsic* (relational). Intrinsic properties are familiar from everyday (classical) world, where attributes are intrinsic. On the other hand, extrinsic, or relational properties depend on the measurement performed on the system (e.g., in quantum experiments).[2]

11.4.1 A Metaphor

Extrinsic properties may seem counterintuitive from a classical perspective. Inspired by the previous quote from Bohr, we aim to make this quantum behaviour a bit less mysterious with the help of a metaphor.

Suppose we have a glass cube – this is a well-defined macroscopic object and has well-defined classical properties. What colour is the cube? If we observe it with a red laser, it appears red; if we examine it with a blue laser, it appears blue. Clearly, colour is not a pre-existing attribute of the cube, but depends on the measuring device (the colour of the laser). The cube is colourless, but this does not make the cube less real. However, this metaphor does not capture very well the behaviour of a quantum spin.

So let's complicate a bit the picture. Assume now we have a cube made of (a hypothetical) quantum-glass, or *qlass*. The quantum-glass has a light-sensitive dye with a peculiar property: if we illuminate it with a laser of colour **c**, the qlass colour randomly becomes either **c** or **c̄** (the complementary color in the RGB space). Thus, if we observe the qlass cube with a red laser, the cube will randomly appear either red or cyan; if we observe it with a blue laser, it will randomly appear either blue or yellow. The probabilities of the two occurrences are determined by the initial state of the qlass cube.

We can extend this metaphor to include, for example, an entangled pair of quantum-glass cubes. In this case, even if the two cubes are spatially separated, when we probe them with lasers of the same color **c**, they will glow in random, but always opposite colours **c** and **c̄**. For space-like separated measurements, Bell-CHSH theorem ensures us that neither pre-existing properties, nor signalling can explain the magnitude of the experimental correlations.

---

[2]Kochen's distinction between intrinsic and extrinsic properties is different from the more well-known one discussed in Stanford Encyclopedia of Philosophy, http://plato.stanford.edu/entries/intrinsic-extrinsic/.

11.4.2   A Realism Revival

The state of a quantum system is fully specified by its wavefunction ψ. In contrast to classical physics, ψ predicts only the probabilities for different measurement results, and not the individual outcomes.

This brings us to a crucial question behind Bohr-Einstein debate, namely how to interpret the wavefunction. Is ψ related merely to our incomplete knowledge of the system (the ψ-epistemic view)? Or does ψ correspond to an objective property of the system (the ψ-ontic view)?

Recently the ψ-ontic vs. ψ-epistemic problem (Harrigan and Spekkens, 2010) became a very active topic in quantum foundations. In a seminal article Pusey, Barrett and Rudolph (PBR) proved that quantum mechanics is incompatible with ψ-epistemic models, if we assume preparation independence (Pusey et al., 2012). The PBR theorem was hailed as the most important result in quantum foundations since Bell's inequality. However, it also generated a heated debate with several articles criticizing its assumptions and defending the ψ-epistemic interpretation.

Significantly, the PBR result reopened the discussion regarding the meaning of the wavefunction. In the wake of the PBR article other no-go theorems for ψ-epistemic models have been proved starting from different assumptions. By assuming free-choice, Colbeck and Renner proved that the wavefunction of a system is in one-to-one correspondence with its elements of reality (Colbeck and Renner, 2012, 2013). Patra et al. (2013a) derived a no-go theorem for ψ-epistemic models starting from continuity and weak separability; this was tested experimentally in Patra et al. (2013b).

Thus the wavefunction cannot be viewed *only* as a state of knowledge, but is directly related to objective attributes of the system (Aharonov et al., 1993). In view of these results, the ψ-epistemic interpretation becomes increasingly difficult to defend. In a way, this mirrors the downfall of hidden-variables theories after Bell-CHSH theorem.

The question now is: what are the implications for realism if the wavefunction is indeed related to objective attributes of a quantum system – or, assuming free-choice, ψ is in one-to-one correspondence with its elements of reality (Colbeck and Renner, 2012)? We address this question in the next section.

11.4.3   Realism, But Not the Way We Know It

In classical physics ontological existence is in one-to-one correspondence with measurable, pre-existing properties: "if I can measure it, it's real; if I can't measure it, it's not". In addition, these pre-existing properties are described by real numbers, or n-tuples of real numbers (vectors, tensors, quaternions etc.).[3]

In quantum mechanics this is no longer the case. The wavefunction ψ determines probabilistically the measurement outcomes (the "properties"), but collapses after measurement. Moreover, Kochen-Specker theorem implies that quantum mechanics is contextual: the measurement results depend on the context, i.e., on the other compatible observables co-measured with it.

Recall the previous discussion about the two meanings of rational/irrational: a mathematical one (ratio of two integers) and a cognitive one. One can detect a similar dynamics here, with *real* having again two distinct meanings: mathematical (real as a number, the power of continuum $\mathfrak{c} = 2^{\aleph_0}$) and existential (real as in *existing-out-there*). In a certain sense, we subconsciously equate *reality* (ontological existence) with attributes possessing real values. I think this preconception prevents us from overcoming the actual crisis of quantum foundations. Thus, in order to comprehend quantum phenomena we need to extend our concept of reality by looking for novel mathematical structures beyond that of real numbers. In essence, we have to stop identifying reality with real numbers, i.e., real-valued pre-existing properties.

The recent theorems of Pusey-Barrett-Rudolph (Pusey et al., 2012) and Colbeck-Renner (Colbeck and Renner, 2012, 2013) imply that, for quantum systems

The wavefunction ψ is real, the properties are not.

Accordingly, only the wavefunction ψ has ontological existence. The measurement results (what classically one would call *properties*), like the spin projection on a given axis, are not pre-existing prior to the observation, and thus are not revealed by the measurement. In contrast to the classical case, for quantum systems the "properties" are created by the act of measurement. However, since ψ is not directly measurable, this is clearly a weaker form of (ontological) existence than the classical one.

The outcome of a quantum measurement depends on two factors: (i) the wavefunction ψ prior to the measurement, and (ii) the measured observable. The outcome itself does not have an objective existence before the measurement. This relates to Kochen's relational (extrinsic) properties discussed above.

Metaphorically, the measurement results are like different shadows of a three-dimensional object (Ananthaswamy, 2013): they reveal information about the underlying reality, but there are not, by themselves, objective properties of this reality (that is, prior to the collapse, see below). The shadow depends both on the object (the wavefunction) and on the viewing angle (the type of measurement we perform). For example, depending on the direction, a cylinder can appear as a disk, as a rectangle, or as anything in between (morphing) (Ionicioiu and Terno, 2011).

---

[3]Due to the finite experimental precision, the outcome of a measurement is always a rational number. However, it is generally assumed that the underlying physical property is continuous and takes values in a subset of $\mathbb{R}$.

Obviously, the cylinder is neither a disk, nor a rectangle – it's a three-dimensional object transcending its shadows.

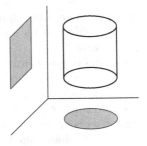

However, a quantum measurement is active, not passive – the measurement changes what we measure. In tune with the previous metaphor, the object transforms into (becomes) its shadow after looking at it – this is the *collapse of the wavefunction.*

To date we don't have a definite answer to the question: *what mathematical structure emerges from quantum experiments and can serve as a basis for a quantum ontology?* The problem is still open, notwithstanding several attempts like quantum logic (Birkhoff and von Neumann, 1936), topos theory (Doering and Isham, 2011), quasisets (French and Krause, 2010), sheaves (Abramsky and Brandenburger, 2011), paraconsistent logic (da Costa and de Ronde, 2013) and many-valued logic (Pykacz, 2014). Quantum logic has been around for more than 70 years, but it yielded few results. Topos theory and quasisets are still in the beginning and until now they produced no breakthroughs.

To end on a more speculative note, I sketch two research directions which can spur the development of a future quantum ontology.

The first is to reconsider the Zermelo-Fraenkel (ZF) axioms of set theory in the context of quantum mechanics. The ZF axioms are based on classical intuitions (distinguishability, intrinsic properties etc.) which appear inappropriate to describe indistinguishable quantum particles without pre-existing properties. A specific example is how to make sense of the Axiom of separation in the context of non-commuting observables and extrinsic properties.

A second line of research is related to Cantor continuum hypothesis: there are no sets with cardinality between $\aleph_0$ and $2^{\aleph_0}$. Gödel and Cohen proved the independence of the continuum hypothesis from the ZFC axioms (ZF with choice). Consequently, one can construct non-Cantorian sets for which the continuum hypothesis is false – this clearly mirrors non-Euclidean geometries in which the parallel postulate does not hold.

Consider now a spin-$\frac{1}{2}$: the wavefunction is continuous, but the measurement outcomes are always discrete. Moreover, the wavefunction collapses after the measurement. In this context it will be interesting to see if a non-Cantorian set (with cardinality between $\aleph_0$ and $2^{\aleph_0}$) is more appropriate to capture this quantum behaviour.

We expect that novel mathematical structures (e.g., non-Cantorian sets, non-classical logic etc.) will provide an ontological basis for quantum foundations. This new ontology will hopefully solve the quantum paradoxes (Aharonov and Rohrlich, 2005; Aharonov et al., 2014), on the one hand, and provide a deeper understanding of quantum mechanics, on the other.

Historically a similar change happened with the advent of general relativity, which replaced the classical structure of Euclidean geometry with Riemannian geometry. Non-Euclidean geometry provided general relativity (GR) with both an adequate *ontology* and a new *metaphor*, the dynamical interplay between spacetime an matter: "spacetime tells matter how to move; matter tells spacetime how to curve" (Ford and Wheeler, 2010). The powerful visual impact of this metaphor generated an intuitive understanding of general relativity. This explains why there are no (deep, unsolved) foundational problems in GR compared to QM.

11.5 Discussion

Like seeds leaving a tree, mathematical concepts have a life of their own, departing from their roots, but still keeping an imprint of their origin. One can thus speak of the *ontological continuity of mathematical concepts*: however abstract or far from reality a mathematical construct might seem, it still has a link, an umbilical cord, to something "existing out-there".

The main intuition behind this article is a conjecture regarding the nature of mathematical concepts and their role in explaining the external world. The ansatz and its consequences can be summarized as follows:

1. Mathematical concepts and structures are distilled from our interaction with the external world.
2. Classical concepts – realism, causality, locality and Aristotelian logic – emerge as structures of a classical universe. Like Euclidean geometry or rational numbers, classical logic is fine per se, as a mathematical structure. However, it is not an adequate model for the quantum world.
3. Paradoxes signal a conflict between reality and our expectations (prejudices) of what reality "should be" (Feynman).
4. Quantum experiments compel us to abandon local realism, causality and non-contextuality. Consequently, we need novel mathematical and logical structures more adequate to describe quantum reality.

We envision that mystifying quantum characteristics (contextuality, nonlocality, the failure of classical realism, collapse of the wavefunction etc.) will emerge naturally as different facets of the same underlying structure, which will form the basis of a new quantum ontology. Despite several attempts in this direction, such a unifying concept is still lacking.

These are interesting times for quantum foundations. We expect the centenary of quantum theory to bring forth not only quantum technologies, but also a genuine understanding and a new quantum ontology.

Acknowledgements I am grateful to Ilie Pârvu, Cristi Stoica and Iulian Toader for discussions and critical comments of the manuscript.

References

Abramsky S, Brandenburger A (2011) The sheaf-theoretic structure of non-locality and contextuality. New J Phys 13:113036

Adler SL (2014) Where is quantum theory headed? arXiv:1401.0896

Aharonov Y, Rohrlich D (2005) Quantum paradoxes. Wiley, Weinheim

Aharonov Y, Anandan J, Vaidman L (1993) Meaning of the wave function. Phys Rev A 47:4616

Aharonov Y, Colombo F, Popescu S, Sabadini I, Struppa DC, Tollaksen J (2014) The quantum pigeonhole principle and the nature of quantum correlations. arXiv:1407.3194

Ananthaswamy A (2013) Quantum shadows. New Scientist 5 Jan 2013, p 36

Bell JS (1964) Physics 1:195

Bell JS (1966) Rev Mod Phys 38:447

Birkhoff G, von Neumann J (1936) The logic of quantum mechanics. Ann Math 37:823

Bohr N (1935) Can quantum-mechanical description of physical reality be considered complete? Phys Rev 48:696

Bohr N (1984) Discussion with Einstein on epistemological problems in atomic physics. In: Wheeler JA, Zurek WH (eds) Quantum theory and measurement. Princeton University Press, Princeton, pp 9–49

Cavalcanti EG, Lal R (2014) On modifications of Reichenbach's principle of common cause in light of Bell's theorem. J Phys A Math Theor 47:424018

Clauser JF, Horne MA, Shimony A, Holt RA (1969) Phys Rev Lett 23:880

Colbeck R, Renner R (2012) Is a system's wave function in one-to-one correspondence with its elements of reality? Phys Rev Lett 108:150402

Colbeck R, Renner R (2013) A system's wave function is uniquely determined by its underlying physical state. arXiv:1312.7353

da Costa N, de Ronde C (2013) The paraconsistent logic of quantum superpositions. Found Phys 43(7):845–858

Doering A, Isham C (2011) What is a thing?: Topos theory in the foundations of physics. In: Bob C (ed) New structures for physics, vol 813, Springer lecture notes in physics. Springer, Heidelberg, pp 753–940

Echenique-Robba P (2013) Shut up and let me think! Or why you should work on the foundations of quantum mechanics as much as you please. arXiv:1308.5619

Einstein A, Podolsky B, Rosen N (1935) Can quantum-mechanical description of physical reality be considered complete? Phys Rev 47:777

Everett H (1957) "Relative State" formulation of quantum mechanics. Rev Mod Phys 29:454

Ford K, Wheeler JA (2010) Geons, black holes, and quantum foam. W.W. Norton, New York, p 235

French S, Krause D (2010) Remarks on the theory of quasi-sets. Studia Logica 95:101

Fuchs CA, Mermin ND, Schack R (2014) An introduction to QBism with an application to the locality of quantum mechanics. Am J Phys 82(8):749–754

Gell-Mann M (1979) In: The nature of the physical universe: the 1976 Nobel conference. Wiley, New York, p 29

Griffiths RB (2014) The new quantum logic. Found Phys 44:610–640

Harrigan N, Spekkens RW (2010) Einstein, incompleteness, and the epistemic view of quantum states. Found Phys 40:125

Ionicioiu R, Terno DR (2011) Proposal for a quantum delayed-choice experiment. Phys Rev Lett 107:230406

Jacques V et al (2007) Experimental realization of Wheeler's delayed-choice Gedanken experiment. Science 315:966

Kastner RE (2013) The transactional interpretation of quantum mechanics. Cambridge University Press, Cambridge

Kochen S (2013) A reconstruction of quantum mechanics. arXiv:1306.3951

Kochen S, Specker EP (1967) J Math Mech 17:59

Laloë F (2012) Do we really understand quantum mechanics? Cambridge University Press, Cambridge. arXiv:quant-ph/0209123

Leggett AJ, Garg A (1985) Phys Rev Lett 54:857

Mac Lane S (1981) Mathematical models: a sketch for the philosophy of mathematics. Am Math Mon 88(7):462–472

Oreshkov O, Costa F, Brukner C (2012) Quantum correlations with no causal order. Nat Commun 3:1092

Patra MK, Pironio S, Massar S (2013a) No-go theorems for ψ-epistemic models based on a continuity assumption. Phys Rev Lett 111:090402

Patra MK, Olislager L, Duport F, Safioui J, Pironio S, Massar S (2013b) Experimental refutation of a class of ψ-epistemic models. Phys Rev A 88:032112

Peres A (1978) Unperformed experiments have no results. Am J Phys 46:745

Pusey MF, Barrett J, Rudolph T (2012) On the reality of the quantum state. Nat Phys 8:475

Putnam H (1968) Is logic empirical? In: Cohen R, Wartofsky M (eds) Boston studies in the philosophy of science, vol 5. Reidel, Dordrecht, pp 216–241

Pykacz J (2014) Can many-valued logic help to comprehend quantum phenomena? arXiv:1408.2697

Rovelli C (1996) Relational quantum mechanics. Int J Theor Phys 35:1637. arXiv:quant-ph/9609002

Schlosshauer M, Kofler J, Zeilinger A (2013) A snapshot of foundational attitudes toward quantum mechanics. Stud Hist Philos Mod Phys 44:222–230

Tang J-S et al (2012) Realization of quantum Wheeler's delayed-choice experiment. Nat Photon 6:600

van Kampen NG (2008) The scandal of quantum mechanics. Am J Phys 76:989

Wheeler JA (1984) Law without Law. In: Wheeler JA, Zurek WH (eds) Quantum theory and measurement. Princeton University Press, Princeton, pp 182–213

Wheeler JA, Zurek WH (eds) (1984) Quantum theory and measurement. Princeton University Press, Princeton

Wigner E (1960) The unreasonable effectiveness of mathematics in the natural sciences. Commun Pure Appl Math 13(1):1–14

Chapter 12
The Quantum Vacuum

Gheorghe S. Paraoanu

12.1 Introduction

The topic discussed in this paper is the vacuum, an entity that has emerged as an object of intense study in physics. Vacuum is what remains when all the matter, or the particles corresponding to all the known fields, are removed from a region of space. The philosophical question that this paper aims at addressing is: in what sense can this remaining entity be said to exist? Does it have any properties – and how can it have properties if nothing is there? If it doesn't have any properties, how can it be described at all – there seems to be nothing to talk about. I will argue that quantum theory offers a radical departure from the classical concept of property as an attribute of an already-existing particle or field. This has testable consequences: measurable effects (e.g. energy shifts, radiation, effects on phase transitions) can be created in a certain region of space without the need of physical objects as carriers of those properties in that region.

In the following, I will use the word "real" or "actual" as referring to classical events – events that leave, somewhere in nature or in the laboratory, a classical record or trace. "Classical" is what both the theory of relativity and quantum mechanics agree upon. Examples of real entities are the results (values) of a measurement in quantum physics, the clicks of particle detectors, the bit of information (either 0 or 1) stored or recorded in a macroscopic register, and the events from the theory of relativity. I take for granted that the classical world is real. For everything else – mathematical constructs, quantum states, structures etc., I will use "to exist" and "to be" in a rather generic way – otherwise I wouldn't know how to refer to these entities. Also I will use of the words "property" and "structure" as

G.S. Paraoanu (✉)
Low Temperature Laboratory, Department of Applied Physics, Aalto University,
P.O. Box 15100, FI-00076 AALTO, Espoo, Finland

© Springer International Publishing Switzerland 2015
I. Pârvu et al. (eds.), *Romanian Studies in Philosophy of Science*, Boston Studies
in the Philosophy and History of Science 313, DOI 10.1007/978-3-319-16655-1_12

referring both to actual and potential properties and structures. Real (actual) entities certainly have real (actual) properties, while structures can be regarded formally as sets of properties with certain relations between them. For example, the momentum and the position at a certain time of a classical particle are real properties, while the Lagrangian, the law of motion, the Poisson bracket, etc. are structures. The difference is that the former are the result of a direct measurement, independent of the law of motion, while the latter describe a specific law of dynamical evolution satisfied by these properties. But also virtual or potential properties, such as the unmeasured value of one observable when the system is in an eigenstate of the canonically conjugate observable, can form structures. Examples of such structures are the modes of the electromagnetic field in a cavity, the geometry, or the non-commutativity of operators in the Hilbert space. As we will see, the main conclusion of this paper is that the quantum vacuum is an entity endowed with plenty of structure, which lies beneath the existential level of "real" matter. Its ontological status as the seed of possibilities is derived from quantum physics. I give first a brief sketch of the historical development leading to the present notion of quantum vacuum, then I present a few interesting open problems and connections between several lines of investigation of this concept, both in physics and philosophy.

12.2 Brief Historical Interlude: The Development of the Concept of Vacuum Up to the Quantum Era

Philosophical reflection about vacuum is as old as philosophy itself. The Greek atomists Leucippus and Democritus were the first to be worried whether the vacuum is a well-defined concept or not. For them this was important, because, after all, what is left between the atoms must be the void. If atoms are to be taken as constituents of the world, then so must be the void. Thus the atomists clearly saw that both the full (the being) and the empty (the non-being) have to be postulated as the primordial elements.

Aristotle, following Plato's thought, devised a number of rather ingenious arguments against the existence of vacuum (Grant, 1981). Some of these arguments have to do with the difficulty of making sense of motion in vacuum, as one would need some reference points with respect to which to describe changes of the position. But in vacuum all points would be equivalent, therefore, worries Aristotle, motion cannot be defined. Moreover, motion in vacuum, if vacuum exists, should continue forever, in flagrant contradiction with Aristotle's own physics that assessed that motion is due to things aiming at reaching their natural place. Besides these physics – based arguments, Aristotle formulated a "logical" argument against the existence of vacuum. Suppose one removes a body from the place it occupies in space. If we were to attribute any reality to the emptiness left behind, we would need to refer to it as a body with the characteristics of existence (being). Imagine now that we put back the body where it was. We are now left with two bodies co-existing in exactly the same region of space. In this case, Aristotle thinks, the vacuum would need to have rather miraculous properties. It should be more like a fluid that perfectly penetrates the initial body, filling exactly the same amount of space.

The Middle Ages did not bring up any significant deviation from Aristotle's arguments. Nature's abhorrence of vacuum (*horror vacui*) was accepted by most thinkers. But what did eventually turn the tables around in favor of vacuum was the experiment: in the seventeenth century a series of experiments due to Torricelli, Pascal, and von Guericke demonstrated that removing the air from an enclosure is technically possible, and from that moment on the vacuum became a legitimate object of study for science. Its ontological status remained however unclear and would change several times during the next centuries. As we will see below, it has remained, until nowadays, tied with the concept of space, and as a result it would go through the reformulations imposed by the Newtonian mechanics, by the theory of relativity, and by the quantum physics.

In Decartes's philosophy the refutation of the reality of absolute space is mostly based on the association between extension and bodies. If bodies are removed, then one cannot talk about extension anymore – thus absolute space is absurd. What we call space is then an ensemble of contiguities: the location of a body is a collection of relations between the body and those immediately contiguous to it. Motion is simply a change in these contiguity relations (Descartes, 1644).

Against this type of relationist thinking, due to Descartes and to Leibniz as well, Newton exposes his conception of absolute space and time in the famous *Scholium* of *Principia* (Newton, 1689),

"Only I must observe, that the common people conceive those quantities under no other notions but from the relation they bear to sensible objects. And thence arise certain prejudices, for the removing of which it will be convenient to distinguish them into absolute and relative, true and apparent, mathematical and common."

Following the great success of Newtonian mechanics, space had become established as the universal receptacle of objects. However, the "common people" ("*vulgus*" in the original) eventually had their way: the corpuscular view of light advocated by Newton had to yield to the wave view of his contemporary Huygens. Later in the nineteenth century the wave theory of light would get experimental confirmation through the work of Young and Fresnel, and will be put on solid mathematical grounds by Maxwell. But light needed a medium into which to propagate as a wave, so it was conjectured that such a medium called "ether", filling the absolute space, would exist. At the end of the nineteenth century, the experiments of Michelson and Morley showed however that there is no motion with respect to the ether.

Finally, the theory of relativity of Einstein made redundant the concept of ether and that of absolute time and space. The conceptual pendulum swang back to the relationists' side (Saunders and Brown, 1991). The special theory of relativity introduced the idea that length and time intervals are not absolute quantities, but, instead, they depend on the state of motion of the observer. The Lorentz transformation and the negative result of the Michelson and Morley experiment were explained as a natural consequence of the postulates of relativity. From now on, spacetime has become defined only in relation to a reference frame, with each object dragging with it its own spacetime as it moves. It is the concept of motion that forces us to attach different vacua to objects moving with respect to each other. Einstein explains it with exquisite clarity (Einstein, 1952),

"When a smaller box s is situated, relatively at rest, inside the hollow space of a larger box S, then the hollow space of s is a part of the hollow space of S, and the same 'space', which contains both of them, belongs to each of the boxes. When s is in motion with respect to S, however, the concept is less simple. One is then inclined to think that s encloses always the same space, but a variable part of the space S. It then becomes necessary to apportion to each box its particular space, not thought of as bounded, and to assume that these two spaces are in motion with respect to each other."

Finally, general relativity puts gravitation and noninertial motion into this picture. In the theory of general relativity the coordinates (space and time) are even more devoid of any physical meaning than in special relativity. The metric is itself a solution of Einstein's equations – if this solution exists, space-time can be rightfully said to exists. If it does not, such as in the singularities of black holes or in the Big Bang, spacetime does not have any meaning. If one somehow removes the metric, as given by solving Einstein's equations, what is left is not the absolute flat spacetime of Newton – nothing is left.

In Einstein's words (Einstein, 1952),

"There is no such thing as an empty space, i.e. a space without field. Space-time does not claim existence on its own, but only as a structural quality of the field. Thus Descartes was not so far from the truth when he believed he must exclude the existence of an empty space."

Here the word field refers to the gravitational field, which in the general theory of relativity can be seen, so to say, almost co-substantial with the metric $g_{\mu\nu}$ (that is why it is also called a metric field). This is, in brief, the great conceptual shift introduced by the general theory of relativity: that spacetime is a field with a dynamics of its own, as determined by the configuration of matter, and not just a fixed background attached to each reference frame, as in the special theory of relativity. In modern mathematical parlance, we say that general relativity is a background-independent theory (due to diffeomorfism invariance), meaning that the theory is not built on a fixed spacetime geometry that exists behind the scenes, unaffected by matter. Einstein's equations tell us explicitly that there can be no such background that is left unbent by the action of matter.

12.3   The Architecture of the Quantum Vacuum

Three major philosophical assumptions about properties can be associated with the Newtonian world-view of the world.

[A1]   Properties are tied to physical objects (particles or non-zero fields).

[A2]   Space is distinct from and exists independently of the objects (carrying properties) one chooses to populate it with. The same is true for time. Spacetime is the immense theater stage where physical processes unfold, the canvas where each dot is an event. One has, in principle, access to any of these points.

[A3]   True randomness does not exist. The observed randomness of the properties of a system is simply a result of our lack of knowledge and imperfect control over the experiment.

The rise of electrodynamics in the mid-nineteenth century did not change much [A1]. It only added fields, mostly through the work of Faraday, as legitimate carriers of properties. Neither did Boltzmann's statistical mechanics present a challenge to [A3], since there the perceived randomness was presumed to be an effect of the motion and collisions of many particles. It was the general theory of relativity that changed [A2] to a large extend: the theory suggests that spacetime itself can be bent due to the presence of matter. A distribution of matter allows one to calculate the metric of spacetime. But accessing any of the points of spacetime is no longer taken for granted – there can exist points where the theory predicts singularities, event horizons prevent the transfer of information from the inside of the region of space which they enclose, and so on.

On this issue, the theory of relativity is not as radical as one can be. As we have seen, for the thinkers before Newton the connection between objects and spacetime was even tighter. Spacetime might not mean anything in the absence of objects. However, even in this conceptual frame it still makes sense to ask what happens when we attempt to remove all the objects from a certain region. There are three possible answers: the first, that the problem is logically ill-defined. This seems to be what Aristotle preferred to believe. Another possibility is to view the objects and their associated spacetime as analogous to fluids: attempting to remove a part of the fluid is hopeless because it will be immediately replaced by another part of the fluid. Finally, the third view could be called a "ceramical" view of spacetime: much like the tiles in a glass mosaic, any attempt to remove one part of the drawing results in the breaking of the glass, extracting objects from the spacetime could result simply in some type of nothingness. In this case, because spacetime is so rigidly attached to the physical objects, it makes sense to wonder if a spacetime structure is useful at all or it is just redundant. This type of conceptual structure might not allow to construct a physical theory in the usual sense: as a story that unfolds in spacetime – simply because there is no spacetime, or it is not clearly distinct from the objects themselves.

But there is another way out. Quantum physics offers a completely different perspective that completely changes [A1] and [A3], and softens the alternatives to [A2] by introducing more conceptual structure. The result is essentially a probabilistic theory in which the evolution is not applied directly to probabilities, but to probability amplitudes (Aaronson, 2013). This automatically means that what evolves are not the properties of the objects, but the possibilities of the objects having certain properties. These properties become actualized (real) only after a measurement of the corresponding observable. The assumption [A2] is to some extend left unscathed: spacetime exists as an independent entity, but it acquires more structure beyond just geometry.

Before proceeding further, it is worth stressing out that so far (Almheiri et al., 2013) there is no prediction of quantum physics that contradicts the general theory of relativity, or the other way around. Of course, we do not know the limits of these theories, and one may reasonably suspect for instance that quantum mechanics will forbid the point-like singularities of general relativity. Still, the fact that the known domain of applicability of both of these theories is so vast – from elementary particles to structures of the size of galaxies – and yet no contradictory result has been obtained is astonishing, especially when one looks at how different are the concepts and assumptions the two theories operate with. Even the combination of the special theory of relativity with quantum physics in the form of relativistic quantum field theory, producing the very successful predictions of the Standard Model of particle physics, is not an easy conceptual marriage (Wigner, 1957). This situation is rather unique in the history of science, and brings a novel twist to the discussion on falsifiability (Popper, 1959), paradigm shifts (Kuhn, 1962), confirmation, etc., which would be worth investigating further in the philosophy of science.

Another remark is that the assumptions [A1]–[A3] are not exhaustive. I left aside for example the very important supposition that interaction is strictly local (there is no action-at-a-distance), a feature which is essential when considering the dynamics of systems. This assumption is maintained in standard quantum field theory – when writing the interaction Hamiltonian between two fields, it is taken for granted that one field couples only to the other field defined at the same point in spacetime. This, however, does not make quantum physics local in the classical sense, because once they have interacted the particles (or fields) can be separated in space and some of the properties that one ascribes to them via measurements cannot result from local probability distributions. This type of quantum non-locality, as famously put first in evidence by Einstein, Podolsky and Rosen, is best expressed by Bell inequalities, but, interestingly, it can also be put in evidence as a purely logical contradiction (Paraoanu, 2011a).

12.3.1 The Ontological Status of the Quantum Vacuum

In quantum physics, vacuum is defined as the ground state of a quantum field. It is a state of minimum energy, corresponding to zero particles. Note that this definition of vacuum employs already the conceptual and formal machinery of quantum field theory. It is justifiable to ask weather it is possible to give a more theory-independent definition with lesser theoretical load. In this situation vacuum would be an entity which is explained – not just defined within and then explored – by quantum field theory. For example, one could attempt an operational definition of vacuum as the state in which no particles are detected. But then we have to specify how to detect the particles, with what efficiency, etc., that is, we need a model for the particle detector. Such a model, known as the Unruh-DeWitt detector, is constructed however from within quantum field theory. Therefore nothing seems to be gained in explanatory power by an operational definition.

The vacuum is simply a special state of the quantum field – implying that quantum physics allows the return of the concept of ether, although in a rather weaker, modified form. This new ether – the quantum vacuum – does not contradict the special theory of relativity because the vacuum of the known fields are constructed to be Lorentz-invariant. In some sense, each particle in motion carries with it its own ether, thus Lorentz transformations act in the same way on the vacuum and on the particle itself. Otherwise, the vacuum state is not that different from any other wavefunction in the Hilbert space. Attaching probability amplitudes to the ground state is allowed to the same degree as attaching probability amplitudes to any other state with nonzero number of particles. In particular, one expects to be able to generate a real property – a value for an observable – in the same way as for any other state: by perturbation, evolution, and measurement. The picture that quantum field theory provides is that both particles and vacuum are now constructed from the same "substance", namely the quantum states of the fields at each point (or, equivalently, that of the modes). What we used to call matter is just another quantum state, and so is the absence of matter – there is no underlying substance that makes up particles as opposed to the absence of this substance when particles are not present. One could even turn around the tables and say that everything is made of vacuum – indeed, the vacuum is just one special combination of states of the quantum field, and so are the particles. In this way, the difference between the two worldviews, the one where everything is a plenum and vacuum does not exist, and the other where the world is empty space (nonbeing) filled with entities that truly have the attribute of being, is completely dissolved. Quantum physics essentially tells us that there is a third option, in which these two pictures of the world are just two complementary aspects. In quantum physics the objects inhabit at the same time the world of the continuum and that of the discrete.

Incidentally, the discussion above has implications for the concept of individuality, a pivotal one both in philosophy and in statistical physics. Two objects are distinguishable if there is at least one property which can be used to make the difference between them. In the classical world, finding this property is not difficult, because any two objects have a large amount of properties that can be analyzed to find a different one. To establish if a painting is fake or it is the original is only a matter of practical difficulty. But, because in quantum field theory objects are only combinations of modes, with no additional properties, it means that one can have objects which cannot be distinguished one from each other even in principle. For example, two electrons are perfectly identical. To use a well-known Aristotelian distinction, they have no accidental properties, they are truly made of the same essence. A very important related problem is that of the distinguishability of non-orthogonal states, which has attracted a lot of attention in quantum information.

Another spectacular application of the idea that properties are detached from objects is quantum computing. Unlike in classical computing, quantum processors do not need to use objects (for example memory elements) as physical support for each of the intermediate result of a calculation (Paraoanu, 2011b). The re-attachment of properties in the form of the result of a calculation is done only at the end of a series of unitary operations, when the registers are measured.

To see in a simple way why quantum physics requires a re-evaluation of the concept of emptiness the following qualitative argument is useful: the Heisenberg uncertainty principle shows that, if a state has a well-defined number of particles (zero) the phase of the corresponding field cannot be well-defined. Thus, quantum fluctuations of the phase appear as an immediate consequence of the very definition of emptiness.

Another argument can be put forward: the classical concept of emptiness assumes the separability of space in distinct volumes. Indeed, to be able to say that nothing exists in a region of space, we implicitly assume that it is possible to delimitate that region of space from the rest of the world. We do this by surrounding it with walls of some sort. In particular, the thickness of the walls is irrelevant in the classical picture, and, as long as the particles do not have enough energy to penetrate the wall, all that matters is the volume cut out from space. Yet, quantum physics teaches us that, due to the phenomenon of tunneling, this is only possible to some extent – there is, in reality, a non-zero probability for a particle to go through the walls even if classically they are prohibited to do so because they do not have enough energy. This already suggests that, even if we start with zero particles in that region, there is no guarantee that the number of particles is conserved if e.g. we change the shape of the enclosure by moving the walls. This is precisely what happens in the case of the dynamical Casimir effect, as described below. Another consequence, which I will not discuss here, is the existence of entanglement between different regions of space in the vacuum state, a somewhat unexpected effect since the concept of entanglement is usually discussed for particles. There is yet another point of view that illustrates that in quantum physics the idea of delimitating a region of space, and taking the particles out of it, is tricky. The very concept of a particle is not a local one in quantum field theory (Colosi and Rovelli, 2009), and defining the number of particle operator in a region of space is not trivial (Redhead, 1994). Particles are extended objects but the operation of removing them is by necessity local – thus when abstractly separating an empty volume of space one needs further care to ensure that no particle leaks in.

All these demonstrate that in quantum field theory the vacuum state is not just an inert background in which fields propagate, but a dynamic entity containing the seeds of multiple possibilities, which are actualized once the vacuum is disturbed in specific ways. This leads to real effects, some of which are discussed in the next subsection: vacuum fluctuations result in shifts in the energy level of electrons (Lamb shift), fast changes in the boundary conditions or in the metric produce particles (dynamical Casimir effect), and accelerated motion and gravitation can create thermal radiation (Unruh and Hawking effects).

12.3.2   Observable Effects Due to the Quantum Vacuum

There are several field-theoretical and many-body effects associated with the existence of vacuum fluctuations (Milonni, 1994; Sciama, 1991).

Measurements showing conclusively that differences in vacuum energies have observable effects provided some of the earliest experimental confirmations of quantum physics. For example, one possibility is to measure the vibrational spectra of molecules and to search for isotope effects (a change in the mass of a nuclei will change the zero-point energy, thus the transition frequencies). The first observation of this effect was done by Mulliken in 1925, using boron monoxide. Since then, the vacuum state has played an important role in countless other experiments. For example, in X-ray scattering on solids, it was shown that the zero-point fluctuations of the phonons produces an additional scattering on top of that due to thermal fluctuations. Other examples are the Lamb shift between the energies of the s and p levels in the hydrogen atom, and the fact that liquid helium does not become solid at normal atmospheric pressure even near zero temperature – the vacuum fluctuations prevent the atoms of coming close enough so that solidification can occur. In nuclear physics, a related problem is that of a fundamental limit of the size of nuclei (Indelicato and Karpov, 2013). As the charge number Z increases beyond approximately $1/\alpha$ (where α is the fine structure constant), the electric fields near the nucleus produce vacuum instability (Rafelski et al., 1978), and particle-antiparticle pairs are generated from vacuum due to the Schwinger effect.

The dynamical Casimir effect was predicted theoretically in 1970 (Moore, 1970) and has been recently observed in two experiments. The first one uses a SQUID terminating a coplanar waveguide (Wilson et al., 2011), creating a fast-moving boundary condition. The other experiment employs an array of SQUIDs, effectively realizing a material with a fast-tunable index of refraction embedded in a cavity (Lähteenmäki et al., 2013). When the boundary condition (in the first setup) or the index of refraction (in the second setup) changes fast enough, one observes real photons emerging from the circuit, even if the system was initially in the vacuum state. Quantum superfluids offer also a rich system to observe vacuum effects: such experiments have been discussed in superfluid He (Volovik, 2003), and recently a thermal analog of the dynamical Casimir effect has been reported in a Bose-Einstein condensate (Jaskula et al., 2012).

In order to understand conceptually the dynamical Casimir effect let us go back to Einstein's *gedankenexperiment* with the two boxes S and s, as presented in the second section. Einstein realized that motion imposes on us the concept of a relativistic, frame-dependent space. This relative space is dragged along by the box (or frame) as it moves. As a result, space is not just a kind of fixed canvas onto which we draw reference frames, but, instead, it is defined by and anchored into the reference frame. With this, we are now ready to push Einstein's thought one step further. Because space is an entity effectively created by some enclosure, this implies that deforming the corresponding box or boundary condition might have an effect on the space inside. For example, we can compress and expand the space itself by operating the box as a piston in a cylinder. The result turns out to the creation of real particles. Einstein would have been amazed: quantum physics brings his own view on space to a very unexpected consequence!

Finally, motion itself has an effect on the vacuum. Let us look again back at Einstein's boxes s and S. Each of them carry their own vacuum. As they move one with respect to each other, is the vacuum of S also seen as a vacuum by s? The principle of relativity guarantees that no phenomenon exists allowing to distinguish the vacua of the two inertial systems, but for non-inertial motion it does not put any restriction. It turns out that if s is moved with respect to S at a constant acceleration, s experiences a thermal background (an environment containing particles in thermal equilibrium). This is the Unruh effect (Unruh, 1976). Now, by the principle of equivalence, gravitation is equivalent to acceleration, so one expects a similar effect to occur in gravitational fields. This is the famous Hawking effect (Hawking, 1975), consisting in emission of radiation at the event horizon of a black hole.

12.3.3 Where Do Properties Come From

We now go back to the main theme of this paper: what is the origin of the properties of physical objects? As we have seen, we have to enlarge the category of entities where properties can originate from, by including the quantum vacuum. To make the difference more clear, suppose that we have a region of space emptied of matter and fields. Nothing real, in the sense defined in the introduction, is there. Classically, the only way to create a property inside that region is to bring in from outside an object carrying that specific property. Note that this simple thought experiment relies on all of the assumptions [A1], [A2], and [A3] listed above. These are not trivial assumptions – although they look very innocuous, it is by no means obvious that nature should obey them. In this sense, Netwonian physics appears as a strongly coerced theory, while relativity and quantum physics introduce different relaxations of these assumptions. Firstly, Newtonian physics needs to have the concept of space as in [A2], existing independently of objects and with all the points easily accessed. General relativity shows that this does not happen if the object carrying the desired property is too massive or if we insist of making it as much as point-like – squeezing too much energy into too little space could result in the formation of a black hole. Secondly, if [A1] and [A3] is not satisfied, then properties could appear spontaneously in vacuum, as they do not require either a real object to be attached to or a causal chain of events that would produce them.

The experiments on generation of particles from the quantum vacuum mentioned above (dynamical Casimir effect) show that there exists another way of generating properties. Note that these experiments still use the classical concept of spacetime background as in [A2], but to explain them one needs to alter dramatically [A1] and [A3] to accommodate the quantum-mechanical account of randomness (there exists pure randomness) and properties (properties are not intrinsically attached to objects, but are created contextually, as shown by the Kochen-Specker theorem). Because in quantum field theory the vacuum has a structure, properties can be generated at a certain point by changes of this structure, and not just by bringing them in from somewhere else. As mentioned above already, one cannot do this classically: if a property were to appear at some point in space, then classical physics would tell us that, according to [A1] there must be a real object that carries this property, and according to [A2] there must be a causal story, enfolding in the region of space-time under consideration, which one must discover in order to have a complete description of the phenomenon. To clarify this point, I can make an analogy with the chairs for the public in a concert hall. The arrangement of chairs in rows and the numbering of the chairs in each row, the association of higher prices to better seats etc., provides a structure for the probability distribution of spectators. For example, if one tries to buy a ticket, the options are limited by the total number of seats, by the number of already-reserved seats, and by the budget of that person. The spectators are here the properties: they might buy a certain seat and show up to the concert – or not. However, to create this arrangement of seats in the concert hall one needs to bring in the chairs from outside: there must be some energy and mass to support this structure, and this energy and mass can be recovered if for example the concert hall is renovated and the chairs are removed. This situation is in contrast to the quantum vacuum, where the structure exists as such, ready to acquire real properties, without being constructed beforehand by energy or mass previously brought in from elsewhere. By definition, the vacuum is the ground state, therefore (unless the system is metastable) there is no other lower-energy state where the system would go to if one attempts to extract energy from it. This feature has experimental applications, for example to verifying that systems such as nanomechanical oscillators have reached the ground state (O'Connell et al., 2010). Note that in the case of the dynamical Casimir experiments mentioned above, the energy of the particles comes from the pump in a two-photon spontaneous downconversion process: the vacuum only provides a structure for this process to occur, and it is not the case that the vacuum energy is converted into photons. In general, deforming, shearing, modifying boundary conditions, and changing the index of refraction of the vacuum results in energy exchange – for example, in the static Casimir effect it costs energy to pull apart the two plates. The quantum vacuum behaves, from this point of view, almost as a real material. Clearly, the ontological status of an entity that is not made of real particles but reacts to external actions does not fall straight into any of the standard philosophical categories of being/non-being.

12.4   An Emptiness Full of Unknowns

A significant number of important open problems in physics are connected to the concept of vacuum. I will briefly discuss here a few of them.

12.4.1   What Lies Beneath the Continuous Spacetime Manifold

If the quantum vacuum displays features that make it resemble a material, albeit a really special one, we can immediately ask: then what is this material made of? Is it a continuum, or are the "atoms" of vacuum? Is vacuum the primordial substance of which everything is made of? Such questions lead us to the very edge of our knowledge. To make these big questions more understandable, we can start by decoupling the concept of vacuum from that of spacetime.

As we have seen, the concept of vacuum as accepted and used in standard quantum field theory is tied with that of spacetime. This is important for the theory of quantum fields, because it leads to observable effects. It is the variation of geometry, either as a change in boundary conditions (Wilson et al., 2011) or as a change in the speed of light (and therefore the metric) (Lähteenmäki et al., 2013) which is responsible for the creation of particles. Now, one can legitimately go further and ask: which one is the fundamental "substance", the space-time or the vacuum? Is the geometry fundamental in any way, or it is just a property of the empty space emerging from a deeper structure?

These questions force us to go back to reexamining the most basic conceptual cornerstones of our physical theories. That geometry and substance can be separated is of course not anything new for philosophers. Aristotle's distinction between form and matter is one example. For Aristotle the "essence" becomes a true reality only when embodied in a form. Otherwise it is just a substratum of potentialities, somewhat similar to what quantum physics suggests. Immanuel Kant was even more radical: the forms, or in general the structures that we think of as either existing in or as being abstracted from the realm of independently-existing reality (the thing-in-itself or the *noumena*) are actually innate categories of the mind, preconditions that make possible our experience of reality as *phenomena*. Structures such as space and time, causality, etc. are a priori forms of intuition – thus by nature very different from anything from the outside reality, and they are used to formulate synthetic a priori judgments. But almost everything that was discovered in modern physics is at odds with Kant's view (Heisenberg, 2000). In modern philosophy perhaps Whitehead's process metaphysics (Whitehead, 1979) provides the closest framework for formulating these problems. For Whitehead, potentialities are continuous, while the actualizations are discrete, much like in the quantum theory the unitary evolution is continuous, while the measurement is non-unitary and in some sense "discrete". An important concept is the "extensive continuum", defined as a "relational complex" containing all the possibilities of objectification. This continuum also contains the potentiality for division; this potentiality is effected in what Whitehead calls "actual entities (occasions)" – the basic blocks of his cosmology. For the pragmatic physicist, since the extensive continuum provides the space of possibilities from which the actual entities arise, it is tempting to identify it with the quantum vacuum (Hättich, 2004). The actual entities are then assimilated with events in spacetime, as resulting from a quantum measurement, or simply with particles. The following caveat is however due: Whitehead's extensive continuum is also devoid of geometrical content, while the quantum vacuum normally carries information about the geometry, be it flat or curved.

It is reasonable to expect that the continuous differentiable manifold that we use as spacetime in physics (and experience in our daily life) is a coarse-grained manifestation of a deeper reality, perhaps also of quantum (probabilistic) nature. This search for the underlying structure of spacetime is part of the wider effort of bringing together quantum physics and the theory of gravitation under the same conceptual umbrella. From various theoretical considerations, it is inferred that this unification should account for physics at the incredibly small scale set by the Planck length, 10^{-35} m, where the effects of gravitation and quantum physics would be comparable. What happens below this scale, which concepts will survive in the new description of the world, is not known. An important point is that, in order to incorporate the main conceptual innovation of general relativity, the theory should be background-independent. This contrasts with the case of the other fields (electromagnetic, Dirac, etc.) that live in the classical background provided by gravitation.

The problem with quantizing gravitation is – if we believe that the general theory of relativity holds in the regime where quantum effects of gravitation would appear, that is, beyond the Planck scale – that there is no underlying background on which the gravitational field lives. There are several suggestions and models for a "pre-geometry" (a term introduced by Wheeler) that are currently actively investigated (see e.g. Meschini et al. 2005 for a non-technical review). This is a question of ongoing investigation and debate, and several research programs in quantum gravity (loops, spinfoams, noncommutative geometry, dynamical triangulations, etc.) have proposed different lines of attack (Boi, 2011). Spacetime would then be an emergent entity, an approximation valid only at scales much larger than the Planck length.

Incidentally, nothing guarantees that background-independence itself is a fundamental concept that will survive in the new theory. For example, string theory is an approach to unifying the Standard Model of particle physics with gravitation which uses quantization in a fixed (non-dynamic) background. In string theory, gravitation is just another force, with the graviton (zero mass and spin 2) obtained as one of the string modes in the perturbative expansion. A background-independent formulation of string theory would be a great achievement, but so far it is not known if it can be achieved.

Models of emergent spacetimes can be constructed by analogy with the low-energy models used in condensed-matter physics (Bain, 2013). One recent particularly simple to understand such construction is the quantum graphity model of Markopoulou and collaborators (Hamma et al., 2010), a model inspired from loop quantum gravity. In this model the geometry emerges from a probabilistic structure which is itself of quantum-mechanical nature: geometrical relations are given by the links between the nodes of a graph, and these links are created and annihilated by standard quantum-mechanical creation and annihilation operators. Two nodes are in a relation of spatial vicinity only if the link between them is in the state "connected", as resulting from the action of the creation operator on the vacuum. Note that the graph does not live in spacetime: it is an abstract lattice describing connection relationships between nodes. The geometry is emergent as the overall connectivity of the graph. The concept of proximity is therefore probabilistic (in the sense of quantum mechanics) and it allows for states that are quantum-mechanical superpositions of connected or disconnected links, yielding also superpositions of geometries.

12.4.2 Time, Gravitation, Energy, and the Origin of the Universe

The relationship between the quantum vacuum and other fundamental concepts in physics such as time, gravitation, and energy is not easy to pin down, but some of these connections are intriguing.

Time is one of the most difficult concepts in physics. It enters in the equations in a rather artificial way – as an external parameter. Although strictly speaking time is a quantity that we measure, it is not possible in quantum physics to define a time-observable in the same way as for the other quantities that we measure (position, momentum, etc.). The intuition that we have about time is that of a uniform flow, as suggested by the regular ticks of clocks. Time flows undisturbed by the variety of events that may occur in an irregular pattern in the world. Similarly, the quantum vacuum is the most regular state one can think of. For example a persistent superconducting current flows at a constant speed – essentially forever. Can then one use the quantum vacuum as a clock? This is a fascinating dispute in condensed-matter physics (Wilczek, 2012), formulated as the problem of existence of time crystals. A time crystal, by analogy with a crystal in space, is a system that displays a time-regularity under measurement, while being in the ground (vacuum) state. These systems might not exist in the form originally proposed (Bruno, 2013), but the research into this new concept will probably bring up unexpected connections between time, the quantum vacuum, and the concept of spontaneoulsy broken symmetry.

Then, if there is an energy (the zero-point energy) associated with empty space, it follows *via* the special theory of relativity that this energy should correspond to an inertial mass. By the principle of equivalence of the general theory of relativity, inertial mass is identical with the gravitational mass. Thus, empty space must gravitate. So, how much does empty space weigh? This question brings us to the frontiers of our knowledge of vacuum – the famous problem of the cosmological constant, a problem that Einstein was wrestling with, and which is still an open issue in modern cosmology (Rugh and Zinkernagel, 2002; Volovik, 2006).

Finally, although we cannot locally extract the zero-point energy of the vacuum fluctuations, the vacuum state of a field can be used to transfer energy from one place to another by using only information. This protocol has been called *quantum energy teleportation* (Hotta, 2008) and uses the fact that different spatial regions of a quantum field in the ground state are entangled. It then becomes possible to extract locally energy from the vacuum by making a measurement in one place, then communicating the result to an experimentalist in a spatially remote region, who would be able then to extract energy by making an appropriate (depending on the result communicated) measurement on her or his local vacuum.

All of the above suggest that the vacuum is the primordial essence, the *ousia* from which everything came into existence. Some models suggests that even the spacetime can be seen as an emergent structure. So does Nature try to tell us something about the grand metaphysical question – why there is something rather than nothing – but what exactly (Albert, 1988)? Does vacuum play the crucial role in the coming into existence of the Universe as we know it (Krauss, 2012; Boi, 2011)?

12.5   Conclusions

To conclude, I describe the concept of quantum vacuum in close relation with the latest experimental results that show how particles can be generated by processes such as the dynamical Casimir effect. I then explore the Newtonian-physics assumptions behind the concept of property and show how these are to be modified by relativity and especially by quantum physics. Quantum physics allows for the vacuum state to have an intrinsic structure that provides the "possibility grid" for events, or for entities that we can call real with full confidence. Potentialities are thus actualized as properties when the vacuum is disturbed or measured in specific ways.

The emergence of properties by this mechanism sheds new light onto the intricate relation between the quantum and the classical, but does not solve the deep clash between these worlds. Fundamentally, it is perhaps the concept of separation that would need revising. Vacuum itself is possible because one can separate things from one region of space into another. In quantum physics we have the separation between the object under study and the observer (the measurement apparatus). The object under study is quantum while the observer is classical – thus each of them is thought of as obeying a different dynamics, the unitary quantum evolution of the wave function for the object, and the classical equations of motion for the observer. The interaction between the two collapses the wavefunction, resulting in a nonunitary evolution of the object. This separation does not exist as such in general relativity – there everything, that is, both the object under study and the observer are part of the same dynamical equation: they experience the curvature of spacetime and, by virtue of having mass, they generate the gravitational field themselves. Yet at the same time, quantum physics allows for the existence of entanglement between objects that are localized at different places in space, a feature that seems difficult to accommodate with the theory of relativity. Merging quantum theory with gravitation will therefore most likely require drastically new concepts, also from the direction of what "emptiness" means. A frontal approach to the problem of merging gravitation and quantum physics – attempting for example to quantize the gravitational field – might not the best way to proceed, since quantum physics assumes (and hides it very well in the formalism) the existence of a spacetime background in which the measuring apparatus is placed. In other words, the distinction between the object to be quantized and this background cannot be maintained when the object is the spacetime itself.

Acknowledgements I wish to thank Prof. I. Pârvu for inviting me to contribute to this volume and for many inspirational discussions during my studies at the Department of Philosophy of the University of Bucharest. The title of the third section of this contribution is inspired from his book, *Arhitectura existentei* (Humanitas, Bucharest, 1990). Financial support from FQXi and from the Academy of Finland, projects 263457 and the Center of Excellence "Low Temperature Quantum Phenomena and Devices" project 250280 is acknowledged.

I am grateful to Grisha Volovik, Gil Jannes, Janne Karimäki, Iulian Toader, and Jonathan Bain for comments on the manuscript.

References

Aaronson S (2013) Quantum computing since democritus, ch. 9. Cambridge University Press, New York

Albert DZ (1988) On the possibility that the present quantum state of the universe is the vacuum. PSA: Proceedings of the biennial meeting of the philosophy of science association 1988. Volume two: symposia and invited papers, p 127. University of Chicago Press

Bain J (2013) The emergence of spacetime in condensed matter approaches to quantum gravity. Stud Hist Philos Mod Phys 44:338

Bruno P (2013) Impossibility of spontaneously rotating time crystals: a no-go theorem. Phys Rev Lett 111:070402

Colosi D, Rovelli C (2009) What is a particle? Class Quantum Gravity 26:025002

Descartes R (1644) Principia philosophiae. Translated V.R. Miller, R.P. Miller. Reidel, Dordrecht (1983)

Einstein A (1952) Relativity and the problem of space. In: Relativity – the special and the general theory. Fifteenth edition (1961). Three Rivers Press, New York

Grant E (1981) Much ado about nothing: theories of space and vacuum from the Middle Ages to the scientific revolution. Cambridge University Press, Cambridge

Hawking SW (1975) Particle creation by black holes. Commun Math Phys 43:199

Heisenberg W (2000) Physics and philosophy – the revolution in modern science, ch. 5. Penguin Books, London

Hotta N (2008) Quantum measurement information as a key to energy extraction from local vacuums. Phys Rev D 78:045006

I leave aside Hawkings' black hole information paradox and the hotly debated issue of "firewalls", recently triggered by Almheiri A, Marolf D, Polchinski J, Sully J (2013) Black holes: complementarity or firewalls? J High Energy Phys 2:62 [arXiv:1207.3123 (2012)]; cf. Braunstein SL, Black hole entropy as entropy of entanglement, or it's curtains for the equivalence principle [arXiv:0907.1190v1]; published as Braunstein SL, Pirandola S, Zyczkowski K (2013) Better late than never: information retrieval from black holes. Phys Rev Lett 110:101301. Which deserve a separate discussion

Indelicato P, Karpov A (2013) Theoretical physics: sizing up atoms. Nature 498:40

Jaskula J-C, Partridge GB, Bonneau M, Lopes R, Ruaudel J, Boiron D, Westbrook CI (2012) An acoustic analog to the dynamical Casimir effect in a Bose-Einstein condensate. Phys Rev Lett 109:220401

Krauss LM (2012) A universe from nothing: why there is something rather than nothing. Free Press, New York. For the dispute that errupted between Lawrence Krauss and philosopher David Albert, see e.g. Albert D (2012) On the origin of everything. The New York Times

Sunday Book Review, 23 Mar 2012; Krauss LM (2012) The consolation of philosophy. Scientific American 27 Apr 2012; Sean Caroll (2012) A Universe from Nothing? Cosmic Variance, Discover Magazine, 28 Apr 2012, etc.

Kuhn TS (1962) The structure of scientific revolutions. University of Chicago Press, Chicago

Lähteenmäki P, Paraoanu GS, Hassel J, Hakonen PJ (2013) Dynamical Casimir effect in a Josephson metamaterial. Proc Natl Acad Sci USA 110:4234

Meschini D, Lehto M, Piilonen J (2005) Geometry, pregeometry and beyond. Stud Hist Philos Mod Phys 36:435

Milonni PW (1994) The quantum vacuum: an introduction to quantum electrodynamics. Academic, San Diego

Moore GT (1970) Quantum theory of the electromagnetic field in a variable-length one-dimensional cavity. J Math Phys 11:2679

Newton I (1689) Scholium to the definitions. In: Philosophiae naturalis principia mathematica. Translated by A. Motte (1729); Revised F. Cajori. University of California Press, Berkeley (1934), pp 6–12

O'Connell AD, Hofheinz M, Ansmann M, Bialczak RC, Lenander M, Lucero E, Neeley M, Sank D, Wang H, Weides M, Wenner J, Martinis JM, Cleland AN (2010) Quantum ground state and single-phonon control of a mechanical resonator. Nature 464:697

Paraoanu GS (2011a) Realism and single-quanta nonlocality. Found Phys 41:734

Paraoanu GS (2011b) Quantum computing: theoretical possibility versus practical possibility. Phys Perspect 13:359

Popper K (1959) The logic of scientific discovery. Basic Books, New York

Rafelski J, Fulcher LP, Klein A (1978) Fermions and bosons interacting with arbitrarily strong external fields. Phys Rep 38:227

Redhead M (1994) The vacuum in relativistic quantum field theory. PSA: proceedings of the biennial meeting of the philosophy of science association 1994. Volume two: Symposia and invited papers, p 77. University of Chicago Press

Rugh SE, Zinkernagel H (2002) The quantum vacuum and the cosmological constant problem. Stud Hist Philos Mod Phys 33:663

Saunders S, Brown HR (1991) Reflections on the ether. In: Saunders S, Brown HR (eds) The philosophy of vacuum. Clarendon Press, Oxford, pp 28–63

Sciama DW (1991) The physical significance of the vacuum state of a quantum field. In: Saunders S, Brown HR (eds) The philosophy of vacuum. Clarendon Press, Oxford, pp 136–158

See especially Boi L (2011) The quantum vacuum: a scientific and philosophical concept, from electrodynamics to string theory and the geometry of the microscopic world. John Hopkins University Press, Baltimore

See Hamma A, Markopoulou F, Lloyd S, Caravelli F, Severini S, Markström K (2010) A quantum Bose-Hubbard model with evolving graph as toy model for emergent spacetime. Phys Rev D 81:104032, and references therein

There exists another interpretation, which associates the vacuum state as defined in algebraic field theory with Whitehead's notion of "universal underlying activity", see Hättich F (2004) Quantum processes: a Whiteheadian interpretation of quantum field theory. Agenda Verlag, Münster; see also Bain J (2005) Quantum processes: a Whiteheadian interpretation of quantum field theory. Stud Hist Philos Mod Phys 36:680

Unruh WG (1976) Notes on black-hole evaporation. Phys Rev D 14:870

Volovik GE (2003) The universe in a Helium droplet. International series of monographs on physics. Clarendon Press, Oxford

Volovik GE (2006) Vacuum energy: myths and reality. Int J Mod Theor Phys A 15:1987

Whitehead AN (1979) Process and reality: an essay in cosmology. Corrected edition, edited by D.R. Griffin, DW. Sherburne, MacMillan Publishing

Wigner EP (1957) Relativistic invariance and quantum phenomena. Rev Mod Phys 29:255

Wilczek F (2012) Quantum time crystals. Phys Rev Lett 109:160401

Wilson CM, Johansson G, Pourkabirian A, Johansson JR, Duty T, Nori F, Delsing P (2011) Observation of the dynamical Casimir effect in a superconducting circuit. Nature 479:376

Chapter 13
Structural Pluralism and S-Dualities: A Project in String Realism

Ioan Muntean

13.1 What Realism for String Dualities?

Duality, as a relation among theories or models, has been known in the twentieth century physics: there was first the electromagnetic E-M duality acknowledged since the late 1800s, then the duality of the Ising model and other statistical systems, discovered in the early 1940s; later on, in the 1970s, two important dualities were recognized: a generalization of the E-M duality in quantum field theory called Olive-Montonen duality (Montonen and Olive, 1977) and the dual resonance model of hadrons. Nevertheless, only in string theory dualities have played the role of powerful "methodological maxims" throughout its history, and especially after the "second string revolution" (which occurred around 1995).[1] What other implications do string dualities have—besides their paramount methodological, mathematical and computational importance?

This paper focuses on the reconceptualization of fundamentalism and realism in the presence of S-dualities (a special and interesting case of string dualities). A terminological clarification is in order here: fundamentalism, similar to realism, is always *about* something: in philosophy of science, it refers primarily to

---

[1]For a comprehensive discussion of the early string theory and its deep connection to S-matrix theory, see Cappelli et al. (2012) and Rickles (2014). The "second string revolution" is marked by E. Witten's conjecture about dualities and the M-theory (Witten, 1995).

I. Muntean (✉)
The Reilly Center for Science, Technology and Values, University of Notre, Dame, IN, USA
e-mail: imuntean@nd.edu

© Springer International Publishing Switzerland 2015   199
I. Pârvu et al. (eds.), *Romanian Studies in Philosophy of Science*, Boston Studies
in the Philosophy and History of Science 313, DOI 10.1007/978-3-319-16655-1_13

laws of nature or theories.[2] In philosophy, fundamentalism can refer likewise to entities, causation, events, facts, etc. Non-fundamental entities are derived, emergent, composed, etc. In philosophy of physics, fundamentalism about entities is more frequently called "fundamentality" (McKenzie, 2011, 2014). It expresses the intuition that there should be a fundamental level of reality, and that the chain of ontological priority and compositionality relations should terminate somewhere, in a "fundamental base". Many philosophers endorse the view that the base has to be physical and would consist of elementary particles, or "whatever our best physics is going to tell us are the basic bits of matter out of which all material things are composed." (Kim, 1998). The string theorist is able to offer, at a first take, the basic bits of matter: the strings. String fundamentalism is probably the simple and somewhat naive view that string are the only components of the fundamental basis the metaphysician asks for.

The question asked here is: can we entertain the same string fundamentalism (i.e. string fundamentality) in the presence of dualities? The conclusion of the present argument is that dualities entail a pluralism, with no or little need for fundamentalism. In a nutshell, starting from a set of assumptions about realism **(1)** and string models **(2)**, the paper infers a conclusions **(3)(a)** and **(3)(b)** about the nature of string realism with dualities:

> **(1)** REALISM WITHOUT SUCCESS: *The realism commitments of a given theory can be assessed independently of theory's empirical success, or maturity.*
>
> **(2)** MODELS AND DUALITIES: *In string theory, taken as a collection of models, S-dualities play a central role.*
>
> **(3)** STRUCTURAL PLURALISM FROM DUALITIES: *With S-dualities, string realism reorients **(a)** from object-oriented (OO) realism, to structure-oriented (SO) realism and **(b)** from fundamentalism (about entities) to pluralism (about structure).*

The first premise **(1)** restricts the type of scientific realism applied to string theory. Probably the most pressing philosophical question about string theory is: can we infer something about the world from it? If not, is it appropriate to inquire the realist commitments of string theory? Isn't it too early to ask? Is it too much to ask from a mere collection of mathematical models with little connection to empirical science? This paper assumes that philosophers can gain insight in a theory in physics by analyzing its realist commitments, at various stages of its development.[3] Standard discussions on scientific realism provide some prerequisites: realists

---

[2]N. Cartwright argued against the fundamentalism about both laws of nature and theories (Cartwright, 1999). L. Sklar and P. Teller offered reactions to Cartwright's attack (Sklar, 2003; Teller, 2004).

[3]The philosophical literature about string theory is still dearth, but is growing rapidly. See Callender and Huggett (2001), Dawid (2006, 2007, 2009, 2013), Matsubara (2013), Muntean (2015), Rickles (2011, 2013, 2014), Taylor (1988), and Weingard (1988). This paper refers more often to work of D. Rickles and R. Dawid and ignores aspects of dualities important for theory-choice and theory-development.

should only commit to those theories which are genuinely successful and to mature theories which have been tested and checked for a significant period of time (Chakravartty, 2007; Psillos, 1999). What is assumed here is that maturity, success and being true are not necessary conditions to inquire the realist commitments of a theory. For example, even if classical electromagnetism (or relativity, or quantum mechanics, for that matter) are known as being more or less false theories, it is nevertheless philosophically and historically relevant to ask: what are its realist commitments? This paper investigates the primitive string ontology theorists are committed to. Hence, a "selective skepticism" about scientific realism is appropriate (Chakravartty, 2007): not *all* the statements of string theory are firmly committed to realism: on the contrary, as one can learn from the arguments marshaled against realism, only some statements of a theory are retained in future developments. At this stage, we pick from the array of statements in string theory those which display a good balance between epistemic risk (novel predictions and/or explanations) and epistemic security (e.g. being better related to existing, successful, mature theories in physics or even to direct experiments). Hence, realism about string theory (or any quantum gravity program for that matter), can be better couched in terms of a counterfactual like this:

> (4) *If one thinks that a string model is 'plausible enough', which ontology is one committed to?*

There are other counterfactual and modal questions about string realism: suppose one were serious about the model M in one world w: what building blocks are needed to build w? Now assume that model M is only possible at w: what is possible and what is not in other worlds, including our own? Other option is to start from the ideology of the theory and ask about its ontology: similar to the project of D. Wallace and C. Timpson for quantum mechanics, we can identify some properties that string theory trades in, and then seek their property bearers (ontology) (Wallace and Timpson, 2010). Or ask: what ontology does string theory prescribe?

All the aforementioned questions can be integrated in "string realism", an inquiry into the realism commitments of string theory. When tackling string realism, ask not whether string theory is true/successful/mature, but reflect upon the implications for realism, including, but not restricted to, a choice for or against fundamentalism, monism, pluralism, etc. Hence, let us agree that, notwithstanding string theory's empirical success or maturity, "string realism" is philosophically a respectable topic.

The next premise (2) acknowledges the non-trivial role of S-dualities in string realism. To ease up the introduction to string dualities, an example of a simpler duality may help: witness that the E-M is an "exact" symmetry between magnetic and electric charges and fields (Castellani, 2009; Rickles, 2011). Magnetic monopoles and electric dipoles both exist, if this duality is real. Magnetic monopoles are too heavy and too hard to produce in normal conditions. Electromagnetism in our universe is a weakly coupled theory, because its "coupling constant" $\alpha = \frac{e^2}{\hbar c}$ has a relatively low value in nature: $1/137$. At this weak coupling, electric charges are simple and fundamental, whereas monopoles, as collective excitations of electric charges, are strongly coupled to the E-M field and therefore "heavy", unstable, and in strong interaction with the field. They are also supervenient on electric charges.

The classical E-M duality has some notable realist consequences. Dirac demonstrated the quantization of electric charge: $q_m q_e = 2\pi n$, for any integer n. It can be read as an conditional: "if there exists somewhere a magnetic monopole of charge in the universe, then all electric charges are quantized." The converse is a realist commitment to the existence of magnetic monopoles inferred exclusively from the duality: "because the electric charge is quantized, there are magnetic monopoles". One can infer that monopoles exist in this world, but they are not fundamental or better, less fundamental than electrical charges. Then, there is a dual model M' at strong electric coupling ($\alpha' \gg 1$), in which monopoles are fundamental and electrical charges non-fundamental. Fundamentality and supervenience are most likely relative to a coupling factor. Last, but not least, the duality points to something outside the theory that integrates the two descriptions: the relativistically invariant tensor $F_{\mu\nu}$ is probably the best candidate, or the quantum electrodynamics formalism in which these two descriptions can be integrated (although with some difficulties).

As an argument by analogy for (3), contrasting and comparing S-dualities in string theory and the E-M duality can offer some help. To anticipate Sect. 13.2.2, string dualities are *correspondence* relations among string models: S-dualities in particular relate the strong coupling of a model with the weak coupling of another (or the same) model (Castellani, 2009; Rickles, 2011, 2013). There are fundamental differences between string dualities and the E-M duality. Unlike the latter, S-dualities relate two models with different formalisms, symmetries, spacetimes (topology), and different primitive ontologies, to the very same set of observable quantities. Moreover, unlike the exact E-M duality, a string duality is an 'almost' one-to-one correspondence among string models. Dirac's proof of magnetic monopoles can be read as an argument from dualities to realism in this model. We learn about magnetic monopoles from experiments with electric charges; and we prove the quantization of electric charges (a fact about our world) from the dual model. But these encouraging results are not echoed in string models. For many, probably for a majority of physicists and philosophers of science, the presence of dualities hastens antirealism: they are mathematical redundancies, or worse, vicissitudes of theories disguised in mysterious, serendipitous connections. For the realist, on the contrary, the mystique of dualities is just a gateway to the discovery of the underlying structure. This paper, it its restricted scope, shows that there are better alternatives than the antirealism pit or the monistic realism, when S-dualities are at stake.

Philosophers of physics (R. Dawid, D. Rickles, E. Castellani) who have discussed dualities relate them to realism. For Rickles and Dawid, string dualities are genuine (and rare!) examples of scientific underdetermination. Dawid claims that dualities entail an "ontological underdetermination", which is a threat to standard realism: the duality principle in string theory clearly "renders obsolete the traditional realist understanding of scientific objects as smaller cousins of visible ones" (Dawid, 2007, p. 25). Dawid's conclusion is that a standard scientific realism position is incompatible with string dualities, which indicates a new form of realism about a "unique consistent structure". For Rickles, dualities are "the structural realists' best 'physics-motivated case' for their position" and "ready-made exemplars for the structural realist position" (Rickles, 2011, p. 66). Nonetheless, Rickles needs E. Witten's mysterious "M-theory" as the universal theory whose various low energy limits in the space of parameters produce all the known models of string theory. Supposedly, M-theory possesses one structure and one ontology that would incorporate every string model, with its structure, ontology and symmetries.

On the contrary, a weaker position is to keep the structure, but assume less about it. This position entails an altogether different ilk of realism, weaker than standard scientific realism, and different than straightforward antirealism. It is not premised on the uniqueness of structure, as in Dawid, or on the underlying theory (M-theory) equipped with this structure, path adopted by Rickles and the majority of theorists. Contrary to the "theory of everything" narrative, the most plausible "string realism", at this stage, is model-based (as opposed to a theory-based), structural, and pluralistic. Structures are here generated from different models, and S-dualities. Strings and branes are not anymore the fundamental entities, due to the presence of these structures. Structure is endowed in this framework with *a relative and relational fundamentality*. One can call this type of realism an "attenuated OSR" (Frigg and Votsis, 2011): each of the string models has its ontological commitments relative to the assumptions, idealizations and abstractions it had been built upon. In this view, similar to S. French's recent account of "structure", it involves the "webs of relations', represented by the relevant assumptions, and "as effectively tied together by higher order symmetry principles representing the invariants in terms of which the 'nodes' in this structure can be described." (French, 2010, p. 92). Structure is in this case multi-layered and multi-aspected because the web itself connects radically different models (French, 2010). Another recent advancement in the same spirit is the 'Rainforest Realism' of J. Ladyman and collaborators (Ladyman et al., 2007), which welcomes the idea of a "scale relative ontology". Moving towards a structural ontology seems more natural than entertaining fundamentalism. String realism is therefore a "coupling relative ontology", a species of "relative ontology".[4]

13.2 A Model-Theoretic View of String Dualities

In order to map different realist commitments of string theory, and especially its "pluralist stance", one can delineate it as a family of "string models".[5] From a philosophy of science point of view, what is proposed here is that the unity of analysis cannot be the "scientific theory" in the syntactic approach to theories: string theory is not a set of statements, closed to logical consequence. Whether "string

---

[4]Coupling, as defined below, is not a energy scale or length/time scale, but a combination of two and many other aspects of strings.

[5]This section overlaps with Muntean (2015).

theory" is a scientific theory *per se* or a 'family of models', in the model-theoretic sense, is not a mere terminological dispute, when it comes to its commitments to realism.

Why is string theory more philosophically enticing as a collection of models? Probably the more obvious reason is historical: the "theory" has evolved through a rapid succession of string revolutions, roughly every decade. We are not lucky to benefit from a Newton, a von Neumann, or a J.S. Bell figure of string theory, somebody to come up with a "systematic formulation" based on a set of few principles or axioms (Weingard, 1988). String theory does not contain laws of nature, or axioms, and has few principles, if any. The theory itself does not generalize smoothly to other areas of physics. Nonetheless, it still makes sense to talk about entity fundamentalism here, as opposed to field fundamentalism, or symmetry fundamentalism, or any form of antirealism that can be inferred from each string models. It is also fair to say that every string revolution reshaped this string realism, its fundamentalism and consequently the answer to (**4**).

Probably for the majority of theorists, such a plurality of models is not a virtue, but a drawback. Since the second string revolution, theorists have hinted towards the M-theory that would reduce all string models to one theory with some principles and a simple ontology. The activity of model-building in string theory outweighs any attempt to provide its rigorous (algebraic, axiomatic, etc.) formulation. This is another reason to prefer the model-theoretic view of string theory, based mainly on its current scientific practice: it is less ambitious, but closer probably to the everyday life of a string theorist.

A closer look to the way string models are built endorses the parallel between string theory and the model-theoretic view about science. Two central features of model building are abstraction and idealization. An abstract model describes a system (or perhaps another model) that cannot be made more realistic simply by adding correction factors. A model is idealized in respect of a property P if it represents the real system as having P (or as not having P) when the system does not have P (or has P). C. Pincock (2005) suggests that a mathematical model is idealized when it meets two conditions: (i) no isomorphism relates it to the represented entity (be it a theory, model or a real situation) and (ii) the relevant agents are aware of this. String theory is the result of both abstraction and idealization as mathematical models: perturbative models are idealized and abstracted. They *represent*, inaccurately, another system (another model, a mathematical structure, information, etc.) without claiming that they are perfect representations. Sometimes this inaccuracy is intentional and serves a particular reason. Sometimes we do not know enough about the system represented. This argument illustrates the philosophical debate around the autonomy of models: are theories reducible to models, or, on the contrary, are models just aspects of theories? Probably it is fair to say that string "theory" is a set of models, in the search for a theory, or it is intentionally depicted sometimes as a theory.

Last but not least, what is the relation among string models? There is a specific type of idealization germane to my discussion: the multiple-model idealization, when inconsistency among models is prevalent (Weisberg, 2007). Building a set of inconsistent models is not a *desideratum,* but a consequence of several factors: the complexity of the problem, our limited abilities to represent the system, the nature of representation and, in the case of string models, the lack of an efficient mathematical formalism. For Weisberg, there are many reasons to use the multiple-models idealization, but the one germane here is building a theory, or finding the set of "idealized models that is maximally useful for creating new structures" (Weisberg, 2007, p. 648). The 'generality' ideal can be accomplished on two dimensions: the 'a-generality'—the number of actual targets to which a model applied and the 'p-generality'—the number of possible, but not necessarily actual, targets a particular model captures. Understanding string models as aspiring to fulfill especially the p-generality ideal is a step towards understanding their realism commitments.

String models are peculiar in yet another respect: it is hard to apply directly a formal definition of scientific models. They are not pure mathematical models, as they do not have phenomena as an intended target. They are mathematical models constrained by some physical theories. The current paper employs the notion of a model in a more informal way: one reason is the difficulty to identify a "shared structure" between models, laws of nature, bridge principles among models and a governing theory. A more promising alternative to Weisberg's approach is the "partial structures" account of Da Costa and French (2003). It may capture better S-dualities as partial structures, but for space consideration this line of thought is not followed here.

13.2.1 Delineating String Theory

There are probably several charitable ways to delineate and describe string theory, but one is relevant to the present discussion:

> **(5)** *String theory is a collection of mathematical models of strings vibrating in various types of spacetimes, on which different symmetries and different fields are postulated. These models may include branes or other objects, and they may or may not represent some aspects of known physics: the Standard Model, gravitation, various gauge theories, black hole thermodynamics, information theory, condensed matter physics, etc. String theory includes a set of conjectures about the relations among the string models.*

This hints towards the idea that even if string models are not directly representing the world, there are unexpected and fecund connections between string models and the physics of our universe. The secret of string models as being related to the world, in any meaningful sense, resides in their mutual relations and not in the model itself. In some cases, such as the AdS/CFT correspondence (Rickles, 2013; Teh, 2013), one explains and calculates unknown aspects of some physical theories (most notably aspects of Quantum Chromodynamics, black hole thermodynamics, condensed matter physics, cosmology, etc.) from string models. Here, the plurality of string models is desirable, each string model being more or less appropriate to a domain in physics. Relations among string models shed new light on entities or structures involved in *building* string models: gauge theories, gravitation, black holes etc. String models may instantiate some *virtues*, usually attributed to scientific theories: unification, explanation, prediction, etc. Following Weisberg (2007) again, these are "representational ideals" of string models.

A string model describes the dynamics of strings in a given spacetime and their modes of vibration. Two types of string are possible: open strings and closed strings. After quantization, different massless modes of vibration can be inferred: some modes of vibrations are related to known theories in physics, some are too exotic or even physically impossible. The gauge bosons of any $U(N)$ theory can be interpreted as vibrations of open strings (Hooft, 1974). Mesons are modeled by open strings, whereas glueballs and gravitation are modeled by closed strings. The existence of a graviton mode for all closed strings was totally unexpected. Why do string models, which are all premised on the idea of Lorentzian invariance and are limited to flat spacetime, contain the vibration modes of the graviton?

One short note about supersymmetry (SUSY) is in order here. The open bosonic model, the simplest of all, includes the tachyon as a mode of vibration. This shows that the theory is unstable and very unrealistic. Furthermore, the model did not include fermionic modes. E. Witten showed that D = 11 was the smallest spacetime to accommodate all the gauge groups of the Standard Model (bosons and fermions), such that the manifold assumed in all string models has to have a dimensionality above four ($D > 4$), and has to be supersymmetric (SUSY). All string models with SUSY are able to cancel the anti-physical tachyonic modes and include fermions.

Here is the situation of string "theory" in the late 1980s, when the existence of following consistent string models was acknowledged (Table 13.1):

In early 1990s, two questions were asked about these models:

(**6**) *How are these models related one to the other?*

(**7**) *What is the strong coupling limit of each model?*

The first answer to (**6**) is to entertain the intuition that string models are independent, given the widely different symmetries, manifolds, fields, charges, etc. they are premised on. Another answer to the multiplicity of models is eliminativism: some theorists conjectured that one of the models is closer to the real world: at least the bosonic model is definitely incomplete and unrealistic (it included tachyons). Another attitude is quite the opposite: a 'democracy' among string models. Barring the bosonic model, all models are born equal, or, alternatively, they are equally unreal. Finally, in the line of an unificatory ideal, one can conjecture, that these models are only partial aspects of an underlying "M-theory" (Townsend, 1995; Witten, 1995). Had it been real, M-theory would be properly speaking a theory, not a model.

Table 13.1 A family of relevant string models

Model	Space-time	Modes	Types of objects	SUSY	Gauge group
Bosonic string	$D = 26$	Bosonic	Open or closed strings	None ($\mathcal{N} = 0$)	$U(1)$
Type I super-string	$D = 10$	Fermionic & bosonic	Not oriented open strings (closed by interaction)	$\mathcal{N} = 1$	$SO(32)$
Type IIA superstring	$D = 10$	Fermionic (non-chiral) & bosonic	Closed oriented strings & D-branes	$\mathcal{N} = 2$	$U(1)$?
Type IIB superstring	$D = 10$	Fermionic (chiral) & bosonic	Closed strings	$\mathcal{N} = 2$	None ?
$SO(32)$ Heterotic	$D = 10$ and $D = 26$	Bosonic & fermionic	Closed strings	$\mathcal{N} = 1$	$SO(32)$
$E_8 \times E_8$ Heterotic	$D = 10$ and $D = 26$	Bosonic & fermionic	Closed strings	$\mathcal{N} = 1$	$E_8 \times E_8$
Supergravity (SUGRA)	$D = 11$	Bosonic & fermionic	No strings	$\mathcal{N} = 2$	

13.2.2 S-Dualities as Inter-model Relations

Surprisingly, **(6)** and **(7)** are not independent questions: one of the revelations of the "second string revolution" was that in order to answer **(6)**, one needs to address firstly **(7)**.

The infamous answer to **(7)**, first conjectured by Witten is: each string model has its strong coupling sector related to other models (or to itself), by a web of S-duality relations. The S-dualities can relate different sectors of the same model, too (the so-called self-dualities).

What is a sector of a model? Every string model has a set of characteristics: background fields needed for the consistency of the model, the size and shape of the compact spaces, symmetries, etc. and, last but not least, a set of coupling constants—similar to other models in physics. Two parameters are very important here: the string coupling constant g_S and the string length ℓ_s. g_S is similar to the fine structure constant in electromagnetism which quantifies the way charges interact with photons: in string models it expresses the dynamics of strings and how they interact with spacetime and other fields postulated in the model. Each string model has a weak coupling sector and a strong coupling sector based on g_S. At low values of g_S ($g_S \ll 1$), strings are weakly coupled, dispersed in space, the number of strings is relatively small and strings vibrate against a flat, fixed background spacetime. At $g_S \gg 1$, a more complete model assumes that strings interact and that enough energy is present in the string vibrations for it to interact with spacetime itself. In the strong coupling regime the "backreaction" with spacetime is assumed and the background spacetime is not fixed anymore. The strong regime is also non-perturbative, in the sense that it would be able to describe interacting strings as well as string that can split and join.[6] Therefore, g_S encodes the probability of a string-string interaction and the probability of a string splitting and emitting a closed string.

The unproven conjecture is that the weak sector of *any* string model is related to the strong coupling model.[7] S-dualities are then inter-model relations, or partial maps between a weak and a strong sector:

> (8) STRING S-DUALITY: *A string S-duality is a conjectured map between (i) the sector of a string model in the weak regime M_{weak} with a set of fundamental entities (possibly one) F_{weak}, a set of non-fundamental, derivative entities d_{weak}, some fields φ_{weak} and a coupling constant g_{weak}, and (ii) the sector of its "dual" model in the strong regime M_{strong} with $g_{strong} = 1/g_{weak}$, another set of fundamental entities F_{strong}, non-fundamental, derivative entities d_{strong} and fields φ_{strong}. S-duality assumes by definition that there is an isomorphism between the consequences of M_{weak} and M_{strong}.*

How do we interpret dualities? Whereas M_{weak} is represented by a perturbative formalism, in most cases there is no formalism for M_{strong}. When there is an attempt to represent non-perturbative models, M_{strong} is more exotic, more complicated and harder to capture. A duality is not a relation of correspondence among theories "by limit", in the way classical mechanics is a limit of special relativity, when $c \to \infty$. The perturbation formalism in string theory is only asymptotic, not convergent, and the weak coupling formalism cannot be extended to describe the strong coupling sector. Mathematically, there should be no relation between M_{weak} and M_{strong}: dualities reveal unexpected and serendipitous relations which go beyond the formalism of the two models. Dualities are based on SUSY states that are present in both sectors. The states in the spectrum of the strong model are called the *BPS* (Bogomol'nyi-Prasad-Sommerfield) states. Once the theorist identifies the common *BPS* states of M_{strong} and M_{weak}, the correspondence relation between fields, topologies, and, last but not least, entities can be inferred.

---

[6] The "intrinsic" string length is $\ell_S = \sqrt{\alpha'}$, where α' is the Regge slope, which at its turn is the inverse of the string "tension" $T = \frac{1}{2\pi\alpha'}$. As one expands the dynamics of a phenomenon in terms of α', one moves from a classical description to a stringy description of the phenomenon. Informally α' is a measure of the quantization of geometry and of its dynamical nature. g_S is related to the topology of spacetime and the length ℓ_s to the "size" of the spacetime.

[7] Technical details about dualities can be found in Sen (1998). The foundational discussions referred here are Rickles (2011, 2013), and (Polchinski, 1996, 1998, esp Ch. 14).

There is an epistemic interpretation of dualities. They are devices used to extend our knowledge from perturbative to non-perturbative models, guided by *BPS* states. Through the duality correspondences, the understanding we have of M_{weak} is "carried over" to M_{strong}. Calculations are performed in the M_{weak} and extrapolated to M_{strong}. Perturbative models are incomplete, highly idealized and abstract, whereas non-perturbative models are closer to a full description of strings in which some idealizations of the perturbative models are eliminated and a more complete description is assumed. Strong coupling means that quantum effects on spacetime are preeminent; the spacetimes of weakly coupled models are "less" quantum and closer to the classical view of spacetime manifolds: they are the stage, the background, with no quantum backreaction. And this is at the crux of the harshest criticisms against string theory.

The aim of the next sections is to show that dualities are not mere epistemic devices. The simplest case of a S-duality relates the strong coupling limit of Heterotic $SO(32)$ model with the Type I model and vice versa. Other interesting case is the duality between the weak IIA model and the strong $SUGRA$ model in D = 11. Here, the 'simpler' physics lives in a spacetime with fewer dimensions than its strong dual model. The strong coupling limits of Type IIA and Heterotic E models are interesting examples because both are ten-dimensional at weak coupling. In their strong coupling limits, an additional 11th dimension "grows up" from the string dynamics! We assume that the extra 11th dimension is compact and its size is related to the ten dimensional coupling constant. At weak coupling, the 11th dimension is small and invisible, but as the coupling increases, this extra dimension unfolds. The theorists expect to encounter completely new, non-perturbative entities in the strong coupling sector: the D-branes. In the perturbation formalism at weak coupling, Dp-branes are geometrical, and do not play a crucial role in the M_{weak} because they are somehow too heavy and too complicated to be captured by the formalism. Closed strings end on D-brane. For $g_S \geq 1$, Dp-branes become more fundamental than the fundamental strings. For example, in some types of string theory, strings end on Dp-branes (a special class of p-branes with Dirichlet conditions). Interestingly, Yang-Mills gauge theories may live on the world volumes of Dp-branes in p dimensions, while gravity extends beyond the branes in bulk space with $D > p$ dimensions.

When only strings are fundamental, fundamentality integrates well with the ideal of "simplicity" in Weisberg (2007). String models are ideally committed to a minimal set of entities that generates a desired set of properties. Accomplishing this ideal is not possible beyond certain point: strong-weak duality unveils a crucial aspect of fundamentality in string theory: it is relative to the position in moduli space, especially to values of g_S. Then, one can ask what is left of the metaphysician's fundamentality when it is relative and admits more types in its base. The following sections exposes the logical map of string realism, once we take dualities 'more seriously' than merely computational devices.

13.3 Strings and Branes as Fundamental Objects in String Realism

We learned from recent work on ontic structural realism that physical theories support typically two or more metaphysical "packages" (French, 2014; French and Krause, 2006). For example, when we have the packages of *non-individual* objects and of classical *individual* objects, the realist faces a "metaphysical underdetermination" and hence closer to being lured into some form of anti-realism. To assess the metaphysical packages available to the string theorist, four forms of string realism are situated between the stronghold of string fundamentalism and a weak form of pluralism. The fundamentalism is represented by two object-oriented (OO) positions: (i) when the fundamental base contains strings only (Sect. 13.3.1), or (ii), the "brane democracy", with strings *and* branes as types in the "fundamental base" (Sect. 13.3.2). Further, the fundamentalism can drop objects from the fundamental base and go for a structure-oriented (SO) fundamentalism. Either (iii) a form of eliminativist ontic structural realism (*OSR*) is available to the string fundamentalism (Sect. 13.3.3) or a model-theoretic structural realism in which structures play the role of fundamental entities (iv). Section 13.4 appraises this latter ilk of realism in the context of S-dualities: it is more pluralistic, and less vulnerable to anti-realist objections. One rebuttal of realism based on the reification of SUSY is shortly discussed in Sect. 13.4.1.

13.3.1 Object-Oriented (OO) Realism

On this simplest, if not naïve, form of string realism, the string theorist is committed to the existence of two types of extended objects: first and foremost strings, and second, Dp-branes. For B. Greene strings are similar to the letters of the alphabet, "A string is simply a string—as there is nothing more fundamental, it can't be described as being composed of any other substance" (Greene, 1999, 141–2).

Then "(spatially) extended simples (without proper parts)" are the fundamental entities. They move against a fixed spacetime background similar to the dynamics of fields in the quantum field theory.[8] They are quantized on the basis of quantum field theory procedures. Although they are objects of quantization, strings are not ordinary quantum objects. The string ontology is only tangentially related to that of quantum mechanics and quantum field theory, because strings are neither fields, not pointlike particles–in any relevant respects. It is productive to take string realism as independent from quantum field realism. Strings are more fundamental than any

---

[8]From a metaphysical point of view "extended simples" may constitute a problem. The metaphysician may need to put some effort in accommodating such new entities: see arguments for and against in Baker (2014), Hudson (2005), McDaniel (2007), and Muntean (2015).

pointlike particles known in physics: graviton, photon, gluon, the Z, W^- and W^+ bosons, as these can, arguably, be inferred as modes of vibration of strings or as their solitonic excitations.

What is the status of fields, symmetries, modes, string states, charges, currents, etc., and ultimately, spacetime, in string realism? Although the OO string realist does not deny that other entities and mathematical objects or geometrical objects are useful to the theory and even endowed with explanatory power, there is no commitment to their existence: the string is an extended object, the only fundamental entity and it exists in all models, but in $SUGRA$ or presumably, in M-theory. Fields, which can be postulated on spacetime or on the worldsheet, symmetries, modes, are all needed for the theory to work properly, probably even indispensable mathematically, but for the OO fundamentalist they are all derived or derivable, not fundamental. Some can be inferred naturally from string theory: "in the context of string theory these symmetries [supersymmetry, gauge symmetries, the symmetries of spacetime], are *consequences*; although their importance is in no way diminished, they are part of the end product of a much larger theoretical structure." (Greene, 2011, p. 333). What are the other theoretical terms of a string model and what reasons do we have to be realists about them? The OO fundamentalist tolerates the *fields* postulated on parts of strings, but insists that her OO string realism commits only to the fundamentality of strings or branes. The worldsheet of one string is interpreted as a theory with fields which are quantized. On each string there is an infinite number of harmonic oscillators and the state of the whole string live in a Fock space. In this space there are creation and annihilation operators that act on the Fock vacua. The excited states of the string represent particles, or fields in the target space-time. Similar to fields, for the OO fundamentalist, the *modes* and the *states* of strings are not objects and lack individuality, as any quantum fields do. They are parts of the string, but they are not fundamental. Here, parts are not taken in an ontological sense. Strings have modes of vibrations and boundary conditions attached to their ends. Modes are attributes of the string, not of points of spacetime.

The tension, although can be predicated about points on a string, is less important to string models: the massless modes, the most interesting vibrations of strings are not determined by the absolute value of the tension (which determines nevertheless the massive modes), but the quantum fluctuations of the tension. The realist has then reasons to think that, as extended entities in one dimension, strings are not localized objects with intrinsic properties. It is indeed an altogether different question to check whether strings are individual and distinguishable entities, just like classical objects. The discussion about Fermi-Dirac and Bose-Einstein statistics that spoils several attempts to realism in quantum field theory, needs to be re-framed in terms of types of strings and their combinatorics: nonetheless, the OO fundamentalist may keep on arguing that string realism is about objects, albeit a new type of objects, metaphysically speaking.

The OO string realist insists that although strings are not pointlike entities, they are countable, and asserts the individuality of strings, at least in the case of the highly idealized models of weak coupling. Historically, one may witness here a return to the OO realism that circumvents the OO realism in quantum field theory where objects have "withered away", while fields or structures (group-theoretical structures) are more fundamental (French, 1998). In string theory we do not have a clear-cut distinction, such as the "particle vs. field" picture in quantum field theory.[9]

First, a suggestion in the literature is that in quantum field theory with interactions, the particle realism is fatally undermined in the light of Haag's theorem.[10] But string interactions are different: they do not encode the information about interaction at the vertices of Feynman diagrams: in string theory there are no vertices, no singularities where interactions occur. The worldsheet of two interacting strings is smooth anywhere: locally, every section of the diagram is a sum of free propagating strings. The diagram reveals the interaction *only* globally. Second, it turns out that unlike quantum field theory, scattering amplitudes in string theory are ultraviolet finite. This achievement of string theory endorses the *OO* realism. In field theory, although Feynman diagrams may be read off in a particle-oriented metaphysics, particles are only a limited component of a much larger perturbative expansion of fields. The creation or annihilation of one particle is not, mereologically speaking, a part of a larger process (Teller, 1997). In field theory, the mechanistic view of processes being divided into parts does not hold. Divergences loom large–smaller parts play a more important role than larger part. A lack of mechanistic metaphysics can undermine fundamentalism as well as object-oriented realism.

Is there any trace of mechanistic fundamentalism in string theory? Given its advantage in respect of divergencies, string *OO* realism is in a much better situation. As it eliminates particles as fundamental entities and the ultraviolet divergences, string theory is not a battlefield between dominant interpretations such as the field and the particle interpretation of quantum field theories. This section suggests that fields, modes, solitonic excitations, tension, cannot be promoted to the status of fundamental entities of string theory. The string fundamentalist needs to do some footwork and enrich the theory of individuals to include extended simples as fundamental entities, but the "objectual" nature of string is granted here, with no credible competitor. In this *OO* interpretation, string theory recedes to an object oriented ontology more or less similar to quantum mechanics.

13.3.2 The 'Brane Democracy' Realism

Although the *OO* string realism denies the reality of modes and fields, of solitonic excitations and of many geometrical objects, it needs to face this question: how do we know that all fundamental objects of all models, at any coupling, have precisely the dimension $p = 1$? Are there fundamental objects with dimension $p > 1$? In

---

[9] An extensive discussion on particle realism in quantum field theory is in Ch. 9 of French and Krause (2006).

[10] For an argument against this, see D. Baker's analysis of the particle picture and the field picture of quantum field theory (Baker, 2009). In his view, none fares better.

the mid-1990s, the theorists envisaged the possibility that Dp-branes are in some cases more fundamental than strings. Townsend compared solitonic solutions of the type IIA model with the solitonic solutions of *SUGRA* model, which is a "super-membrane":

> But this difference disappears once one identifies the fundamental string or membrane with the solitonic ones; both theories then have exactly the same spectrum of extended objects. In fact, it becomes a matter of convention whether one calls the theory a string theory, a membrane theory, or a *p*-brane theory for any of the other values of *p* for which there is a soliton solution; all are equal partners in a p-brane democracy. (Townsend, 1995)

This form of 'p-brane democracy' opens the door to underdetermination and ultimately to anti-realism. If both models are equally successful, and yet so different in their realist commitments, which is the best? One needs to find a way to break the underdetermination between the two models. Before giving up fundamentalism and succumb to a string anti-realism, the *OO* realist may look for some implicit discrimination in the 'p-brane democracy'. Here is a recipe against democracy in string realism: acknowledge that closed strings are more fundamental than open strings. In string theory, closed strings are necessary objects: there is no model without closed strings, but some string models may have only closed strings.[11] Dp-branes are p-dimensional hyperplanes on which open strings with Dirichlet boundary conditions end, so their existence may be deemed as dependent on open strings. Closed strings always exist in the space between Dp-branes, therefore they are more fundamental than Dp-branes.

But there is a fly in the ointment: in some models, at different couplings, strings are not stable and can split, as a probabilistic process. An analysis of strings reveals that although some fundamental strings are stable (the type IIB model), others are not (the type I string). The probability of breaking is proportional to $\sqrt{g_s}$. In this case, the *OO* fundamentalism ends up being relative to the coupling factor. At stronger coupling, some fundamental entities split, and some fundamental entities can join and form another fundamental entity (this is the time-reversal process of string splitting). As the argument goes, in some strong coupling models, Dp-branes are more fundamental than strings because they are stable. A reformed fundamentalist would insist that strings need to end on a physical object and take Dp-branes as entity of the fundamental base, together with strings (Becker et al., 2007, p. 194). Dp-branes are always dynamical hyperplanes that fluctuate and curve, and they are quantum geometric entities. It is natural to assume that non-perturbative representations are always more fundamental and more realistic than any perturbative model, despite us not being able to represent them through the mathematical formalism. Formally, S-dualities may look like symmetric isomorphisms, but when it comes to realist commitments, we cannot treat them on equal footing, despite Townsend's promising 'p-brane democracy'. In other words, the string realist needs

---

[11]As closed strings always contain in their spectrum the graviton, this choice is not physically innocent: it corresponds to the fact that gravitational interaction is always present in the physical world.

to commit to non-perturbative models and a literal string fundamentalism does not survive the transition to non-perturbative models. Non-perturbative, strong coupling models are *always* closer to the ideal of a background-independent string theory.

The democracy can be supported by another argument: one can take Dp-branes as more fundamental because they provide a simple and unified way of introducing gauge symmetries in string models on multiple coincident Dp-branes. Non-Abelian gauge symmetries in particular are extremely useful in representing Yang-Mills theories in string theory. So a connection to "real" physics is not always possible without Dp-branes. Strings are not enough to connect string models to known physics! In non-perturbative models they are back-staged and hidden and presumably will disappear from the subsequent formalism.

Finally, a string "pragmatist" may use the perturbative models as epistemic guidance to non-perturbative models. The string pragmatist progresses and calculates from weak coupling models, and is silent about non-perturbative models. On the contrary, for the realist, the string models in which we should believe, and potentially M-theory itself, have to be at strong coupling. Although Dp-branes are ultimately on equal footing with strings, they are landmarks on the path to non-perturbative and more realistic models. For the M-theory optimist, the hope of a ultimate theory with no strings is real.

13.3.3   Structure-Oriented (SO) Realism: Uniqueness and Fundamentality

A radical change of strategy comes from ontic structural realism (*OSR*). The structure-oriented (*SO*) string realist rejects the *OO* realism and adopts the structure as the fundamental entity. This is how the proposals of Rickles and (partially) Dawid can be integrated into *SO* realism: they consider the structure as being unique and fundamental. For the *SO* realist, there is *one* structure underlying all models, more fundamental than strings and Dp-branes, at *all* regimes. This section adopts a more skeptical perspective about the "all" and the "one" quantifiers in the previous statement. The *SO* realist argues that for the elimination of strings and D-branes or, weaker, for "thin" concepts of "string" or "brane" with their individuality reworded in structural terms. *SO* string realism should be aware of some differences with the *OSR* that is so well-fitted with quantum field theory. As discussed in Sect. 13.3.1, the main source of inspiration for *OSR* is a metaphysical underdetermination between the "field-as-substance" and "field-at-spacetime-points" metaphysical packages (French, 2010). The present paper does not aim to balance the two metaphysical packages because in string theory we may have a different packaging of the structure. A motivation for *OSR* stems from quantum (field) statistics. The *SO* realist can keep the spirit of *OSR*, if not its letter. Nevertheless, the lack of a string statistics is not a major blow for *SO* realism.

These reasons are enough to convince the *SO* string realist to ground her argument against *OO* on different assumptions than the *SO* realist in quantum field theory. First, the *SO* realism can use M-theory and argue for fundamentalism and uniqueness about its structure; second, another *SO* realism can drop fundamentalism and the ontological uniqueness of structure and adopt a model-centered structural realism (see Sect. 13.4).

The *SO* who keeps fundamentalism and uniqueness needs a theory to 'present' the structure. *OSR* favors a theory-dependence of models: lacking a theory, string models need the uniqueness of structure from elsewhere.[12] If the *SO* realist wants to keep fundamentalism, then the easiest interpretation is to postulate that M-theory is the seat of a unique structure. For Rickles and many string theorists, all dualities point to a deeper, unificatory and more fundamental *theory*, the "M-theory", of which we know very little. Such a theory is purportedly without strings, so its can be the ideal candidate for the SO realism that replaces the *OO* ontological assumption. Second, one may wish that M-theory will have a *unique* structure in which all string models can be embedded:

> All string theories are connected [...] to something new and surprising: an eleven-dimensional super-gravity theory with no strings, known provisionally as 'M-theory.' (Polchinski, 1996)

> The web of dualities is taken to restore the uniqueness that was thought to characterize the earliest incarnation of string theory. (Rickles, 2011)

Consequently, different *models* are just perturbative expansions of a unique underlying *theory* about five different, consistent quantum vacua. *SUGRA* is one of the low-energy effective descriptions of the M-theory. Dp-branes do not exist in M-theory, only the M2-brane and the M5-brane, and "end-of-the-world" 9-branes. As these are not defined anymore through the end points of strings as Dp-branes, the SO realist may well speculate that they are not objects, but object-like aspects of the underlying structure, or putative 'objects'. Probably the *OSR* in string theory would reduce objects such as strings and Dp-branes to representations of the M-theory structure. More than that, the suggestion is that there is strictly *one* structure which is presented in M-theory, and no strings: indeed, M-theory is touted in recent years as a string-free theory. For the M-theory enthusiast, all we know about string theory now will be inferred from, or better, reduced to, M-theory. The *SO* string realism may *ideally* keep both the fundamentalism and the uniqueness of structure and that would be the best metaphysical package.

13.4 The Structural Pluralism of String Realism

The last version of realism under scrutiny meet only partially the general requirements of the *SO* realism: what if one drops the uniqueness of structure and tackles differently the fundamentality in string models? In this form of *SO* realism,

---

[12]This is probably grist to the mill for "autonomous models" view of M. Morrison, N. Cartwright, etc. See Morgan and Morrison (1999).

pluralism is a consequence of dualities and contributes to a reconceptualization of fundamentality. The fundamental base is now relative to models and coupling factors, as includes Dp-branes that can replace strings as the fundamental entity of the model. It does not assume a unique structure for all models at all couplings.

We witness inconsistencies and incompatibilities between string models, which is acceptable to *OSR* or to the "rainforest" realism: string theory as a collection of models is still evolving, both by discovering new dualities and by the analysis of new sectors, especially the non-perturbative ones. The *SO* string realism defended here is based on dualities and on SUSY. The structures have a pure relational nature: they connect fundamental and non-fundamental objects in different models, without being themselves entities. The plurality of the structure is granted by the very diverse nature of the S-dualities (and other dualities for that matter). But a skeptic may ask: is this diversity just apparent? What underlays the "web of relations"?

13.4.1 Supersymmetry (SUSY), Reification and String Realism

One can adopt a simplified form of skepticism about string realism based on SUSY. In S-dualities, the two models display usually very different structures and symmetries, but SUSY is always present. As a common assumption of all "realistic" string models, SUSY can shape the prospect of a pluralistic structural realism with S-dualities. A partial isomorphism between several string models is the consequence of SUSY. Stability as part of fundamentality is always related to SUSY: for example, the type I F-string is unstable and breaks because there are no enough invariants to keep it together, and because it does not have enough supersymmetry. SUSY helps with the transition from weak to strong coupling models: it acts as a protection of extrapolating unwanted features of weak coupling to strong coupling models, and generates the invariants that are conserved in the S-duality transformation, the *BPS* state. The skeptic about fundamental entities rejects realism and shows how it reifies symmetries (here, mostly SUSY). In a slogan: "String realism? Go back to symmetries in the models and be quietist about entities".

13.4.2 Structural Pluralism and String Realism

One way to retort this skepticism is to admit a multi-faceted structure which is not identical to SUSY. The *SO* realist drops fundamentality "as we know it" from previous physics, and accepts its plurality and its relativity to coupling. The new relative and pluralistic fundamentality is better couched in terms of structure than in terms of entities. The *SO* pluralism entails that the fundamental base is not unique for different models, in stark contrast to the assumptions in Sect. 13.3.1. The structure needed here comes in various forms, from various models, and brings in a multiplicity of 'presentations': it also implies a "model-relative ontology" (more exactly a sector-relative ontology, in the moduli space). Some dualities are presenting a more explicit common structure. In T-dualities, geometry is deformed by the scalar field that represents the radius of compactification. Rickles emphasizes that in the moduli space different models are in fact opposite ends to a continuum of geometries. Each sector presents the structure. Can we state something similar about S-dualities which are not self-dualities? Do we have a shared structure among string models presented in dualities?

This paper indicates the contrary, and offers an alternative: structural pluralism. In its *weak* form, the structural pluralism emphasizes that there is more than one *type* of structures in string realism. Structure is here multifaceted and multilayered (French, 2010). The pluralism of structure comes from the way our theories present it. The weak pluralism may remain silent about the uniqueness of structure but admits it is presented in a variety of ways. The strong pluralism denies the identity of structure belonging to different models and postulates that each model comes with its own structure, and that they are different both in their nature and in their presentation. Asking for the identity of such a structure over models and sectors, as the weak pluralist claims, would ignore the idea that string models are impoverished models. There is some degree of similarity of structure based on analogy, but nothing involving realist commitments.

For the weak structural pluralism, the common structure is not the symmetry of spacetime alone, and not SUSY alone. As a model-centered realism, the pluralism acknowledge the important role that g_S plays in shaping the structure and the fundamentality of entities. A similar argument can be developed for other coupling factors, such as the length of strings or subspaces of spacetime. Fundamentality can change drastically, depending on the spacetime which entities live in: move a system of strings and branes on a different manifold with different compactification, and the fundamentality relation changes drastically. Even within one model, the structure is multi-faceted and multilayered.

The other option is strong pluralism: the structure is "multi-aspected" because there is no single structure, but structures, each being part of a model. To each presentation, layer or aspect, one can assign a structure. The string realist asks whether moving from *one* "multi-aspected" structure to a proper *multiplicity* of structure is granted by the physics of string theory.

By way of conclusion, both forms of structural pluralism have advantages over the monist fundamentalism. Do we have the formal resources to put all the aspects and layers of structure under one concept? For such a question, a more prudent attitude is to "wait-and-see": the M-theory, a new field string theory, a string theory without SUSY, or, why not?, a completely new quantum gravity program may prove the argument of the present section wrong.

Acknowledgements Many thanks go to Jeremy Butterfield, Craig Callender, Antonio Cancio, Richard Dawid, Steven French, Jeffrey Harvey, Don Howard, Nick Huggett, Ken Intrilligator, James Maxin, Dean Rickles, for suggestions, comments and ideas that have morphed finally in this paper. I am indebted to the audience, the organizers and the referees of the *Philosophy of Science Association* meeting (Montreal, 2010), *Structure and Identity* workshop, (Bristol,

2012), the *Bucharest Colloquium in Analytic Philosophy. New Directions in the Philosophy of Physics* (2013). Last but not least, special acknowledgments to Ilie Pârvu and Iulian Toader for throughout suggestions and corrections that improved the quality of the argument. This material is complemented by a similar work on "string metaphysics" (Muntean, 2015), with several overlaps in form and content: thanks to the referee and editors of the *Poznan Studies*.

References

Baker DJ (2009) Against field interpretations of quantum field theory. Br J Philos Sci 60(3):585–609

Baker DJ (2014) Does string theory posit extended simples?. Preprint available at: http://philsci-archive.pitt.edu/11053/

Becker K, Becker M, Schwarz JH (2007) String theory and M-theory. Cambridge University Press, New York

Callender C, Huggett N (eds) (2001) Physics meets philosophy at the Planck scale: contemporary theories in quantum gravity. Cambridge University Press, Cambridge

Cappelli A, Colomo F, Di Vecchia P, Castellani E (eds) (2012) The birth of string theory. Cambridge University Press, Cambridge

Cartwright N (1999) The Dappled World: a study of the boundaries of science. Cambridge University Press, Cambridge

Castellani E (2009) Dualities and intertheoretic relations. In: Suarez M, Dorato M, Redei M (eds) Launch of the European Philosophy of Science Association. Springer, Dordrecht

Chakravartty A (2007) A metaphysics for scientific realism: knowing the unobservable. Cambridge University Press, Cambridge

Da Costa NCA, French S (2003) Science and partial truth: a unitary approach to models and scientific reasoning. Oxford University Press, New York

Dawid R (2006) Underdetermination and theory succession from the perspective of string theory. Philos Sci 73(3):298

Dawid R (2007) Scientific realism in the age of string theory. Phys & Philos 11:1–35

Dawid R (2009) On the conflicting assessments of the current status of string theory. Philos Sci 76(5):984–996

Dawid R (2013) String theory and the scientific method. Cambridge University Press, Cambridge

French S (1998) On the withering away of physical objects. In: Castellani E (ed) Interpreting bodies. Princeton University Press, Princeton, pp 93–113

French S (2010) The interdependence of structure, objects and dependence. Synthese 175:89–109

French S (2014) The structure of the world: metaphysics and representation. Oxford University Press, Oxford

French S, Krause D (2006) Identity in physics: a historical, philosophical, and formal analysis. Oxford University Press, Oxford

Frigg R, Votsis I (2011) Everything you always wanted to know about structural realism but were afraid to ask. Eur J Philos Sci 1(2):227–276

Greene B (1999) The elegant universe: superstrings, hidden dimensions, and the quest for the ultimate theory. W. W. Norton, New York.

Greene B (2011) The hidden reality: parallel universes and the deep laws of the cosmos. Knopf, New York

Hooft G (1974) Magnetic monopoles in unified gauge theories. Nucl Phys B 79(2):276–284

Hudson H (2005) The metaphysics of hyperspace. Oxford University Press, New York

Kim J (1998) Mind in a physical world an essay on the mind-body problem and mental causation. MIT, Cambridge

Ladyman J, Ross D, Spurrett D, Collier JG (2007) Every thing must go: metaphysics naturalized. Oxford University Press, Oxford

Matsubara K (2013) Realism, underdetermination and string theory dualities. Synthese 190(3):471–489

McDaniel K (2007) Extended simples. Philos Stud 133(1):131–141

McKenzie K (2011) Arguing against fundamentality. Stud Hist Philos Sci B: Stud Hist Philos Mod Phys 42(4):244–255

McKenzie K (2014) Priority and particle physics: ontic structural realism as a fundamentality thesis. Br J Philos Sci 65(2):353–380

Montonen C, Olive D (1977) Magnetic monopoles as gauge particles? Phys Lett B 72(1):117–120

Morgan MS, Morrison M (eds) (1999) Models as mediators: perspectives on natural and social science. Number 52 in ideas in context. Cambridge University Press, Cambridge

Muntean I (2015, forthcoming) A metaphysics from string dualities: pluralism, fundamentalism, modality. In: Bigaj T, Wüthrich C (eds) Metaphysics in contemporary physics. Poznan studies in the philosophy of the sciences and the humanities. Brill/Rodopi Amsterdam

Pincock C (2005) Overextending partial structures: idealization and abstraction. Philos Sci 72(5):1248–1259

Polchinski J (1996) String duality. Rev Mod Phys 68(4):1245–1258

Polchinski J (1998) String theory. Volume 2: superstring theory and beyond, 1st edn. Cambridge University Press, New York

Psillos S (1999) Scientific realism: how science tracks truth. Routledge, London

Rickles D (2011) A philosopher looks at string dualities. Stud Hist Philos Sci B: Stud Hist Philos Mod Phys 42(1):54–67

Rickles D (2013) AdS/CFT duality and the emergence of spacetime. Stud Hist Philos Sci B: Stud Hist Philos Mod Phys 44(3):312–320

Rickles DP (2014) A brief history of string theory. From dual models to M-theory. Springer, Berlin/Heidelberg

Sen A (1998) An introduction to non-perturbative string theory. arXiv:hep-th/9802051

Sklar L (2003) Dappled theories in a uniform world. Philos Sci 70(2):424–441

Taylor CC (1988) String theory, quantum gravity and locality. PSA: Proc Bienn Meet Philos Sci Assoc 1988:107–111

Teh NJ (2013) Holography and emergence. Stud Hist Philos Sci B: Stud Hist Philos Mod Phys 44(3):300–311

Teller P (1997) A metaphysics for contemporary field theories. Stud Hist Philos Mod Phys 28B(4):507–522

Teller P (2004) How we dapple the world. Philos Sci 71(4):425–447

Townsend P (1995) The eleven-dimensional supermembrane revisited. Phys Lett B 350(2):184–188

Wallace D, Timpson CG (2010) Quantum mechanics on spacetime i: spacetime state realism. Br J Philos Sci 61(4):697–727

Weingard R (1988) A philosopher looks at string theory. PSA: Proc Bienn Meet Philos Sci Assoc 2:95–106

Weisberg M (2007) Three kinds of idealization. J Philos 104(12):639–659

Witten E (1995) String theory dynamics in various dimensions. Nucl Phys B 443(1–2):85–126

Chapter 14
The Prospects for Fusion Emergence

Alexandru Manafu

14.1 Introduction

In a number of articles, Humphreys (1996, 1997a,b, 2008) has offered an account of emergence which aims to provide the grounds for an ontology of the special sciences. Humphreys' account (called *fusion emergence*) presents a series of challenges to at least three widely accepted assumptions about ontology: (i) that the right way to represent the relation between lower-level and higher-level properties is supervenience, (ii) that our world's ontology is wholly compositional, and (iii) that the physical domain is causally closed.[1]

Humphreys has argued not only that fusion emergence can be consistently described (1997b), but also that our own world exhibits cases of this kind of emergence (2008). According to Humphreys, covalent bonding is a "core example of fusion emergence" (2008, p. 7). The purpose of this paper is to raise some concerns about Humphreys' account in general and about his core example of fusion emergence in particular. It will be suggested that the extent to which covalent bonding undermines the second assumption mentioned above has been overstated.

---

[1] (i) Has been discussed extensively by Kim and Lewis, (ii) by Lewis, (iii) by Papineau.

A. Manafu (✉)
Institute for History and Philosophy of Science and Technology, University of Paris-1 Pantheon Sorbonne, 13 rue du Four, 75006, Paris, France
e-mail: Alex.Manafu@univ-paris1.fr; http://alexmanafu.com

© Springer International Publishing Switzerland 2015
I. Pârvu et al. (eds.), *Romanian Studies in Philosophy of Science*, Boston Studies in the Philosophy and History of Science 313, DOI 10.1007/978-3-319-16655-1_14

14.2   Humphreys' Fusion Emergence

Humphreys' account of emergence was motivated by the desire to avoid the exclusion argument or a generalized version thereof, whose conclusion is that higher-level emergent properties are excluded from affecting lower-level properties, since all the causal work is done by the latter (see Kim 1992, 1999, 2006). The exclusion argument has unwelcome consequences for the ontology of the special sciences. If one thinks of the special science properties (e.g., chemical or biological) as occupying higher levels than do physical properties, then the exclusion argument entails that no event involving a special science property could ever causally influence a physical event. The idea of special science causation is thus threatened. Also, the exclusion argument challenges the idea that special science properties deserve a place in our ontology: if special science properties are causally idle, what is the point of having them in our ontology? The exclusion argument has unwelcome consequences for physics, too. If one thinks of physics itself as stratified (e.g., with high energy physics, solid state physics and thermodynamics occupying different strata), the exclusion argument entails that only the most basic physical properties can be causally efficacious, and – as a result – all other causal claims within contemporary physics are false.

While the exclusion argument denies that the higher-level properties that special sciences are concerned with are capable of downward causation, emergence seems to require it explicitly. It has been argued that the only way to cause an emergent property to be instantiated is by causing its emergence base property to be instantiated (Kim, 1992, p. 136). This is known as the downward causation argument, and it shares with the exclusion argument the assumption that the right way to represent the relation between lower-level and higher-level properties is supervenience.

In his work, Paul Humphreys challenges both the exclusion argument and the downward causation argument by explicitly denying their common assumption, namely that supervenience is the right way to represent the relation between lower-level and higher-level properties (1997a). He also argues that thinking of higher-level emergent properties in terms of supervenience is mistaken. Instead, he links the possibility of emergence with the existence of a *fusion operation* that operates on i-level properties and outputs $i + 1$-level properties, which have novel causal powers.[2]

The process of fusion is formally represented as follows. Let $P_m^i(x_r^i)t_1$ represent an i-level entity, x_r, instantiating an i-level property, P_m, at time t_1. $P_n^i(x_s^i)t_1$ will denote another i-level entity, x_s, instantiating another i-level property, P_n, at time t_1.

---

[2]For the sake of brevity, sometimes I will use "property" instead of "property instance". It should be noted however that for Humphreys the arguments of the fusion operation are property instances.

Humphreys introduces the *fusion operation* symbolized by [. $*$.], which takes as arguments the two property instances $P_m^i(x_r^i)t_1$ and $P_n^i(x_s^i)t_1$ and fuses them: $[P_m^i(x_r^i)t_1 * P_n^i(x_s^i)t_1]$. The fusion operation is a *i*-level operation, i.e., an operation of the same level as its arguments. The result of the fusion operation is the fused property $[P_m^i * P_n^i][(x_r^i)+(x_s^i)](t_2)$ at the *i*+1-level, which can also be written as $[P_l^{i+1}][x_l^{i+1}](t_2)$. The fused property is a unified whole in the sense that its causal effects cannot be represented in terms of the separate causal effects of the original property instances. Also, within the fused property instance $[P_m^i(x_r^i)t_1 * P_n^i(x_s^i)t_1]$, the original property instances $P_m^i(x_r^i)t_1$ and $P_n^i(x_s^i)t_1$ no longer exist as separate entities and they do not have all of their *i*-level causal powers available for use at the *i*+1-level (Humphreys, 1997b, p. 10).

Humphreys argues that this particularity of fusion emergence is what enables this brand of emergentism to avoid the threats of the exclusion and downward causation arguments. At the time when the fused property instance $[P_m^i(x_r^i)t_1 * P_n^i(x_s^i)t_1]$ comes into existence, the original property instances $P_m^i(x_r^i)t_1$ and $P_n^i(x_s^i)t_1$ go out of existence. Therefore, it is a fortiori the case that they cannot compete as causes with the emergent property instance. On Humphreys account, emergents don't coexist with their bases, and this feature prevents the exclusion argument to get off the ground.

Humphreys' fusion emergence also deals with the downward causation argument. This argument is also committed to the idea that emergent properties supervene on lower-level properties. The argument assumes that the *only* way to bring about an emergent property instance at time *t* is by bringing about its subvenience base at time *t*. But if fusion emergents are not synchronous with their bases, this assumption is unwarranted. There is no reason to suppose that an $i + 1$-level property instance could not *directly* produce another $i + 1$-level property instance e.g., by directly transforming into it or by transforming another, already existing, $i + 1$-level property instance – in both cases, other property instances may contribute (1997b, p. 13; 2008, p. 8).

By avoiding the threats to the ontology of the special sciences posed by the exclusion and downward causation arguments, Humphreys' emergentist account attempts to rescue the autonomy of the special sciences and to depict an ontologically antireductionist image of the world in which the subject matters of the various special sciences correspond to irreducible ontological strata.[3] For Humphreys, there is a hierarchy of levels of properties $L_0, L_1, \ldots L_n \ldots$ of which at least one distinct level is associated with the subject matter of each special science, and L_j cannot be reduced to L_i for any $i < j$ (Humphreys, 1997a, p. 5).

---

[3]Humphreys admits however that the boundary between the physical level and other levels is not sharp (Humphreys, 1997a, p. S345).

14.3   Previous Criticisms of Fusion Emergence

Humphreys remarks that philosophers have long thought of the ontology of the special sciences in terms of supervenience. On this view, the higher-level properties are "composed of" or "supervenient upon" lower-level properties.[4] But Humphreys finds supervenience unsatisfactory. He complains that supervenience does not provide any understanding of ontological relationships holding between levels.[5] If these levels are emergent, they contain emergent properties. According to Humphreys, an important characteristic of emergent properties is that they result from the interaction between their constituents.[6] However, the level of detail that emergent properties demand makes the use of supervenience relations seem simplistic. This is one of the reasons why Humphreys argues that emergence should not be understood in terms of supervenience. Add to this the threats posed by the exclusion and downward causation arguments, and supervenience seems completely inappropriate for providing the grounds for an ontology of the special sciences.

As mentioned in the previous section, on Humphreys' account, emergents are not co-instantiated with their bases. Wong (2006) has called this the *basal loss* feature of fusion emergentism, and he has claimed that it is both problematic and unmotivated (2006, p. 346). According to Wong, the dissapearance of the lower-level properties of an entity is problematic for two reasons. First, because it threatenes the structural properties crucial to the proper functioning of that entity. The basal properties that fuse to become emergents may also constitute nonemergent, structural properties which may be indispensable to the proper functioning of the system. However, if basal properties are destroyed by the fusion process, then so would the structural properties. Second, the dissapearance of the lower-level properties generates what Wong calls "the correlation problem". It is empirically established that many special science properties have lower-level correlates with which they are copresent (e.g., mental properties are synchronously correlated with neurophysiological properties). However, if we are to treat the special science properties as fusion emergents, then we deny the copresence of their lower-level correlates, which Wong sees as empirically implausible.

Wong considers the basal loss feature of fusion emergentism as unmotivated for the reason that on Humphreys' account, basal and emergent properties don't have causal profiles that overlap significantly and thus cannot compete as overdeterminers of their effects. According to Wong, emergents supplement the underlying dynamics rather than merely overdetermine physical effects (2006, p. 361).

---

[4]The notion of supervenience that Humphreys uses is Kim's strong supervenience: "A family of properties M strongly supervenes on a family N of properties iff, necessarily, for each x and each property F in M, if F(x) then there is a property G in N such that G(x) and necessarily if any y has G it has F" (Kim, 1993, p. 65).

[5]This worry is in fact shared with Kim.

[6]Humphreys sees this interaction as nomologically necessary for the existence of emergent properties (1997a, p. S342)

In his response to Wong's criticisms, Humphreys (2008) argues that most systems possess multiple properties, some of which are essential to carrying out the system's function, whereas others are not. In general, the fusion process will affect only the latter. If a system's state is given by <P,Q,R,...Z> (x), the fusion between P(x) and Q(x), will leave R...Z unchanged and able to sustain the proper functioning of the system. Also, given that most properties are quantitative, part of P and part of Q will fuse, leaving the remainder to maintain the state. Wong's challenge to Humphreys is to show that this is will *always* be the case (Wong, 2006, p. 357). However, Wong's demand is unreasonable. If Humphreys can show that at least some of the special science properties are examples of fusion emergence, then this is enough to challenge the three assumptions mentioned at the beginning of the paper. But are Humphreys' examples able to do this? Before addressing this question, a couple of quick general points about Humphreys' account of emergence are in order.

14.4 The Division of Labor Between Properties and the Notion of a Physical Operation

Although Humphreys does not say it, his distinction between properties which are able to undergo fusion and those which are essential in the functioning of the system does in fact rely on two other dichotomies: first, between properties that are able to undergo fusion (PAUF) and properties that are not (PNAUF); and second, between properties which are essential in the functioning of the system (PEFS) and those that are not (PNEFS). Thus, Humphreys' distinction results from crossing two criteria: first, whether the properties are able to undergo fusion; second, whether the properties are essential in the functioning of the system. Humphreys assumes that the application of these two criteria delivers co-extensive subsets of properties, so that the properties which are able to undergo fusion will also be the ones that are not essential in the functioning of the system. Humphreys can, of course, maintain his distinction between properties without threatening the coherency of his account. That is, he can maintain that in any given entity there will be a "division of labor" between properties: some will undergo fusion, while others will preserve the functioning of the system. In this case, the properties to which the emergent character of an entity is due will not also be structural properties with a role in the in the functioning of that entity. Conversely, the structural properties that are crucial to the proper functioning of an entity won't participate in the fusion processes that that entity may undergo. However, these two last claims are far from trivial. Unless we have an independent justification of why the two dichotomies overlap, one may worry that Humphreys' division of labor between properties constitutes an ad hoc response to the problem of basal loss (Fig. 14.1).

The other point has to do with the nature of the fusion operation. On Humphreys' account, the fusion operation is not necessarily causal. However, fusion is supposed

Fig. 14.1 The dichotomy between the properties that are able to undergo fusion (PAUF) and those that are not (PNAUF) overlaps with the dichotomy between the properties that are not essential in the functioning of the system (PNEFS) and those that are (PEFS)

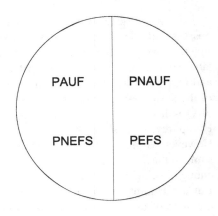

to operate on real properties in the world, not on their representations. Humphreys claims that the fusion operation is "a real physical operation" as opposed to a merely logical one like conjunction or disjunction (1997b, p. 10).

At this point, one may pause and ask what a *physical* operation is. We know what a *logical* operation like conjunction or disjunction is because there are logical/mathematical theories in which such operations are defined (e.g., sentential logic, predicate logic, Boolean logic, etc.). In the absence of these theories, our understanding of the logical operations will be greatly impoverished. What is the corresponding theory for the fusion operation? Humphreys claims that fusion is a *physical* operation. However, what it means for something to be a physical operation is not entirely clear. For example, in physics textbooks one does not find such an operation being defined. Humphreys may be taken as being uncommitted to the exact nature of the fusion operation pending futher empirical work (Wong, 2006, p. 352). It may turn out that fusion is implemented by single physical process (already discovered or yet to be discovered), or by a host of physical processes. In any case, an understanding of fusion as a physical operation depends on how well one understands its physical implementation. In order to achieve this, one needs to engage with empirical issues. It is to these empirical issues that I will now turn.

14.5 Humphreys' Examples

Whether the theory of fusion emergence can be coherently formulated is one thing; whether it applies to anything in the world is quite another. The former is a theoretical aspect that can be addressed largely on a priori grounds, while the latter is a an empirical issue. To argue that fusion emergence is not a metaphysician's fiction but a real phenomenon, one needs more than appeals to imagined scenarios; one needs concrete examples taken from the sciences. Humphreys presents such examples. According to Humphreys, "the clearest cases of fusion emergence is the entangled states of quantum systems" (2008, p. 4).[7]

According to Humphreys, the existence of such cases of emergence entails that our world's ontology is not wholly compositional. By a compositional ontology Humphreys means an ontology in which "all non-fundamental entities are aggregated or structured collections of other entities that can be generated by the use of explicitly stated rules of combination, where the constituent entities retain their identities within the structure" (2008, p. 2).

Humphreys thinks that the entangled state of a composite quantum system does not conform to the requirements of a compositional ontology because it is non-separable – the state of the system cannot be written as a tensor product of the states of its parts. Although there may be worries that a theory whose physical interpretation is still heavily debated might not be our best guide to ontology, let's grant that the entangled state in quantum mechanics is a *bona fide* example of fusion emergence. The question then becomes whether there are other examples of fusion emergence in our world, preferably in the special sciences.[8] Humphreys' answer is affirmative. The example of fusion emergence that is discussed in most detail by Humphreys is that of the covalent chemical bond. As mentioned, according to Humphreys, covalent bonding is a "core example of fusion emergence"(2008, p. 7).

Why does Humphreys think that the covalent bond exemplifies fusion? Humphreys notes that a covalent bond occurs when a pair of electrons is shared by two atoms; he also notes that the electron density of the electrons which participate in the covalent bond is distributed over the entire molecule rather than the individual atoms. Humphreys also claims that while some properties remain unchanged after the fusion (e.g., the charge and mass of the nucleons, the total charge of the molecule), others are affected by it; for example, there is a slight lowering of the energy of the combined molecular arrangement compared to the energies of the atoms before fusion. According to Humphreys, this energy that emerges upon fusion is responsible for the characteristic properties of the molecule.

Humphreys contrasts the covalent bond with the ionic bond. He suggests that ionic compounds are the result of electrostatic forces between positively and negatively charged ions and can be understood within the framework of a compositional ontology. On the other hand, molecules (resulting from covalent bonding) exemplify fusion and therefore are non-compositional in the sense explicated above. Humphreys does not elaborate much on why ionic bonding is compositional and covalent bonding isn't. He only claims that there is a contrast between the two types

---

[7] Humphreys' suggestion that the entangled state is an example of fusion emergence has been developed in more detail by Kronz and Tiehen (2002), who also discussed its ramifications and limitations.

[8] The exclusion argument to which fusion emergentism is an objection threatens the special sciences to a greater extent than physics.

of bonding and that the fact that "the molecule is not simply a spatial arrangement of the two atoms (. . .) is one of the things that distinguishes fusion from composition." (2008, p. 7).

14.6   Questioning the Ionic-Covalent Dichotomy

Insofar as Humphreys takes the two types of bonding as having different ontological requirements (and thus supporting incompatible ontologies), he is committed to a contrast between them that is not simply a matter of degree. However, the sharp contrast between ionic and covalent bonding that Humphreys' example assumes does not receive as much support from physical chemistry as one may think.

Ionic and covalent bonding are viewed as two extreme models of the chemical bond (Atkins and Jones, 2002, p. 92). With the exception of the bonds of homonuclear diatomic molecules, all chemical bonds lie somewhere between purely ionic and purely covalent. If the electronegativity difference $\Delta\chi$ increases, so does the ionic character of the bond.[9] Generally, if $\Delta\chi > 1.6$, the bond is considered ionic. If $\Delta\chi < 0.5$, the bond is considered covalent non-polar. And if $\Delta\chi$ is between 0.5 and 1.6, the bond is considered covalent polar. However, there is no principled way to choose these values and they may vary slightly from one chemistry textbook to another. There is no sharp distinction between an ionic and a polar covalent bond; rather, the difference between them is a matter of degree. If the difference between the two types of bonding is only gradual, then how can they be accomodated within different ontological frameworks? Where should the boundary between compositional and non-compositional be placed?

One may argue that as long as there exist clear cases of covalent and ionic bonding, this should be enough to justify the requirement of different ontological frameworks. However, while pure covalent bonding exists (between the atoms of homonuclear diatomic molecules such as Cl_2, H_2, O_2), pure ionic bonding cannot exist, since it would require that the electronegativity difference $\Delta\chi$ between the atoms be infinite or at least exceedingly large (Carter, 1979, p. 124). Therefore, all bonds have some covalent character. Does non-compositionality characterize only those pure cases of covalent bonding, or should all types of bonding be accountable within a single (non-compositional) ontology? If neither, how should the discrete border between two distinct ontological frameworks be superimposed onto the covalent-ionic continuum? These questions are not in themselves sufficient to show that Humphreys' account fails, but they are certainly indicative of a lack of harmony

---

[9]On the Pauling scale, the difference in electronegativity between atoms A and B is a dimensionless quantity: $\chi_A - \chi_B = (eV)^{-1/2}\sqrt{E_d(AB) - [E_d(AA) + E_d(BB)]/2}$, where $E_d(XY)$ represents the dissociation energy between atoms X and Y in electronvolts. Pauling defined the amount of ionic character of a chemical bond as $1 - e^{-1/4(\chi_A - \chi_B)}$ (Pauling, 1960, p. 98).

between the sharp character of the boundary between a compositional and a non-compositional ontology and the non-sharp character of the boundary between ionic and covalent bonding.

There is another problem with viewing chemical compounds through the ionic-covalent dichotomy. These two types of chemical bonding are models, i.e., they are idealizations which have their virtues but distort reality in some respect. For example, they represent the pair of electrons participating in a covalent bond as being shared by just one pair of atoms, even when the molecule is polyatomic. Chemical bonds between atoms can be described more accurately using the concept of resonance. Resonance refers to the representation of the electronic structure of a molecular entity in terms of distinct contributing structures (also called resonance structures). Electrons involved in resonance structures are said to be delocalized: for example, in the case of a polyatomic molecule the sharing of an electron pair is distributed over several pairs of atoms and cannot be identified with just one pair of atoms. A resonance hybrid is a blend of the contributing structures.

All compounds, regardless of whether they are considered ionic or covalent, can be viewed as resonance hybrids of purely covalent and purely ionic *resonance structures*. For example, the structure of a homonuclear diatomic molecule, in which two atoms of the same element are covalently bonded to each other, can be described as a resonance hybrid of two ionic structures (Atkins and Jones, 2002, p. 93).

$$A-A \longleftrightarrow A^{\delta-}A^{\delta+} \longleftrightarrow A^{\delta+}A^{\delta-}$$

In the case of homonuclear diatomic molecules, the ionic structures make only a small contribution to the resonance hybrid. Also, the two ionic structures have the same energy and make equal contributions to the hybrid, so the average charge on each atom is zero. In a heteronuclear molecule, the resonance hybrid has unequal contributions from the two ionic structures – the structure with the negative charge on the atom that has a greater electron affinity will make a bigger contribution to the resonance hybrid.

The representation of chemical compounds in terms of resonance structures is more accurate than the ionic-covalent representation but it is strictly speaking incompatible with it. The resonance model challenges the view of chemical compounds as either ionic or covalent because resonance hybrids are a blend of resonance structures rather than the flickering of a compound between different structures, just as a mule is a blend of a horse and a donkey, not a creature that flickers between the two (Atkins and Jones, 2002, p. 80).

14.7 The Level-Relativeness of Fusion

According to Humphreys, the covalent bonding exemplifies a kind of ontological emergence which shows that the ontology of our world is not exclusively compositional. Why does Humphreys think that molecules cannot be understood in the framework of a compositional ontology? Molecules consist of atoms, so at a first glance, the compositionality condition would seem to be satisfied. However, at a closer look, one realizes that molecules are not simply the result of the combination or spatial juxtaposition of atoms. A molecule is the *sharing* of electrons between two or more atoms. Because of this, Humphreys is justified in claiming that the molecule can be described as the *fusion* of two or more atoms, not as a combination or aggregation of atoms.

However, if one thinks of molecules not as collections of atoms but as collections of nuclei and electrons, what looks like fusion between two atoms can be described as composition of nuclei and electrons. The Aufbau principle consists in a number of explicitly stated rules that allow us to understand the atom (any atom) as a physical system that is built by successively adding electrons around the nucleus.

1. The principle of the minimum energy: the electrons occupy atomic orbitals in such a way that the total energy of the atom is a minimum; they fill orbitals starting at the lowest possible energy states before filling higher states.
2. The Pauli exclusion principle: every electron in an atom is described by its own distinct set of four quantum numbers, not shared with any other electron. This entails that a given orbital is to be occupied by no more than two electrons, case in which their spins, denoted by the m_s quantum number, are paired.
3. The Madelung rule: orbitals with a lower $n + l$ value are filled before those with higher $n + l$ values.
4. Hund's rule of maximum multiplicity: electron pairing will not take place in orbitals of the same sub-shell until orbitals are singly filled by electrons with parallel spin.

It should be recognized that the Madelung rule and Hund's rule of maximum multiplicity are not exceptionless. They are rules of thumb, but they are helpful. There must be some deeper reason of why these rules work (when they do), although deriving these rules from deeper physical principles has proven to be not an easy task. What these four rules show is that the atom is a complex physical system in which the nucleus and the electrons are subject to a number of physical constraints and interact with each other according to physical laws. It is these physical laws and constraints that are the more basic rules of composition in the multi-electron atom. The atom appears to be more than just a collection of individual particles because of the complexity of the interactions between these particles.

In contrast with the entangled state, which is non-decomposable into separate states of each of the two electrons and thus cannot be written as a tensor product of the states of the individual electrons, the wavefunction of a multi-electron atom can be thought of as resulting from the separate contributions of each electron wavefunctions, and it can be written as a product of individual atomic orbitals: $\psi(r_1, r_2, \ldots r_n) = \phi_1(r_1)\phi_2(r_2)\ldots\phi_n(r_n)$. This strategy of learning about the wavefunction of a multi-electron atom on the basis of the individual electrons is known as the *orbital approximation* and is a remarkably useful tool in the attempts at solving the Schrödinger equation for atoms that have more than one electron. When applied to multi-electronic atoms, the (atomic) orbital approximation assumes that each electron behaves independently of the others, and thus the electronic Hamiltonian can be separated into as many components as there are electrons: $\hat{H}_e = \hat{H}_1 + \hat{H}_2 + \ldots + \hat{H}_n$.

The treatment of the multi-electron atom in physical chemistry is, I think, an illustration of compositionality. Admittedly, the orbital approximation is an approximation – the inter-electronic repulsion forces which are due to the Coulomb potential are deliberately ignored, to make the Schrödinger equation more tractable. However, the existence of such forces does not show that the atom cannot be understood compositionally; after all, the electron-electron repulsion itself obeys compositionality – the repulsive force depends on the charge and and distance between the individual electrons.

The presence of compositionality principles seems to be abundant in the physical chemistry of molecules, too. In contrast to older theories such as the valence shell electron pair repulsion theory (VSEPR), the molecular orbital theory describes the electrons in a molecule as delocalized; they are not confined to pairs of atoms, but are spread over the whole molecule. The central claim of the molecular orbital theory is that molecular orbitals are obtained from summing up atomic orbitals. More rigorously, each one-electron molecular orbital ϕ_i is expressed as a linear combination of atomic orbitals (LCAO): $\phi_i = c_{1i}\sigma_1 + c_{2i}\sigma_2 + c_{3i}\sigma_3 + \ldots + c_{ni}\sigma_n$, where the coefficients represent the weights of the contributions of each atomic orbital to the molecular orbital and are found using the Hartree-Fock method. The wavefunction for the molecule is then written as a product of one-electron wavefunctions. This is the molecular orbital approximation: the wavefunction of a multi-electron molecule is approximated as the product of individual molecular orbitals: $\Psi(r_1, r_2, \ldots r_n) = \Phi_1(r_1)\Phi_2(r_2)\ldots\Phi_n(r_n)$. The electron configuration of molecules is obtained from the same set of rules that yielded the electron configuration of multi-electron atoms.

In some sense, Humphreys is justified in thinking of the molecule non-compositionally, for a molecule is not simply the result of the spatial arrangement of atoms. If one descends one ontological level (e.g., from the level of the molecule to the level of atoms), the molecule cannot be described compositionally, in terms of separate but interacting atoms. However, if one descends *two* ontological levels (e.g., from the level of the molecule the level of nuclei and electrons), the molecule *can* be described in terms of separate but interacting components. What what looks like fusion at the *i*-level (molecular level) can be represented as composition at the *i*-2-level (level of electrons and nuclei). For example, in the case of a simple molecule such as the dihydrogen molecule, what looks like fusion between two hydrogen atoms could be understood as composition between two nuclei and two electrons.

14.8 Entanglement to the Rescue?

An argument that challenges this conclusion may in fact be available to the defender of non-compositionality. The argument is based on the remark that the electrons which participate in the covalent bond have opposite spins (are paired), and thus they are entangled (i.e., they form a singlet state, or a state in which their total spin is zero). If the entangled state is a *bona fide* case of fusion and hence it does not conform to the requirements of a compositional ontology, then the molecule must also be an example of fusion. On this view, the fact that the electrons participating in the covalent bond cease to possess separate states is sufficient grounds for concluding that the molecule is a non-separable whole which defies a compositional ontology. The defender of non-compositionality could argue that the covalent bond (and hence the molecule) owes its existence to the entanglement of the electrons constituting the bond. On this view, once two electrons belonging to different atoms have become entangled, a covalent bond occurs between the atoms and a new entity emerges: the molecule.

The problem with the argument above is that it does not give an accurate characterization of the origin and nature of the chemical bond. Chemical bonds are due to the interplay of four sets of forces: the attraction of each electron to the nucleus of its own atom, the attraction of each electron to the nucleus of the other atom, the electron-electron repulsion, and the nucleus-nucleus repulsion. The fact that the electrons participating in a covalent bond are paired is a consequence of their obeying the Pauli exclusion principle. However, the Pauli exclusion principle is not a force, but a constraint that the electrons must satisfy if a covalent bond is to be formed. Consider two hydrogen atoms whose electrons have parallel spins. If the atoms are brought together, the charge density from each electron is accumulated in the antibonding region (i.e., at the extremities of the system), rather than in the bonding region between the nuclei. Therefore, they will not form a dihydrogen molecule. The role of the Pauli exclusion principle is to veto those systems that cannot form a molecule by imposing a constraint on the spin of the electrons participating in the covalent bond. The Pauli exclusion principle tells us that only those hydrogen atoms whose electrons have opposite spins are eligible for forming a bond. The spin entanglement that can be found in a molecule does not play the role of a force holding the molecule together.[10]

In fact, there are molecules in which not all of the electrons are entangled, such as molecules with unpaired electrons or an open shell configuration. Although usually the unpaired electrons are found in the antibonding orbitals and they are expected to

---

[10]The so-called "exchange force" or "exchange interaction" which decreases the expectation value of the distance between two electrons (or fermions, more generally) with identical quantum numbers when their wave functions overlap is not a true force and should not be confused with the exchange forces produced by the exchange of force carriers, such as the electromagnetic force produced between two electrons by the exchange of a photon, or the strong force between two quarks produced by the exchange of a gluon.

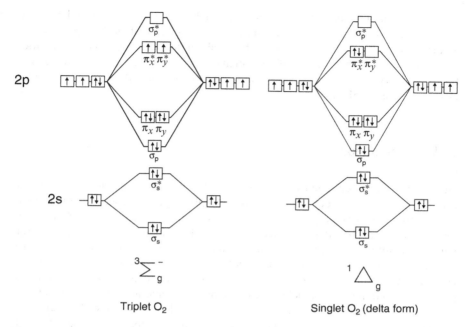

Fig. 14.2 Triplet O_2 vs. singlet O_2

lower the bond order and thus decrease the bond energy, there are cases such as the oxygen molecule, O_2, in which the unpaired electrons actually increase the strength of the bond. The ground state of the oxygen molecule is also known as the triplet oxygen because the total spin of the molecule is 1: the electrons occupy two different $2p\pi^*$ molecular orbitals singly and, according to Hund's rule, their spins are parallel (this can also be deduced empirically, from oxygen's paramagnetism) (See Fig. 14.2 below). The triplet oxygen is known to be more stable than the singlet oxygen – a diamagnetic form in which the electrons are paired in the same $2p\pi^*$ orbital (Wiberg et al., 2001, p. 476).[11] Another example is that of the molecular hydrogen ion, H_2^+, in which there is no entanglement since there is only one electron. The bond holding together the dihydrogen cation is described as a "one-electron bond", and has a formal bond order of $\frac{1}{2}$ (Pauling and Wilson, 1963, p. 362). For this simple system the Schrödinger equation can be solved accurately and the calculations show clearly that the molecular entity possesses a bound state, i.e., it possesses a ground state whose energy is less than that of a hydrogen atom and a free electron. In the case of dilithium, the binding energy is greater for the one-electron Li_2^+ than for the two-electron Li_2, although in the Li_2^+ there is no entanglement involved in bonding, while in the Li_2 there is (James, 1935).

---

[11]The specific form referenced here is $O_2 a^1 \Delta_g$.

These examples show that the connection between the chemical bond and entanglement is not as strong as the argument that is available to Humphreys may assume it to be. They show that (i) the chemical bonding of some molecular entities is possible even in the absence of entanglement, as in the case of the one-electron bonds, and (ii) the chemical bond of some molecules is actually stronger if not all electrons are entangled.

14.9 Conclusion

Humphreys' fusion emergence is an elegant solution to the exclusion problem, but it is not without its difficulties. There are a couple of general concerns. First, there is the worry that Humphreys' division of labor between properties may be an ad hoc response to the problem of basal loss if we don't have independent justification for why the dichotomy between the properties that are able to undergo fusion and those that are not should overlap with the dichotomy between the properties that are not essential in the functioning of the system and those that are. Second, the notion of a physical operation plays a crucial role in Humphreys' emergentist account, but it is not clear what this operation means, and how it is implemented at the physical level.

There are also more specific concerns which regard Humphreys' core example of fusion emergence. In developing this example, Humphreys assumes a deep contrast between ionic and covalent bonding that is not warranted by physical chemistry. It is not clear how the the fuzzy boundary between ionic and covalent bonding maps onto the discrete boundary between a compositional and a non-compositional ontology.

Finally, Humphreys' claim that chemistry gives us reasons to reject a compositional ontology is problematic. While it is true that the molecule can be described as the fusion of atoms, if one thinks of the molecule not as a collection of atoms but as a collection of electrons and nuclei, what looks like fusion between two atoms could perhaps be described as composition of electrons and nuclei. In fact, chemistry is full of compositional principles: in the molecular orbital theory, each one-electron molecular orbital is expressed as a linear combination of atomic orbitals; the orbital approximation gives us a way of learning about the wavefunction of a multi-electron atom or molecule on the basis of the wavefunctions of the individual electrons; and the Aufbau principle gives us explicit rules of composition for obtaining the electronic structure of atoms and molecules. Given the effectiveness of these rules and principles and the fact that they are compositional *par excellence*, it is premature to conclude that the entities forming the subject matter of chemistry cannot be accommodated within the framework of a compositional ontology.

References

Atkins P, Jones L (2002) Chemical principles. The quest for insight, 2nd edn. W.H. Freeman and Company, New York

Carter GF (1979) Principles of physical and chemical metallurgy. American Society for Metals, Metals Park

Humphreys P (1996) Aspects of emergence. Philos Top 24:53

Humphreys P (1997a) Emergence, not supervenience. Philos Sci 64:S337–S345

Humphreys P (1997b) How properties emerge. Philos Sci 64(1):1–17

Humphreys P (2008) A defence of ontological emergence. Paper delivered at the International School on Complexity 9: Emergence in the Physical and Biological Worlds, Ettore Majorana Foundation, Sicily

James HM (1935) Wave-mechanical treatment of the Li2 molecule. J Chem Phys 3(9):9–14

Kim J (1992) Multiple realization and the metaphysics of reduction. Philos Phenomenol Res 52(1):1–26

Kim J (1993) Supervenience and mind. Cambridge University Press, New York

Kim J (1999) Making sense of emergence. Philos Stud 95(1–2):3–36

Kim J (2006) Emergence: core ideas and issues. Synthese 151(3):547–559

Kronz FM, Tiehen JT (2002) Emergence and quantum mechanics. Philos Sci 69:324–347

Pauling L (1960) The nature of the chemical bond and the structure of molecules and crystals; an introduction to modern structural chemistry, 3rd edn. Cornell University Press, Ithaca

Pauling L, Wilson EB (1963) Introduction to quantum mechanics: with applications to chemistry. Dover, New York

Wiberg E, Wiberg N, Holleman AF (2001) Inorganic chemistry, 1st English edn. Academic, San Diego/De Gruyter, Berlin/New York

Wong H (2006) Emergents from fusion. Philos Sci 73(3):345–367

Part V
Explanation, Models, and Mechanisms

Chapter 15
Scientific Progress, Understanding and Unification

Sorin Bangu

15.1 Introduction

Undoubtedly, the astronomical theories of Copernicus and Kepler were better than that of Ptolemy; Kepler's and Galileo's theories were superseded by the Newtonian physics, which in turn yielded to Einstein's theories of relativity and to quantum mechanics. Such successions are not peculiar to physics, but can be documented everywhere in science. Contemplating them leads to the formulation of the central question of this paper: if the present science is better than the past one, how can we characterize this improvement? The question is of course old; it has been debated by both scientists (e.g., Bragg 1936) and philosophers, and most recently revived by Alexander Bird in an insightful paper 'What is Scientific Progress?' (2007) (See also Bird 2008).

I begin with a critical presentation of the various answers on the table (including Bird's); then, since I am not entirely satisfied with them, I shall sketch an alternative proposal. In essence, I argue that scientific progress is best characterized as an increase in scientists' *understanding* of the world. The point seems rather obvious but, as far as I can tell, has not yet been made.[1] I would speculate that it went unnoticed, at least in the form in which I attempt to it articulate here, because not much attention has been paid to connecting two rather large philosophical themes:

---

[1]Bird himself alludes to this idea in his (2007, 84): 'I will however leave a detailed discussion of the important question of what contributions to knowledge contribute most to progress (and in particular the role of understanding) for another occasion—not least because it is a much more difficult question.'

S. Bangu (✉)
Department of Philosophy, University of Bergen, Bergen, Norway
e-mail: sorin.bangu@fof.uib.no

© Springer International Publishing Switzerland 2015
I. Pârvu et al. (eds.), *Romanian Studies in Philosophy of Science*, Boston Studies
in the Philosophy and History of Science 313, DOI 10.1007/978-3-319-16655-1_15

on one hand, the question regarding scientific progress; on the other, the nature of scientific understanding – which I spell out here in terms emerging from the debate on explanation and unification originating in the works of M. Friedman (1974, 1983) and P. Kitcher (1981, 1985, 1989).[2]

15.2 Accounts of Scientific Progress, and Their Shortcomings

Confronted with our question, five (on my count) answers can be documented.[3] One is developed by building up on the idea that progress is a success term, and thus should be distinguished from the mere descriptive term 'scientific change' (Niiniluoto 1995; Laudan et al. 1986). If a branch of science makes progress, then this is to be understood as a move forward, toward a goal: the closer to the goal we are, the more we have progressed. And, what this goal can be other than the truth? But this view, which I will call the 'truth-as-goal' account of progress, is problematic, for the well-known reason that the kind of truth needed to substantiate this view – the pure, objective truth, the 'view from nowhere', uncontaminated theoretically – is a chimera (or, as is sometimes put, this kind of truth is 'transcendent'[4]). One way out of this impasse was to think about truth differently. Thus, a reaction to the problems raised by the elusive, ultimate truth ('Truth') was to go for approximate truth instead (Popper 1959, 1963, 1972; Dilworth 1981; Niiniluoto 1987, 2010; Psillos 1999; Balzer 2000; Barrett 2008). The main motivation behind this idea is the natural intuition that some theories offer more accurate descriptions of the world than others. Hence, in this sense, the more a theory approximates the truth, the better it is. But, just like the truth-as-goal account, this 'truth-approximation' account is not satisfactory, as we'll see below.

Thus, a new proposal was to ascertain progress retrospectively (Stegmüller 1976). Using the journey metaphor once again, the idea is that we should not evaluate how far we are from its end-point, but rather how far we are from the point of departure. One way to spell out this insight was in terms of accumulation of true beliefs: if the starting-point can be described as a collection of (true) beliefs, then the more such beliefs we pile up as time goes by, the more progress we have made. We may never reach the (ultimate) truth, but we can claim that we have made progress when we have accumulated more true beliefs than those we began with. An appropriate label for this idea is thus the 'truth-accumulation' account of

---

[2]To clarify: I shall be drawing on these views here, as I agree with the core idea. However, there are aspects of unificationism which I find problematic, but I can't discuss them here.

[3]Needless to say, I might have missed certain answers; the ones I discuss here are among the most prominent. My presentation and discussion of these accounts of progress, in this section and the next, owes a lot to Niiniluoto (2011) and Bird (2007).

[4]For more on this, in the context of a discussion of scientific progress in relation to Thomas Kuhn, see Bird (2000).

progress. This account, however, has raised many (antirealist) eyebrows in the post-Kuhnian philosophy of science and, consequently, a different way to make sense of progress has been advanced. The key-move was to set the issue of truth (and approximate truth) aside, and think of a theory in more instrumentalist terms – as a problem-solving device (Kuhn 1970, 1977; Laudan 1977). Then, we can regard a theory T as better than other theory T' when T solves more problems than T'. (For critical discussion, see Scheibe 1976; Rescher 1978; Kleiner 1993; Jonkisz 2000).

Because of the difficulties encountered by these accounts (including the 'problem-solving' one, as I will refer to the view just described), an alternative proposal is currently on the table. This more recent view, due to Bird (2007), builds on a fundamental distinction, between true scientific beliefs and scientific *knowledge* – where what separates the two is that having knowledge is having true beliefs which are also well-justified (conclusively, reliably justified).[5] Generally speaking, knowing demands more from a subject than possessing true beliefs, namely believing true things for the right reasons, not just accidentally. Thus, once the distinction between true belief and knowledge is in place, one can formulate another (the fifth) account of progress, which I will call the 'knowledge-accumulation' view, as follows: progress consists in the accumulation of scientific knowledge (as opposed to the mere accumulation of scientific truths.) As Bird demonstrates, this view has many virtues, and clear advantages over its rivals. I am sympathetic to it, but I will argue that it is incomplete, as it leaves out an important aspect of the notion of scientific progress: the central role of *understanding*.

Yet, before we get to the details of this account, an important preliminary task is to review the reasons for which the abovementioned accounts have been regarded as unsatisfactory. As I hinted above, the strongest motivation for developing a new way to conceive of scientific progress is that the accounts listed above are fraught with problems.

To begin with, recall the reaction to the difficulty raised by thinking of progress in terms of the elusive ultimate truth: just go for approximate truth instead. The idea is attractive, but it faces unexpected difficulties, which can be appreciated by reflecting on the following point due to Bird (2007). It is natural to compare the accuracy of individual beliefs: if it is 12 o'clock, Alice's belief that it is 12.05 is incorrect, but more precise, or a better approximation of the truth, than Bob's belief that it is 12.15. Yet scientific theories are of course more complicated than that, in the sense that they typically involve conjunctions of claims. Hence a problem arises. Suppose it is 12:00 and the temperature is 20C, then which claim is a better approximation of the truth: Alice's, that it is 12:05 and 17C, or Bob's, that it is 11:57 and 25C?

We encounter another subtle difficulty when we move on to considering the problem-solving account. Here we have to distinguish between genuine scientific

---

[5]Note that the proviso here is that the beliefs constituting knowledge are not only justified, but *well*-justified, or reliably, conclusively justified; that is, justified in such a way as to preclude the famous Gettier-type counterexamples (Gettier 1963).

problems and only *perceived* problems. Consider the seventeenth century chemical theories of combustion, and their 'problem' of making sense of the negative mass (Lakatos and Musgrave 1970; Howson 1976). If, as observed in experiments, magnesium (for instance) gained mass upon burning, and phlogiston was supposed to be the substance released by combustion, then one wonders, how is it that phlogiston has negative mass? From the current scientific standpoint this is a pseudo, or only perceived, problem – there is no such thing as phlogiston, and combustion is explained through the role of oxygen fixing and energy transformation. Hence, someone skeptical about this account may ask: should we judge as progress the theories' ability to solve the perceived problems, or only the 'real' problems? If so, how do we single out the latter? In connection with this point, serious difficulties ensue: are scientific problems self-standing, framework-independent puzzles or, on the contrary, we can only make sense of them from a certain philosophical-scientific perspective? (Feyerabend 1962)

The truth-accumulation view doesn't fare better either. It stumbles upon a rather simple difficulty: collecting truths after truths about the world is just not enough, since they are less valuable than collecting knowledge, or, as we recall, true beliefs for which we have a good justification to believe they are true. This is surely a fair point, but this more recent 'knowledge-accumulation' account is not, I argue, entirely satisfactory. For one thing, what is missing from it (and from the other accounts as well) is an elucidation of the natural and fundamental relation between knowledge and *understanding*. It is the examination of this relation that suggests the formulation of the first principle of the new conception I propose: we want science to deepen our understanding of how the universe works, and thus scientific progress is achieved when an increase in understanding is obtained.

Scientific understanding emerges as the central notion here, and at this point we should ask what understanding *is*, after all. However, as is the case with many other central questions in philosophy of science, the consensus on the answer still eludes us. The issue is currently hotly debated,[6] and an overview of the many positions and arguments recently advanced is not possible here. What I shall do, however, is sketch my preferred approach, which spells out scientific understanding in terms of unification.

15.3 Scientific Understanding, Explanation and Unification

One of the main motivations for the revisions undergone by the deductive-nomological (DN) model of scientific explanation was the desire to increase the ability of this model to account for the strong intuition that explanation and

---

[6]Grimm (2013, forthcoming), Hindricks (2013), Khalifa and Gadomski (2013), Khalifa (2013), Newman (2013), Strevens (forthcoming). Earlier discussions include Trout (2002), de Regt and Dieks (2005), de Regt (2009), Elgin (2009).

understanding are related. The very point of giving scientific explanations is, it is often said, to advance our understanding of the world.[7] Although understanding is a multifaceted faculty (featuring subjective, psychological connotations) a basic assumption of these revisions is that an objective sense of this notion can, and ought to, be captured by any adequate theory of scientific explanation.

In light of this desideratum, Friedman (1974, 1983) and Kitcher (1981, 1989) proposed a reinterpretation of the DN model, the core idea of their proposal being to construe explanation as unification. Both these unificationist versions of the DN model preserve its central characteristic, that explanation is a form of derivation,[8] and also aim to do justice to other strong intuitions we have about what constitutes a good explanation.[9] Notwithstanding their disagreement over the details about how to characterize unification, both Friedman and Kitcher submit that 'unification is the essence of scientific explanation' (Friedman 1974, 15) and – directly relevant for my purpose here – that unificatory power is the property of scientific theories able to help us attain 'the ideal of scientific understanding' (Kitcher 1985, 638). Although attempts have been made to question the coupling of unification, explanation and understanding (e.g., Barnes 1992a, b; Humphreys 1993; Halonen and Hintikka 1999; Morrison 2000; Woodward 2003; Strevens 2004, 2008), unificationism continues to be a live option in contemporary philosophy of science.[10]

The central unificationist idea holds that the explanation of a phenomenon advances our understanding when the phenomenon is integrated into a more comprehensive system of ideas. This insight is not new; in fact, the question 'What is understanding in physics?' has been asked ever since the birth of the first field theory, Maxwell's electromagnetism,[11] and one of the earliest approaches to this question attempted to connect explanation and understanding to unification. Let me briefly trace the development of this connection. In the modern era, Heisenberg's essay 'Understanding in Modern Physics' (1971) is particularly useful in learning about the views of the creators of quantum mechanics on this matter. This essay

---

[7]Not everybody agrees that the main and only way to increase understanding is by resorting to scientific explanations; see, for instance, van Fraassen (1985, 642).

[8]W. Salmon's causal (ontic) conception of scientific explanation (Salmon 1984, 1998) is also motivated by the various failures of DN model. However, Salmon maintains that the core idea of this model (namely that explanations are derivations, or arguments) is inadequate.

[9]The difficulties of the DN model are various (and notorious). One of the most important, whose solution was attempted by Friedman in his 1974 paper, stems from the so called 'conjunction problem': given two laws L and K, we can formally derive L from L&K but this cannot intuitively count as an explanation of L. Obviously, our understanding as to why L holds is not enhanced by this sort of derivation – explanation.

[10]For endorsements and elaborations of unificationism, see for instance Weber (1999), Schurz (1999), Bartelborth (2002).

[11]Suggestively, this is the context in which Lord Kelvin's famously remarked 'It seems to me that the test of 'Do we or not understand a particular subject in physics?' is 'Can we make a mechanical model of it?' (Kargon and Achinstein 1987, 3; 111).

consists of free transcriptions of a number of conversations taking place between 1920 and 1922 between Pauli, Heisenberg, their less famous fellow, the physicist O. Laporte (all three students of A. Sommerfeld at that time), and their mentor, Niels Bohr. As we'll see, one curious thing about these debates is that the ideas advanced there anticipated, with remarkable precision, the main conceptions to be elaborated in the philosophy of scientific explanation in the next 70-years or so.

The starting point of the discussions focusing on understanding in physics is the profound impression that the Special Theory of Relativity made on these then-young physicists. Answering Pauli's inquiry about his understanding of Einstein's theory, Heisenberg begins with a number of remarks on the relation between understanding, mathematics and prediction. Heisenberg writes:

> Thus Wolfgang [Pauli] asked me (. . .) whether I at long last understood Einstein's relativity theory, on which Sommerfeld laid so much stress. I could only say I did not really know what was meant by 'understanding' in physics. The mathematical framework of relativity theory caused me no difficulties, but that did not necessarily mean that I had 'understood' why a moving observer means something different by 'time' than an observer at rest. The whole thing baffled me, and struck me as being quite 'incomprehensible'. (1971, 29)

Pauli insists, and tentatively proposes a certain view – to be developed by N. R. Hanson about 40 years later (Hanson 1963) – namely the identification of understanding with the capacity to make predictions:

> But once you have grasped the mathematical framework (. . .) you can surely predict what an observer at rest and a moving observer ought to observe or measure. And we have good reason to assume that a real experiment will bear out these predictions. (1971, 29)

Pauli goes on by challenging Heisenberg with a new riddle: once we have this 'good reason', notes Pauli, 'what more can you ask?' about understanding. Heisenberg confesses his perplexity once again, and steers the discussion toward another important aspect, the psychological-pragmatical connotations of the notion of understanding:

> This is precisely my problem, (. . .) that I do not know what more to ask. I feel somewhat cheated by the logic of the new mathematical framework. You might even say that I have grasped the theory with my brain, but not yet with my heart. (1971, 29)

After mentioning a causal conception of explanation and understanding (to be developed in detail by Salmon (1984)), according to which what explains and imparts understanding is the discovery of the causes of phenomena,[12] the discussants return to, and dismiss, cashing out understanding in terms of prediction:

> The ability to predict is often the consequence of understanding, of having the right concepts, but is not identical with understanding. (Heisenberg 1971, 33)

(Note that earlier on, Heisenberg agreed with Pauli: '(. . .) correct predictions were not a sign of true understanding.' (Heisenberg 1971, 31))

---

[12]Pauli says: 'I, for one, see a basic distinction between Newton's astronomy and Ptolemy's. (. . .) To begin with, Newton posed the whole problem quite differently; he inquired into the causes of planetary motions not into the motions themselves. These causes, he discovered, were forces (. . .)' (Heisenberg 1971, 32).

Pauli then makes a final attempt and construes understanding in terms of unification:

> 'Understanding' probably means nothing more than having whatever ideas and concepts are needed to recognize that a great many different phenomena are part of a coherent whole. Our mind becomes less puzzled once we have recognized that a special, apparently confused situation is merely a special case of a something wider, that as a result it can be formulated much more simply. The reduction of a colorful variety of phenomena to a general and simple principle, or, as the Greeks would have put it, the reduction of the many to the one, is precisely what we mean by 'understanding'. (Heisenberg 1971, 33)

While mentioned in passing by some of the early positivists (Hempel 1965, 345; 444; Feigl 1970, 12), Pauli's unificationist conception of understanding owes its current philosophically sophisticated form mainly to the work of Friedman (1974, 1983) and Kitcher (1981, 1985, 1989). Their work is usually discussed together, but, given my focus on understanding, Friedman is the more relevant author, so I will confine my analysis to his view. As is easy to notice, Kitcher's main concern is explanation, not understanding. In fact, he doesn't even address this issue until the end of his seminal 1981 paper, where he acknowledges his agreement with Friedman in this matter:

> In conclusion, let me indicate very briefly how my view of explanation as unification suggests how scientific explanation yields understanding. By using a few patterns of argument in the derivation of many beliefs we minimize the number of types of premises we must take as underived. That is, we reduce, in so far as possible, the number of types of facts we must accept as brute. Hence we can endorse something close to Friedman's view of the merits of explanatory unification. (Kitcher 1981, 540)

Friedman (1974) begins by formulating his version of 'the central problem' for a theory of scientific explanation. In essence, such a theory has to answer the following questions: (1) What kind of relation between two phenomena has to hold to say that one phenomenon is the explanation of the other? and (2) What is about this relation that advances our understanding? (1974, 6) After an examination (followed by rejection) of several answers available in the literature, Friedman develops his own view, a descendant of Hempel's DN view. With regard to (1), Friedman suggests that the relation we have to look for is the 'derivation' of the phenomenon[13] to be explained (i.e., the *explanandum*) from other phenomena (*explanans*). Friedman's contribution consists in specifying a supplementary requirement that this derivation has to meet: the premises of the explanatory derivations have to be *more comprehensive* than the conclusions (1974, 19).

This requirement reflects a change in the approach to explanation and understanding, from a 'local' approach to a 'global' one. Friedman notes that a look at the structure of the debate on the nature of scientific explanation reveals that it focused on the derivation relation holding locally, usually between two phenomena.

---

[13]Strictly speaking, we derive *descriptions* of phenomenona, not the phenomena themselves. In this paper I use the word 'phenomenon' liberally, referring to any kind of thing explained in science (laws, events, facts, etc.)

Yet, urges Friedman, we should enlarge the perspective, and come to see the global aspects of explanation, namely the relation between the phenomenon to be explained and the total set of accepted phenomena.

On the local approach, to explain amounts to simply replace one puzzling phenomenon with another. Thus the problem of understanding can still rear its head: even if one may agree that the (local) derivation of X from Y offered a good explanation of X ('why X? because Y'), one is still entitled to ask how is our overall understanding enhanced by this explanation, since the number of puzzles has not been reduced.[14] Although phenomenon X has been explained, phenomenon Y is still in need of explanation.[15]

In contrast, on the global approach, we seek to derive the phenomenon to be explained from a more comprehensive phenomenon. In schematic terms, the situation can be presented as follows. If x and w are some distinct explananda and y, z are explanans, the situation represented as

$$y \rightarrow x$$
$$z \rightarrow w$$

is less desirable than the one represented as

$$u \Big\langle {}^{\nearrow \, x}_{\searrow \, w}$$

(The arrow stands for 'is derived from', and u is another explanans.)

The point of this change of approach is to allow Friedman to answer question (2). If the relation he was searching for was identified as 'derivation', then to say that the premises must be more comprehensive is to say that they must be such that as to allow us not only the derivation of the phenomenon we wanted to explain initially, but the derivation of several other, apparently unrelated phenomena. This, in turn, leads to a more unified account of the world, reflected in a reduction of the number (and presumably of variety) of phenomena we have to ultimately accept as brute, or unexplained. It is precisely this reduction of the number of basic, brute phenomena which amounts to an increase in understanding:

> Science increases our understanding of the world by reducing the total number of independent phenomena that we have to accept as ultimate or given. A world with fewer independent phenomena is, other things equal, more comprehensible than one with more. (1974, 15)[16]

---

[14] As Salmon (2002, 94) points out, Friedman is too quick in assuming that counting the number of fundamental laws (what serves as basis of derivation) is possible.

[15] Of course, one can reply that the *explanans* (Y) may be easier to understand, more 'familiar' than the phenomenon to be explained (X). Friedman counters this objection by arguing that familiarity (and other related notions) should not to be confused with intelligibility (1974, 10).

[16] Friedman's proposal prompted a number of insightful objections (Kitcher 1976; Barnes 1992b), but none of them will concern me here.

On Friedman's account understanding is a form of conceptual economy, in an almost literal sense: the point of unification is the point of any functional economy, to get 'more' from 'less'. The reduction of the number of brute facts we must ultimately accept is the *only* thing that matters when it comes to estimating an increase in understanding. Note also a major gain of this account of understanding: the subjective, or psychological, aspects associated with this notion, i.e., how we *feel* about the phenomena under scrutiny, are rendered completely irrelevant.

Friedman's own example of unification, the kinetic theory of gases, illustrates this approach very well. Within the framework of this theory, instead of two brute facts – the inverse proportionality of pressure and volume of a gas at constant temperature (captured by Boyle's ideal gas law) and the dependence of the rate of diffusion of gases on volume (expressed in Graham's law of diffusion) – we have to accept only one brute fact, that molecules obey the laws of Newtonian mechanics.[17] This one fact is enough to derive the two facts or phenomena we wanted to explain; and so it does look like we get more from less.

15.4 Scientific Progress as an Increase in Objective Global Understanding

The present attempt to explicate scientific progress distinguishes itself from what has been said until now by giving due consideration to a neglected aspect, the role of understanding in articulating our conception of scientific progress. Two important features of this account – henceforth the 'understanding-accumulation' account – deserve highlighting. The first is an intrinsic property of it; the second has a comparative nature, namely the ability of this approach to complement the knowledge-accumulation approach. I will close this paper by discussing them in turn.

To begin with the first point, the view of understanding I endorse here can be characterized as public-objective. When I claim that gaining more scientific understanding constitutes a significant part of scientific progress, I don't have in mind the kind of understanding manifested as a private episode, experienced by individual scientists (the 'aha!' exclamation moment, the feeling of having understood, the illumination sentiment, etc.) I surely agree that such episodes are crucial for science as practiced, since they increase the scientists' personal confidence in their methods and in the direction of research they pursue. However, what I take to be relevant for the present account is the public-objective aspect of understanding. It is this shift in emphasis, from private-subjective to pubic-objective, which motivates the preference for the unificationist conception of understanding – or, to be more precise, for what has been called above the 'global' notion of understanding. When compared to other ways of characterizing understanding, this one seems to me most

---

[17]Friedman (1974, 14–5) refers to this as a fact (and I followed his usage) although this is not, strictly speaking, a fact; it is rather something that Kitcher (1981, 1989) calls a 'type of fact'.

clearly divorced from the private-subjective notion. An increase in understanding does not occur when one scientist, or a group, reports ('feels') that they have understood a certain issue, but when the following situation takes place: they realize that instead of having to answer two (or more) why-questions, they have to answer only one.

Now someone may of course point out that this way of looking at things is open to the objection that although a decrease in the number, or (so to speak) in the *quantity* of questions provides prima facie an objective measure of the epistemic situation, it is not immediately clear that this is a measure of understanding – and, as I propose here, of progress. This is so for the simple reason that the one, or the fewer, question(s) we are left with may be harder than those they replaced! In other words, the reduction in the quantity of questions may lead to an increase in their *quality* – so to speak. More directly put, what happens is that we may trade two moderate mysteries for a bigger one, and this is hardly an advantage.[18]

This is a fair point, but the suggestion should be resisted. In fact, upon deeper reflection, the exact opposite is the case: a typical sign of understanding something is that one is in the position to ask fewer and more profound, indeed harder, questions about it. One typically only *begins* one's inquiry by having many, direct and simple questions; it is a sign of lack of progress and depth of one's investigation when one *ends* it with questions of the same nature. It is the novice, not the expert, who asks many ('easy') questions, circling around the gist of the matter. Indeed, it is the beginner who doesn't know *what* to ask. If deeper answers are a sign of achieving understanding and making progress, they can only be answers to deeper, harder questions.

The second point I will elaborate on is the complementarity between the understanding-accumulation approach sketched here and the knowledge-accumulation account. I mentioned above (fn. 1) Bird's own acknowledgment that his account should have included a discussion of scientific understanding. He also provides, in the same paragraph, a good example of accumulation of knowledge without accumulation of understanding:

> Imagine a team of researchers engaged in the process of counting, measuring, and classifying geologically the billions of grains of sand on a beach between two points. Grant that this may add to scientific knowledge. But it does not add much to understanding. Correspondingly it adds little to scientific progress. (2007, 84)

This is indeed so, and a related, though less intuitive, point can be added: an increase in unificationist understanding leads to an increase in knowledge. Let me explain, using a toy-model.

Suppose that r and s are true and well-justified scientific propositions, so we can say that they are straightforward, unproblematic, 'proper' items of scientific knowledge. Moreover, their justification is given by derivation from some basic scientific truths (let us call them P and Q; their status is marked by using capital letters).

---

[18]There is a vast literature criticizing the unificationist approach (part of it mentioned here), but I'm not sure this objection has been advanced before.

We can illustrate this relation by using arrows: P → r, Q → s. Propositions P and Q are basic, foundational ('self-evident') truths any science has to assume. In being so, there is something deficient about their epistemic status; the very labels 'self-evident', or 'self-justified' ('not in need of justification', etc.) indicate that they are not justified in the ordinary way, or 'properly' justified. Now, as the knowledge-accumulation account requires, to make progress we should ensure that more true and properly justified propositions are identified and added to our body of beliefs. So far so good; but note that nothing is being said about *how* these additions can, or should, be done. So let us compare several ways in which they can be done.

One possibility is that a true proposition t is justified on the basis of a basic truth K (as different from P and Q), so the diagram looks like this:

(1) P → r, Q → s, K → t

The net result is that one more proper item of knowledge, t, has been added to our body of knowledge.

Another way is to make an addition is by identifying a proposition t which is also true and not basic, but which is justified by derivation from either P, or Q, or r, or s (or some combination of them). Without affecting the generality of the argument, let's say that t is justified on the basis of s, so the diagram looks like this:

(2) P → r, Q → s → t

Proposition t is true and justified, and thus a proper item of knowledge; hence the condition for progress required by the knowledge-accumulation account has been satisfied.

Yet there is a third way in which the addition can be made. Suppose that proposition U is found, such that U is a basic truth, and such that U → p and U → q. Thus, the diagram looks like this:

(3) U ⟨ p → r
        q → s

As is clear, proposition U acts as a *unifier* for our body of beliefs and, in so far as the number of brute phenomena has been reduced (to one instead of three, or two, as above), we can claim, according to the unificationist view, an increase in understanding. Furthermore, this third alternative also reveals something else: the addition of U turns the initially basic truths P and Q into unproblematic (proper) items of knowledge p and q, since now they are not only true, but also properly justified by derivation from U (note that they don't appear as capital letters in the last diagram). Thus, the net result of bringing U into the picture is that *two* new unproblematic items of knowledge (p and q) have been added to our body of beliefs, and this is more than the one item (t) we have added previously.[19]

---

[19]Note that even if we count the new basic truths added as items of knowledge, diagram (3) is still depicting a more desirable situation than diagrams (1) and (2): in (3) we have three new items of knowledge (U, p and q), as compared to only two (K and t) in (1), or only one (t) in (2).

The lesson of this comparison is easy to grasp: the quest for (unificationist) understanding has positive consequences for acquiring knowledge; increasing understanding leads to increasing knowledge. Thus, understanding, in the sense adopted here, is more fundamental. While we do make progress when we add up more pieces of knowledge to our list (Bird got this right), a somewhat counterintuitive effect arises: the addition of a proposition which is prima facie *not* a proper item of knowledge (U), can lead, if it has a global-unifying character, to even better consequences for the amount of knowledge we possess – better, that is, than the (local) addition of a proposition that is a proper item of knowledge. The goal of achieving more knowledge is served to an even higher degree when we concern ourselves with strengthening the interrelations holding within the body of beliefs – that is, when considerations having to do with the unification and the systematization of our body of beliefs take center stage.

15.5   Conclusion

Even in this sketchy form, the proposed understanding-accumulation account of scientific progress is in the position to offer some conceptual advantages over the previous existing accounts, and all this while acknowledging their naturalness. It not only essentially complements the knowledge-accumulation account, but also neatly integrates some of the ideas defining some of the other accounts. For instance, it is just natural to think that being good at solving many genuine problems must be a result of understanding how the world is put together. Also, some other ideas we have examined, that progress involves steps forward toward a goal, and that it can be thought of in quantitative-measurable terms (either cumulatively, or as 'distance-from-a-starting-point'), are easily incorporated. There is a lot of plausibility in the proposal that understanding is the goal of science, and that progress means moving closer to this goal or, equivalently, moving farther away from the epistemic state in which everything we knew was just a collection of basic, brute facts. Moreover, this goal is less elusive than others, such as the Truth – since, as we recall, the unificationist insight can capture this goal in relatively clear, objective-quantitative terms, as the decrease in the amount of basic beliefs we have to assume.

Summing up, having more true beliefs, even more justified true beliefs (i.e., more knowledge), or being able to solve more (genuine) problems are all important steps in the right direction. Yet a complete characterization of what it is to advance science must include a role for how all these items of scientific knowledge are systematized within a coherent picture of the world, thus increasing its comprehensibility – that is, our understanding.

Acknowledgments I thank the editors for soliciting my contribution, an anonymous referee for suggestions and Alexander Bird for reading an early draft. I presented sections of this paper at Aarhus University and I'm grateful to Sam Schindler, Sara Green and Helge Kragh for comments. All responsibility for possible errors remains mine. I dedicate this paper to the memory of M-R. Solcan.

Bibliography

Balzer W (2000) On approximate reduction. In Jonkisz and Koj (2000), pp 153–170
Barnes E (1992a) Explanatory unification and scientific understanding. In: Hull D, Forbes M, Okruhlik K (eds) PSA 1992, vol 1. Philosophy of Science Association, East Lansing, pp 3–12
Barnes E (1992b) Explanatory unification and the problem of asymmetry. Philos Sci 59:558–571
Barrett JA (2008) Approximate truth and descriptive nesting. Erkenntnis 68:213–224
Bartelborth T (2002) Explanatory unification. Synthese 130:91–107
Bird A (2000) Thomas Kuhn. Acumen, Chesham
Bird A (2007) What is scientific progress? Noûs 41:92–117
Bird A (2008) Scientific progress as accumulation of knowledge: a reply to Rowbottom. Stud Hist Philos Sci 39:279–281
Bragg W (1936) The progress of physical science. In: Jeans J et al (eds) 1936 scientific progress. George Allen and Unwin, London
De Regt HW (2009) Understanding and scientific explanation. In: de Regt HW, Leonelli S, Eigner K (eds) Scientific understanding: philosophical perspectives. University of Pittsburgh Press, Pittsburgh
De Regt HW, Dieks D (2005) A contextual approach to scientific understanding. Synthese 144:137–170
Dilworth C (1981) Scientific progress: a study concerning the nature of the relation between successive scientific theories. Reidel, Dordrecht
Elgin C (2009) Exemplification, idealization, and scientific understanding. In: Suarez M (ed) Fictions in sciences: philosophical essays on modeling and idealization. Routledge, New York
Feigl H (1970) The 'orthodox' view of theories: remarks in defense as well as critique. In: Radner M, Winokur S (eds) Minnesota studies in philosophy of science, vol IV. University of Minnesota Press, Minneapolis
Feyerabend P (1962) Explanation, reduction, and empiricism. In: Feigl H, Maxwell G (eds) Minnesota studies in the philosophy of science, vol II. University of Minnesota Press, Minneapolis, pp 28–97
Friedman M (1974) Explanation and scientific understanding. J Philos 71:5–19
Friedman M (1983) Foundations of space-time theories. Princeton University Press, Princeton
Gettier E (1963) Is justified true belied knowledge? Analysis 23:121–123
Grimm S (2013) Understanding. In: Pritchard D, Berneker S (eds) The Routledge companion to epistemology. Routledge, New York
Grimm S (forthcoming) Understanding as knowledge of causes. In: Fairweather A (ed) Virtue Scientia: essays in philosophy of science and virtue epistemology. Special issue of *Synthese*
Halonen I, Hintikka J (1999) Unification – it's magnificent but is it explanation? Synthese 120:27–47
Hanson P (1963) The concept of the positron. Cambridge University Press, Cambridge
Heisenberg W (1971) Physics and beyond. Encounters and conversations. Translation from German by Pomerans AJ. Harper and Row, New York
Hempel C (1965) Aspects of scientific explanation. Free Press, New York
Hindricks F (2013) Explanation, understanding, and unrealistic models. Stud Hist Philos Sci 44(3):523–531
Howson C (ed) (1976) Method and appraisal in the physical sciences: the critical background to modern science, 1800–1905. Cambridge University Press, Cambridge
Humphreys P (1993) Greater unification equals greater understanding? Analysis 53:183–188
Jonkisz A (2000) On relative progress in science. In: Jonkisz and Koj (2000), pp 199–234
Jonkisz A, Koj L (eds) (2000) On comparing and evaluating scientific theories. Rodopi, Amsterdam
Kargon R, Achinstein P (eds) (1987) Kelvin's Baltimore lectures and modern theoretical physics. MIT Press, Cambridge, MA
Khalifa K (2013) The role of explanation in understanding. Brit J Philos Sci 64(1):161–187

Khalifa K, Gadomski M (2013) Understanding as explanatory knowledge: the case of bjorken scaling. Stud Hist Philos Sci 44:384–392

Kitcher P (1976) Explanation, conjunction, and unification. J Philos 73:207–212

Kitcher P (1981) Explanatory unification. Philos Sci 48:507–531

Kitcher P (1985) Two approaches to explanation. J Philos 82:632–639

Kitcher P (1989) Explanatory unification and the causal structure of the world. In: Kitcher P, Salmon W (eds) Scientific explanation, vol 13, Minnesota studies in the philosophy of science. University of Minnesota Press, Minneapolis, pp 410–505

Kleiner SA (1993) The logic of discovery: a theory of the rationality of scientific research. Kluwer, Dordrecht

Kuhn TS (1970) The structure of scientific revolutions. University of Chicago Press, Chicago (1st ed. 1962)

Kuhn TS (1977) The essential tension. University of Chicago Press, Chicago

Lakatos I, Musgrave A (eds) (1970) Criticism and the growth of knowledge. Cambridge University Press, Cambridge

Laudan L (1977) Progress and its problems: toward a theory of scientific growth. Routledge and Kegan Paul, London

Laudan L et al (1986) Scientific change: philosophical models and historical research. Synthese 69:141–224

Morrison M (2000) Unifying scientific theories. Physical concepts and mathematical structures. Cambridge University Press, Cambridge

Newman M (2013) Refining the inferential model of scientific understanding. Int Stud Philos Sci 27:173–197

Niiniluoto I (1987) Truthlikeness. Reidel, Dordrecht

Niiniluoto I (1995) Is there progress in science? In: Stachowiak H (ed) Pragmatik, Handbuch pragmatischen Denkens, Band V. Felix Meiner Verlag, Hamburg, pp 30–58

Niiniluoto I (2010) Theory change, truthlikeness, and belief revision. In: Suárez M et al (eds) EPSA epistemology and methodology of science. Springer, Dordrecht, pp 189–199

Niiniluoto I (2011) Scientific progress. In: The Stanford encyclopedia of philosophy (Summer 2011 Edition), Zalta EN (ed), http://plato.stanford.edu/archives/sum2011/entries/scientific-progress/

Popper K (1959) The logic of scientific discovery. Hutchinson, London

Popper K (1963) Conjectures and refutations: the growth of scientific knowledge. Hutchinson, London

Popper K (1972) Objective knowledge: an evolutionary approach. Oxford University Press, Oxford. 2nd enlarged ed. 1979

Psillos S (1999) Scientific realism: how science tracks truth. Routledge, London

Rescher N (1978) Scientific progress: a philosophical essay on the economics of research in natural science. Blackwell, Oxford

Salmon W (1984) Scientific explanation and the causal structure of the world. Princeton University Press, Princeton

Salmon W (1998) Causality and explanation. University Press, Oxford

Salmon W (2002) Scientific explanation: causation and unification. In: Balashov Y, Rosenberg A (eds) Philosophy of science. Contemporary readings. Routledge, London, pp 92–105, First published in Critica. Revista Hispanoamericana de Filosofia, 1990, 22(66): 3–21

Scheibe E (1976) Conditions of progress and comparability of theories. In: Cohen RS et al (eds) Essays on memory of Imre Lakatos. Reidel, Dordrecht, pp 547–568

Schurz G (1999) Explanation as unification. Synthese 120:95–114

Stegmüller W (1976) The structure and dynamics of theories. Springer, New York/Heidelberg/Berlin

Strevens M (2004) The causal and unification approaches to explanation unified – causally. Nous 38(1):154–176

Strevens M (2008) Depth: an account of scientific explanation. Harvard University Press, Cambridge, MA

Strevens M (forthcoming) No understanding without explanation. In: Studies in history and philosophy of science Part A

Trout JD (2002) Scientific explanation and the sense of understanding. Philos Sci 69:212–233

van Fraassen BC (1985) Salmon on explanation. J Philos 82:639–651

Weber E (1999) Unification: what is it, how do we reach it and why do we want it? Synthese 118:479–499

Woodward J (2003) Making things happen: a theory of causal explanation. Oxford University Press, Oxford

Chapter 16
When Is a Mechanistic Explanation Satisfactory? Reductionism and Antireductionism in the Context of Mechanistic Explanations

Tudor M. Băetu

16.1 Introduction

Some of the most successful and influential explanations in the life sciences amount to descriptions of mechanisms, where mechanisms are characterized as organized systems of parts that operate in such a way as to produce phenomena (Bechtel and Abrahamsen 2005; Glennan 2002; Machamer et al. 2000; McKay Illari and Williamson 2012). There is no mystery, however, that the entities of a biological mechanism can be further decomposed into subparts, activities into sub-activities, and mechanisms into more fine grained sub-mechanisms (Bechtel and Richardson 2010; Craver 2007). Nor there is any doubt that biological mechanisms are parts of progressively more comprehensive systems of mechanisms, ranging from molecular networks to planetary ecosystems, where more systemic mechanisms can both depend on the functioning of the sub-mechanisms of which they are composed and impose constraints on their mode of operation (Hooker 2011). Thus, in the realm of mechanistic explanations, the issue of reductionism in biology[1] can be

---

[1]In biology, the reductionism debate is primarily about the relationship between molecular biology and other branches of biology, such as classical genetics [e.g., (Waters 1990) vs. (Kitcher 1984)] and developmental biology [e.g., (Rosenberg 2006) vs. (Oyama 1985)]. In the contemporary literature, reductionists agree that a mechanistic explanation does not need to bottom down at the most fundamental building blocks of physical reality, and that a satisfactory explanation can be articulated at the level of molecular interactions. Likewise, even the most fervent proponents of antireductionism agree that some, but not all contexts, and certainly not the totality of the universe, are important for understanding biological phenomena. If there is a resistance to molecular or genetic reductionism, the concern is that certain features of the cell, organism or the direct

T.M. Băetu (✉)
Programa de Filosofia, Universidade do Vale do Rio dos Sinos, São Leopoldo, Brazil
e-mail: tudormb@unisnos.br

© Springer International Publishing Switzerland 2015
I. Pârvu et al. (eds.), *Romanian Studies in Philosophy of Science*, Boston Studies in the Philosophy and History of Science 313, DOI 10.1007/978-3-319-16655-1_16

reformulated as a combo of questions, one about the level of composition at which mechanistic descriptions bottom out, and the second, about whether mechanisms act as independent modules that can continue to function when separated from the systems in which they are embedded.

The goal of this paper is to provide an answer to these questions. I argue that the solution lies in the elaboration of norms for evaluating the completeness of mechanistic explanations. According to current accounts (Craver 2006, 2007; Machamer et al. 2000), a satisfactory mechanistic explanation should describe the mechanism actually producing the phenomenon of interest and include all of the relevant features of the mechanism, its component entities and activities, their properties and their organization, as well as exhibit productive continuity. It is not specified, however, how this kind of mechanistic completeness can be demonstrated in scientific practice. Current accounts emphasize the role of experimental interventions demonstrating that various components of a mechanism are actually involved in the production of the phenomenon (Baetu 2012; Craver 2006, 2007). However, a strictly interventionist approach is not enough. I argue that an increasingly popular strategy for determining whether all the relevant mechanistic components and information about these components have been taken into consideration relies on mathematical modeling. Once it is possible to demonstrate that a given mechanism is actually involved in the production of a phenomenon and that it can produce that phenomenon solely in virtue of its identified components, their known properties, organization and activities, then there is no need to further elaborate the description of the mechanism by bottoming out at deeper levels of composition or to expand it in order to include a more systemic perspective, thus providing a principled way of determining when a mechanistic explanation is satisfactorily complete for the purposes of accounting for the phenomenon of interest.

The paper is organized as follows. In Sect. 16.2, I discuss currently elaborated guidelines for developing norms of mechanistic explanation. In Sect. 16.3, I discuss the role of experimental interventions in demonstrating that a mechanism and its components are necessary, and actually involved in the production of phenomena, as well as the limitations of a strictly interventionist approach. In Sect. 16.4, I elaborate the notions of quantitative and parameter sufficiency inferences from mathematical models and show how they can provide a principled way of determining where a mechanistic explanation can safely bottom out and what is the cutoff point beyond which external factors can be ignored. Finally, some broader-interest implications are discussed in Sect. 16.5.

---

environment of an organism have been neglected. It is within these boundaries that the issue of reductionism is considered here.

16.2   Guidelines for Developing Norms of Mechanistic Explanation

A mechanistic explanation is analogous to a recipe for producing a phenomenon starting from a list of ingredients, where the ingredients are mechanistic entities and their properties, and the recipe amounts to the organization and sequence of activities these entities perform. The mechanistic explanation is deemed satisfactory when (1) it is known by means of which particular 'mechanistic recipe' the phenomenon of interest is actually produced in the biological system of interest, and (2) there are no missing ingredients and no missing lines in the description of the 'recipe' for producing the phenomenon. In more technical terms, it is important "(1) to distinguish how-possibly explanations from how-actually explanations, and (2) to distinguish mechanism sketches from mechanism schemata" (Craver 2007, 111). Aim (1) refers to the distinction between conjectures about possible mechanisms that might be able to produce the phenomenon and descriptions of the actual components, activities, and organizational features of the mechanism that in fact produce the phenomenon (Craver 2007, 112). Aim (2) alludes to the completeness of the description of a mechanism. A mechanism schema is a "truncated abstract description of a mechanism that can be filled with descriptions of known component parts and activities. [. . .] When instantiated, mechanism schemata yield mechanistic explanations of the phenomenon that the mechanism produces" (Machamer et al. 2000, 15, 17). A satisfactory mechanistic explanation should "include all of the relevant features of the mechanism, its component entities and activities, their properties, and their organization" (Craver 2006, 367); and "exhibit productive continuity without gaps from the set up to termination conditions" (Machamer et al. 2000, 3). By contrast, a mechanism sketch is an incomplete explanation "for which bottom out entities and activities cannot (yet) be supplied or which contains gaps in its stages" (2000, 18).

If it were possible to demonstrate that the 'mechanistic recipe' for producing a phenomenon is actual and complete, the explanation of the phenomenon could safely be reduced to this 'recipe' in the sense that adding further ingredients or lines to the 'recipe' would either not cause any changes in the phenomenon, meaning that such additions are causally and explanatorily irrelevant, neutral or redundant, or interfere with the functioning of the mechanism causing a failure to produce the phenomenon as it is measured in the biological system of interest, in which case the explanation would fail. The task, therefore, is to determine what kind of evidence is necessary to support the claim that the 'mechanistic recipe' is actual and complete.

16.3   The Role of Experimental Interventions in the Elucidation of Biological Mechanisms

In the life sciences, mechanisms are usually elucidated experimentally, by carefully circumscribing a putative mechanism within the boundaries of a well characterized experimental setup (Baetu 2013); by means of decomposition strategies (Bechtel and Richardson 2010); by conducting exploratory interventions aimed at identifying correlating factors providing an initial pool of putative mechanistic components (Baetu 2012); by performing specific interventions aimed at demonstrating the causal relevance of the entities, activities, and organizational features of a hypothesized mechanism (Craver 2007; Woodward 2002, 2003) and elucidating their causal roles relative to the operation of the mechanism (Craver 2001); and by tracking causal pathways (Craver 2007; Darden 2006).

Currently elaborated norms for evaluating mechanistic explanations are inspired from the experimental practice of the life sciences. By intervening on the components of a mechanism, it is possible to demonstrate that the mechanism is necessary to produce the phenomenon, as well as that the mechanism in question is actually involved in the production of the phenomenon (Craver 2007). Given a suitable experimental design (e.g., standardized quantitative measurements, multivariable intervention experiments), experimental interventions can provide further evidence that no parallel or convergent causal pathways are actually involved in the production of a phenomenon in a particular experimental setup (Baetu 2012). For example, in a typical knockout experiment, two factors, the initial conditions and a mechanistic component, are simultaneously manipulated on an independent basis and the effects on the output conditions are observed. If the knocking out of the component results in a complete inhibition of the output, one can infer that the mechanism is necessary and sufficient for producing the phenomenon of interest, in the sense that there are no other mechanisms that produce the phenomenon via alternate causal pathways that do not involve the knocked out component (Fig. 16.1).

Experimental interventions are used to demonstrate that mechanistic components are necessary and actually involved in the production of phenomena, thus providing methodological criteria for distinguishing how-possibly explanations from how-actually explanations. However, interventions don't tell us if and when all the explanatorily relevant details have been filled in or whether there are gaps in the productive continuity of a mechanism. One way of framing the problem is in terms of the ability to physically construct biological mechanisms: if the mechanism described in the proposed explanation were to be artificially synthesized from components organized, acting, and having the properties described in the mechanistic explanation, would it succeed in producing the phenomenon of interest as it was originally measured? To clarify, it is not question here of further explaining why the components have the properties they have, why they are organized the way they are, or why they are doing whatever they are doing. Nor is there any doubt about the fact the identified mechanistic components, along with their experimentally demonstrated properties, organization and activities are necessary for and actually involved in the production of the phenomenon. Rather, the issue under scrutiny is whether entities, properties of entities, activities or organizational features have been omitted, such that the mechanistic explanation amounts to an incomplete recipe missing some ingredient or step in the sequence of events necessary for the production of the phenomenon of interest.

Consider, for instance, the following example. When exposed to certain stimulants, such as pathogens, white blood cells, and T-cells in particular express a

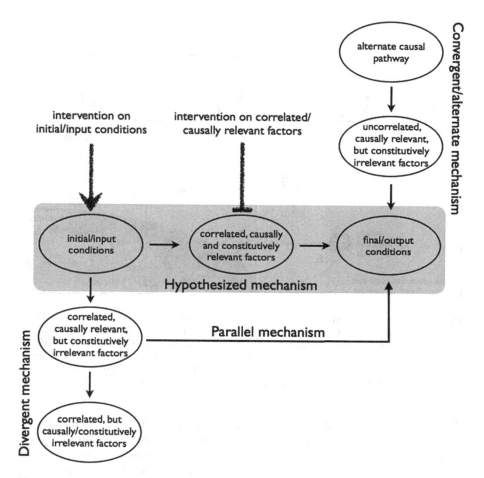

Fig. 16.1 Sorting putative mechanistic components by a two-variable knockout experiment

variety of genes required for mounting an immune response, after which they automatically return to their initial resting state. This spike of gene expression following stimulation is explained by a negative feedback regulatory mechanism whereby a transcriptional factor (nuclear factor κB, or NF-κB) is initially activated, then subsequently inactivated by an inhibitory protein (inhibitor of κB, or IκB) coded by a gene under its transcriptional control (Fig. 16.2).

There are many details missing from the above mechanistic description. The mechanistic description can be further elaborated by bottoming down at the deeper level of biochemical details rather than the lower resolution level of molecular interactions depicted in Fig. 16.2, most notably by including additional information about the tridimensional conformations of the proteins involved and their role vis-à-vis molecular function, such as structural motifs involved in specific binding

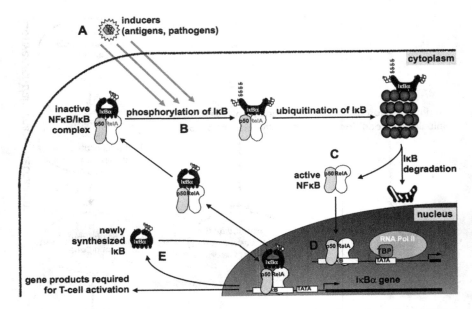

Fig. 16.2 The NF-κB negative feedback loop regulatory mechanism. In resting cells, the NF-κB transcriptional activator is held in the cytoplasm by the IκB inhibitor. When cells are stimulated (**a**), a chain of protein-protein interactions leads to the degradation of IκB (**b**); NF-κB is freed (**c**), translocates to the nucleus (**d**) where it binds specific sequences in the promoter regions of target genes drastically enhancing their transcription. NF-κB also binds the promoter of the IκB gene (**e**), and the newly synthesized IκB binds NF-κB, trapping it back in the cytoplasm

(Fig. 16.3, panel C). By digging deeper, researchers typically hope to gain a better understanding of why and how mechanistic components are able to do what they are doing, as well as discover new ways in which mechanistic components can be manipulated for experimental and medical purposes. This kind of knowledge and the interventions it renders possible play a crucial role in elucidating mechanisms. At the same time, the mechanistic description can also be expanded by taking into account other molecular mechanisms, most notably upstream signaling pathways and downstream mechanisms triggered via the expression of new genes (Fig. 16.3, Panel A). In this particular case, the negative feedback loop mechanism is known to be involved in a number of rather diverse biological phenomena, ranging from development and cell differentiation to immune responses and cell death. By adopting a more systemic viewpoint, one may hope to gain a better understanding of how immunity relates to other biological activities. This is particularly important for understanding possible side effects of therapies designed to enhance desirable immune responses or inhibit deleterious ones.

While both a more fine grained description bottoming out at deeper levels of composition and taking into consideration a more systemic perspective amount to a net gain of knowledge, it is not obvious how this additional information can support the conclusion that the mechanism described in Fig. 16.2 generates the phenomenon

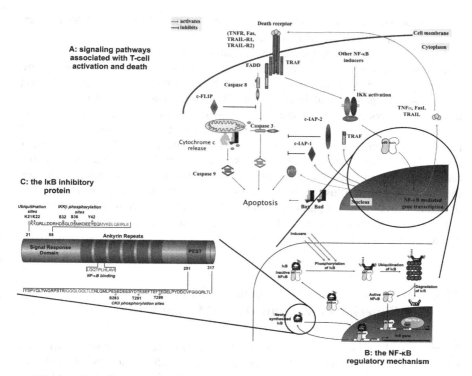

Fig. 16.3 Molecular levels [Panels adapted from (Baetu and Hiscott 2002; Baetu et al. 2001)]

of interest in virtue of its identified components, their properties, organization and activities. Higher resolution structural details of the NF-κB transcriptional activator and the IκB inhibitor are crucial for understanding how these two proteins bind each other, and which alterations (e.g., mutations) result in a loss in binding ability. Nevertheless, given experimentally gained knowledge that the two bind, further knowing how and why they bind does not tell us whether it is possible to artificially synthesize the feedback regulatory mechanism starting from a pool of NF-κB transcriptional activators, IκB inhibitor proteins and other molecular components organized as described in Fig. 16.2. There is, therefore, a worry that the mere fact that the various components of a mechanism can be analyzed at progressively lower levels of composition creates the reductionist illusion that biological phenomena are ultimately explainable by and reducible to the theories of particle physics, while in truth this analysis does not necessarily contribute to the mechanistic explanation, which should be a story about how an organized system of parts succeeds in producing a phenomenon. From the standpoint of mechanistic thinking, the goal is to figure out that precise level of composition at which parts, organized and acting as described in the proposed explanation, can generate the phenomenon in need of an explanation, and not to explain how or why the parts have the properties they have, which is a different research question.

Likewise, if a more systemic understanding of how this regulatory mechanism contributes to a variety of biological activities is crucial for grasping the physiological and evolutionary relevance of the mechanism, as well as evaluating the therapeutic potential of interventions, this knowledge does not tell us if the mechanism's contribution to the regulation of immune responses is mediated solely by means of the feedback loop regulation of the expression of the genes required for mounting an immune response, and independently of the mechanism's involvement in other biological activities.[2] The worry here is that the fact that biological mechanisms are embedded in or connected to other mechanisms creates the antireductionist illusion that everything is inextricably interconnected and ultimately irreducible to parts or sums of parts, while in truth the mere fact of connectedness does not allow us to determine whether or not a given mechanism can be treated as an independent module.

16.4   Inferences from Mathematical Models of Experimentally Elucidated Mechanisms

Specifying where an explanation can safely bottom out and when the mechanism can be considered an independent module requires a different kind of evidence, which is not likely to emerge from the accumulation of information bought about the further decomposition of mechanistic components or by taking into account progressively more systemic contexts. What is needed, is a reconstruction of the mechanism starting from a set of parts having the properties and organization specified by a proposed mechanistic explanation, in order to determine if, thus reconstructed, the mechanism can indeed produce the phenomenon it is supposed to produce. While the physical reconstruction of mechanisms is documented in contemporary biology (Morange 2009; Weber 2005, Chap. 5), a much more common and accessible alternative relies on the mathematical modeling of experimentally elucidated mechanisms.[3]

---

[2] The organicist debate that raged in the nineteenth century biology centered on the claim that living things are organic wholes that cannot be decomposed into a set of independent mechanisms. Critics of molecular biology and its methods often appeal organicist arguments to defend more holistic approaches, and part of the manifesto of systems biology is precisely to provide a more holistic understanding of life. Contemporary echoes of this debate can be found in Nicholson (2013).

[3] Mathematical modeling is by no means a novel practice in biology. The Hodgkin-Huxley model of the action potential, the Michaelis-Menten model of enzyme kinetics, and Knudson's two-hit model of cancer development made use of theoretical tools in order to demonstrate that biological and biochemical phenomena can be accounted for as consequences of laws or rules governing the behavior of certain systems. These same models played an important role in guiding the subsequent elucidation of the molecular mechanisms. More recently, mathematical models have been used

Commenting on a study by Hoffmann et al. (2002), where the authors constructed and tested a mathematical model of the NF-κB negative feedback regulatory mechanism described earlier (Fig. 16.2 above), Alice Ting and Drew Endy make the following point:

> A limitation of computational modeling is that, in the absence of complete information about cell parts and interconnections, it is easy to omit critical parameters that might influence the state of a cell or signaling pathway. This is illustrated in the Hoffmann et al. work. [. . .] When they used this model to predict the behavior of wild-type cells, the outcome was very different from what was actually measured, even though many of the parameters were empirically obtained. Such discrepancies could be due to compensatory changes in expression and signaling state from one cell line to the next, or to additional pathway components and regulatory mechanisms beyond the current model (2002, 1190).

The limitation of computational modeling to which they allude is not one due to abstraction, idealizations or the instrumental nature of the models used, but rather the concern that, even when constructing detailed and highly realistic mathematical models of previously elucidated molecular mechanisms, and even when the values of the parameters of model are based on empirical measurements, these models can only be as complete as our knowledge of the modeled mechanisms is. However, as the authors quickly point out, there is a bright side to this limitation. If the output of the model fails to closely match the phenomenon known to be produced by the modeled mechanism, then this can be an indication that something is missing from the mechanistic explanation. That is, the mechanistic explanation might be incomplete because not all the components of the mechanism have been identified, or other mechanisms are needed to produce the phenomenon of interest.

It should be immediately noted that the kind of explanatory completeness evaluated by mathematical models has nothing to do with an ultimate understanding of how everything works at the level of systemic interactions between the most fundamental building blocks of physical reality. Rather, it is an engineer's understanding of completeness, framed in terms of information required to reconstruct *in silico* a mechanism capable of producing the phenomenon of interest starting from components organized, acting, and having the properties described in the mechanistic explanation.

If the output of the mathematical model of the proposed mechanism matches experimental measurements of the phenomenon, this is taken as evidence supporting the claim that the proposed mechanism is *quantitatively sufficient* for generating that phenomenon. This is an important piece of information. Qualitative descriptions associated with traditional mechanistic explanations usually suffice to provide an intuitive understanding of how a mechanism may produce something roughly resembling the phenomenon to be explained. For instance, by contemplating Fig. 16.2, one can intuitively understand how a negative feedback loop switching

---

to account for quantitative-dynamic features of phenomena meant to complement traditional qualitative descriptions of mechanisms (Baetu 2015; Bechtel 2012; Bechtel and Abrahamsen 2010; Brigandt 2013). In such cases, mathematical models act as *in silico* surrogates for investigating the properties of systems they model (Baetu 2014).

gene expression 'on' and 'off' in response to persistent exposure to triggering conditions can generate oscillating peaks of gene expression. Nevertheless, the question remains whether the feedback loop mechanism described in Fig. 16.2 can generate oscillations matching the amplitude, frequency, dampening and other minute quantitative-dynamic quirks of experimentally measured NF-κB mediated peaks of gene expression. Mathematical modeling provides the means to address this question.

When quantitative sufficiency is demonstrated by means of a detailed and realistic model, *parameter sufficiency* is further inferred. If the model simulations and predictions match experimental data, it can be argued that a more complex model, including additional parameters, is not needed. In as much as all the parameters have a clear physical interpretation, meaning that they describe known physical properties of the components of the mechanism, and at least some values of these parameters are based on independent empirical measurements, a close match between simulation and experimental measurements of the phenomenon of interest is taken as evidence in support of the claim that a more complex mechanism, including additional components, or additional mechanisms are not likely to be needed to produce the phenomenon.[4]

Parameter sufficiency plays an important role in guiding the design of artificial molecular mechanisms aimed at producing a desired phenomenon. Most famously, the repressilator (Elowitz and Leibler 2000), an artificial molecular oscillator, was designed on the basis of mathematical models predicting that sustained oscillations, the desired outcome, are favored by transcriptional regulation mechanisms constructed from molecular components organized in a certain way (in this case, negative feedback loops) and having a particular set of properties (strong promoters, low leakiness, etc.). Even though this first attempt to construct a synthetic mechanism turned out to be only a partial success – the mechanism did produce oscillations, but lacked the desired degree of robustness – , it did demonstrate that mathematical models can be in principle used to evaluate and predict whether a mechanism synthesized from the components described in the designed mechanism can generate the phenomenon of interest down to minute quantitative-dynamic aspects.

Beyond the specific needs of synthetic biology, parameter sufficiency also provides the means to figure out whether it is safe to bottom out at the level of composition at which the mechanism is described, in the sense that a more

---

[4]Klipp (2005, 8–9) makes a clear distinction between 'black-box' input-output correlations and realistic models in which known mechanistic details are taken into account: "It must be noted that different system structures may produce similar system behavior (output). The structure determines the behavior, not the other way around. Therefore the system output is often not sufficient to predict the internal organization [...] The intention of modeling is to answer particular questions. Modeling is, therefore, a subjective and selective procedure. It may, for example, aim at predicting the system output. In this case it might be sufficient to obtain precise input-output relation, while the system internals can be regarded as black box. However, if the function of an object is to be elucidated, then its structure and the relations between its parts must be described realistically".

detailed description is not required for the immediate purpose of explaining how the components of the mechanism produce the phenomenon in virtue of their properties, organization and activities; and whether it is safe to treat the mechanism as an independent module that can be separated from the system in which it is embedded and yet continue to produce the phenomenon for which it is responsible.

In the NF-κB regulatory mechanism example (Fig. 16.2 above), the key finding amounted to the realization that the initial negative feedback loop mechanism needs to be augmented to include a parallel pathway of activation not subjected to negative feedback, and that it takes the combined activity of both pathways in order to produce peaks of gene expression matching experimental observations.[5] The bottoming out argument here is that in order to produce the phenomenon of interest, the key requirement is that of a double activation pathway involving experimentally identified molecular components shown to be necessary for the production of the phenomenon and shown to interact in such a way as to make possible the double activation pathway. For the immediate purpose of explaining the phenomenon of interest, it is not essential to further understand why and how these molecular components interact the way they do, how these components were produced in the cell or how they evolved. The expectation here is that certain changes would not influence in any way the ability of the mechanism to produce its target phenomenon. Most notably, the NF-κB activator, its DNA binding motifs and the IκB inhibitor could tolerate significant changes in sequence and structure, yet the mechanism would continue to function undisturbed on condition that some key features are preserved, such as the dual activation pathway and the affinity and kinetics of chemical interactions.[6] There is therefore a clear sense in which certain lower-level details can be ignored and the phenomenon of interest can be satisfactorily explained in terms of higher-level description of mechanistic components, their properties, organization and activities.

Likewise, a tight quantitative match between the predictions of the model and experimental measurements support the claim that, at least relative to the timeframe in which the phenomenon is characterized, other mechanisms at work in the cell, as well as effects triggered downstream as a result of the functioning of the mechanism are not required to produce the phenomenon of interest or interfere with the ability to produce it. It is expected therefore that an in vitro reconstituted NF-κB regulatory mechanism should produce peaks of gene activation closely resembling those produced in vivo, thus acting as an independent module. Again, this specifies a sense in which a more systemic context can be ignored such that a satisfactory explanation can be focused on a local mechanism.

---

[5]For a more detailed discussion, see (Baetu 2015).

[6]This occurs, for example, when complementary mutations in several components rescue the wild-type phenotype.

16.5 Conclusion

Mathematical modeling provides an accessible substitute for something which is missing in biology: a rich theoretical apparatus from which one could derive detailed hypotheses capable of guiding experimental research from an initial description of the phenomenon of interest to the final explanation. In the absence of such theories, experimental research is bound to remain largely exploratory, and exploration implies a fundamental incertitude about how much is known and how much remains to be investigated. While it cannot rival with the all encompassing theories of physics, mathematical models can nevertheless provide a useful workaround by providing a principled way of evaluating the completeness of the information included in a mechanistic explanation, thus specifying where a mechanistic explanation can safely bottom out and what is the cutoff point beyond which external factors can be ignored.

Beyond the philosophical interest relative to the problem of reductionism, there are practical implications to be considered as well. During the discovery process, evidence that an explanation is satisfactory is an indication that the research project is on the right path. Before worrying about the countless ways in which a mechanistic explanation could be further detailed and expanded, it is crucial to gather at least some evidence that the proposed mechanism, at the level of composition at which it is described, can and does produce the phenomenon of interest. It would be misguided to try to understand how and why the components of a mechanism do what they are doing, how the mechanism and its organizational features came into being, and how the mechanism as a whole integrates the greater whole which the living organism, in the absence of evidence that the mechanism described in the proposed explanation can produce the phenomenon to be explained. At various points in project, researchers can stop, recompose the many bits and pieces of experimental results into mechanistic descriptions and then model these descriptions in order to gain at least a rough estimate of whether, thus far, they 'got things right' and the proposed mechanisms, at the level of composition at which they are described, can indeed produce the phenomena which they are supposed to explain. Furthermore, since mechanistic explanations often provide the rationale for developing technologies for gaining control over phenomena and medical treatments, evidence that the explanation is satisfactory is key for making an enlightened decision about how much trust to put on the probability of a successful outcome, especially when there is a little room for trial and error.

Acknowledgments This work was supported by a generous fellowship from the KLI Institute. I thank Stuart Glennan, Mathieu Charbonneau, Dan Nicholson, Maarten Boudry, Argyris Arnellos, Laura Nuño de la Rosa and Michael Rammerstorfer for their much appreciated input.

Bibliography

Baetu TM (2012) Filling in the mechanistic details: two-variable experiments as tests for constitutive relevance. Eur J Philos Sci 2(3):337–353

Baetu TM (2013) Chance, experimental reproducibility, and mechanistic regularity. Int Stud Hist Philos Sci 27(3):255–273

Baetu TM (2014) Models and the mosaic of scientific knowledge. The case of immunology. Stud Hist Philos Biol Biomed Sci 45:49–56

Baetu TM (2015) From mechanisms to mathematical models and back to mechanisms: quantitative mechanistic explanations. In: Braillard P-A, Malaterre C (eds) Explanation in biology. an enquiry into the diversity of explanatory patterns in the life sciences. Springer, Dordrecht

Baetu TM, Hiscott J (2002) On the TRAIL to apoptosis. Cytokine Growth Factor Rev 13:199–207

Baetu TM, Kwon H, Sharma S, Grandveaux N, Hiscott J (2001) Disruption of NF-kB signalling reveals a novel role for NF-kB in the regulation of TNF-related apoptosis-inducing ligand expression. J Immunol 167:3164–3173

Bechtel W (2012) Understanding endogenously active mechanisms: a scientific and philosophical challenge. Eur J Philos Sci 2:233–248

Bechtel W, Abrahamsen A (2005) Explanation: a mechanist alternative. Stud Hist Philos Biol Biomed Sci 36:421–441

Bechtel W, Abrahamsen A (2010) Dynamic mechanistic explanation: computational modeling of circadian rhythms as an exemplar for cognitive science. Stud Hist Philos Sci Part A 41:321–333

Bechtel W, Richardson R (2010) Discovering complexity: decomposition and localization as strategies in scientific research. MIT Press, Cambridge, MA

Brigandt I (2013) Systems biology and the integration of mechanistic explanation and mathematical explanation. Stud Hist Philos Biol Biomed Sci. doi:10.1016/j.shpsc.2013.06.002

Craver C (2001) Role functions, mechanisms, and hierarchy. Philos Sci 68:53–74

Craver C (2006) When mechanistic models explain. Synthese 153:355–376

Craver C (2007) Explaining the brain: mechanisms and the mosaic unity of neuroscience. Clarendon, Oxford

Darden L (2006) Reasoning in biological discoveries: essays on mechanisms, interfield relations, and anomaly resolution. Cambridge University Press, Cambridge

Elowitz M, Leibler S (2000) Synthetic Gene oscillatory network of transcriptional regulators. Nature 403:335–338

Glennan S (2002) Rethinking mechanistic explanation. Philos Sci 69:S342–S353

Hoffmann A, Levchenko A, Scott M, Baltimore D (2002) The I κB – NF-κB signaling module: temporal control and selective gene activation. Science 298:1241–1245

Hooker C (2011) Philosophy of complex systems. Elsevier, New York

Kitcher P (1984) 1953 and All that: a tale of two sciences. Philos Rev 93:335–373

Klipp E (2005) Systems biology in practice: concepts, implementation and application. Wiley, Weinheim

Machamer P, Darden L, Craver C (2000) Thinking about mechanisms. Philos Sci 67:1–25

McKay Illari P, Williamson J (2012) What is a mechanism? thinking about mechanisms *across* the sciences. Eur J Philos Sci 2(1):119–135

Morange M (2009) Synthetic biology: a bridge between functional and evolutionary biology. Biol Theory 4(4):368–377

Nicholson D (2013) Organisms ≠ machines. Stud Hist Philos Biol Biomed Sci. 44 (4):669–78

Oyama S (1985) The Ontogeny of information: developmental systems and evolution. Cambridge University Press, New York

Rosenberg A (2006) Darwinian reductionism, or, How to stop worrying and love molecular biology. University of Chicago Press, Chicago

Ting AY, Endy D (2002) Decoding NF-κB signaling. Science 298:1189–1190

Waters CK (1990) Why the Anti-reductionist consensus won't survive: the case of classical Mendelian genetics. In: Proceedings to the biennial meeting of the philosophy of science association, vol 1990, Volume One: Contributed Papers, pp 125–139

Weber M (2005) Philosophy of experimental biology. Cambridge University Press, Cambridge

Woodward J (2002) What is a mechanism? a counterfactual account. Philos Sci 69:S366–S377

Woodward J (2003) Making things happen: a theory of causal explanation. Oxford University Press, Oxford

Chapter 17
Causal and Mechanistic Explanations, and a Lesson from Ecology

Viorel Pâslaru

17.1 Introduction

The mechanistic perspective on scientific explanation is typically described as a reaction to the deductive-nomological view (Machamer et al. 2000; Bechtel and Abrahamsen 2005). The mechanistic perspective also defines itself by contrast to causal conceptions, and mechanisms are opposed to causes. Glennan (1996) seeks mechanisms to explain causation, while Machamer, Darden, and Craver (2000) – henceforth MDC – argue that the term "cause" is abstract and has to be specified in terms of more specific activities, such as push, carry, scrape, if it is to become meaningful. Subsequent developments integrate causation, in particular as conceived along counterfactual lines by Woodward (2000, 2003) to account for mechanisms. The work of Glennan (2002) and Craver (2007) illustrates this approach, while Woodward (2002) himself proposes a counterfactual account of mechanisms, conceiving of them as networks of causal relations understood counterfactually. He continues this approach in a recent response (Woodward 2011) to Waskan's (2011) arguments for maintaining mechanistic explanation distinct from counterfactual theories of explanation and causation.

This article contributes to the aforementioned debate by scrutinizing two recent examinations of the relationship between causal claims and description of mechanisms in scientific explanations. Jani Raerinne (2011) examines representative cases of research by ecologists, and argues that in ecology many causal explanations are "phenomenological" invariant generalizations that do not offer satisfactory explanations. Mechanistic explanations of ecological phenomena could prove satisfactory, but such explanations in ecology are undetermined by data, and hence, fall short of

V. Pâslaru (✉)
Department of Philosophy, University of Dayton, 300 College Park,
Dayton, OH 45469-1546, USA
e-mail: vpaslaru1@udayton.edu

© Springer International Publishing Switzerland 2015
I. Pârvu et al. (eds.), *Romanian Studies in Philosophy of Science*, Boston Studies
in the Philosophy and History of Science 313, DOI 10.1007/978-3-319-16655-1_17

expectations. Lindley Darden (2013) examines this issue too, but in the context of biology and medicine, with a special focus on the case of cystic fibrosis. Although she does not discuss ecological cases, it is important to examine her view, since she intends her conception of mechanisms and its relationship to causation to be general and applicable beyond the area of molecular and cell biology. Darden argues that causal claims of the kind "C causes E" do not offer satisfactory explanations, but descriptions of mechanisms do. Accordingly, causal talk has to be replaced with mechanistic talk in the sense of MDC.

I question these claims of Raerinne and Darden and argue that both causal and mechanistic perspectives are necessary to formulate scientific explanations and to account for the explanatory practice of scientists. My reasoning is based on examination of examples from ecology and it proceeds as follows: In Sect. 17.2, I outline the views of Raerinne and Darden on mechanisms and causality. I formulate four theses that I think summarize their stances on these topics. After that, in Sect. 17.3, I describe the use of structural equation modeling and of causal models for the study of causal structures. For illustration, I look at a study on pollination by Randal Mitchell and at a study on competition by Eric G. Lamb and James F. Cahill. In light of this examination, I show in Sect. 17.4 that the four theses that I take to express the views of Raerinne and Darden do not characterize adequately the nature and use of causal claims in ecology and that mechanistic talk cannot replace causal talk. Instead, it has to incorporate it for successful explanations and to account for the explanatory practice of ecologists.

17.2 Raerinne and Darden on Mechanisms and Causation

Jani Raerinne takes up a distinction made in philosophical literature between two types of causal explanation that he calls "simple causal claims" and "mechanistic causal explanations":

> A simple causal claim describes the causal connection between the phenomenon-to-be-explained and the thing that does the explaining. It refers to a 'phenomenological' or superficial causal explanation in which one has an invariant relation between variables, but no account — or mechanistic explanation — as to why or how the relation holds between the variables (Raerinne 2011, p. 264).

He understands causal claims and the corresponding causal relations in terms of Woodward's (2003) manipulationist account of causal explanation: causal claims describe causal dependency relationships between changes in the values of independent variables and changes in the values of dependent variables. To be explanatory, a causal claim has to be invariant under interventions on independent variables that bring about changes in the dependent variable.

As the name suggests, mechanistic explanations consist of descriptions of mechanisms of phenomena; they are causal and bottom-up explanations. A mechanistic explanation "describes the underlying mechanism *within* the system by showing how the system is constituted and how this produces the phenomenon-to-be-explained" (Raerinne 2011, p. 264). Just like causal explanations based on simple causal dependencies, mechanistic explanations are causal and rely on invariance and modularity. As for the relationship between the two types of explanation, Raerinne claims that mechanistic explanations complement causal explanations formulated in terms of simple causal claims, for the former describes how the dependency relation between the thing that does the explaining and the phenomenon-to-be-explained produces the latter. That is, a mechanism underlies a simple causal relationship, and a mechanistic explanation accounts for a simple causal relationship. Raerinne seems to conceive of mechanisms and mechanistic explanations as elaborate sets of simple causal relations and explanations, respectively, just like Woodward (2002) does with his counterfactual account of mechanisms.

Raerinne views the situation of explanation in ecology as wanting because "many causal explanations in ecology are simple causal claims in the sense that there are no known or confirmed mechanistic explanations, for how the causes of these explanations produce their effects" (2011, p. 267). He offers the several rules of the equilibrium theory of island biogeography — the area rule, the distance rule, the diversity-stability, the endemicity rule, and the intermediate disturbance rule — as illustrations of causal explanations in ecology. The rules of island biogeography are simple in the sense that they link two variables, where one variable is independent and represents the cause, while the other one is dependent and stands for the effect. That is, C causes E. For example, the area rule states that species numbers tend to increase with island area. Put in causal terms, the rule states that an increase in island area causes an increase in species numbers.

The following two theses summarize the foregoing:

R (1): Causal explanations in ecology consist of simple causal claims that offer "phenomenological" or superficial accounts of invariant dependency relations between variables, but no account of why or how the relationships hold.

R (2): Mechanistic explanations ought to describe invariant and modular causal structures that underlie the phenomenon-to-be-explained.

In light of important cases of ecological research, I show in Sect. 17.4 that R (1) does not apply to those cases and, hence, does not adequately characterize explanation in ecology. I agree with R (2), but Raerinne thinks that ecological mechanisms are undermined by data and poorly known. Ecologists have yet to offer accurate accounts of mechanisms. I disagree, given the examples of ecological research.

Lindley Darden argues that claims such as "C causes E" are impoverished compared to the claim that "this mechanism produces this phenomenon." The claim "A mutation in the CFTR gene causes cystic fibrosis" is impoverished by comparison to an account of the very large number of mechanisms involved in the production of the disease (2013). She conceives of mechanisms in terms of a characterization that has become a *locus classicus* in the literature and is known as the MDC view: "Mechanisms are entities and activities organized such that they are productive of regular changes from start or set up to finish or termination conditions" (Machamer et al. 2000, p. 3).

The problem with the talk in terms of "cause" and "effect" is that these words are general and they have to be specified by more specific causal terms. By doing so, one offers a more accurate account of scientific explanation. Darden shows how talk of "causes" and "effect" could be linked to "mechanism" and "phenomenon." In a mechanism, causes could be specified as activities, a mechanism at a lower level, an earlier stage of the mechanism, or "start or setup conditions." The effect is the phenomenon of interest that the mechanism produces (Darden 2013, pp. 24–26). Moreover, mechanisms are sought for three reasons: explanation, prediction, and control. Description of mechanisms goes through a process of recharacterization.

The same cases of ecological research that I use to challenge Raerinne's theses help me dispute two key claims that I think summarize Darden's view on the explanatory value of causal claims and of descriptions of mechanisms:

D (1): Causal claims of the form "C causes E" are impoverished claims about phenomena under scrutiny, while descriptions of underlying mechanisms *sensu* MDC offer satisfactory explanations.

D (2): Causal talk has to be replaced with talk of activities, sub-mechanisms, stages and setup conditions.

In Sect. 17.4, I show that D (1) does not accurately characterize ecologists' use of causal claims and their role in formulating causal explanations. D (2) cannot be applied to some population-level causal relationships and that renders causal-talk unavoidable. Note that R (1) and D (1) are similar in their assertion that causal claims do not offer satisfactory explanations. Only mechanistic explanations do. The two theses differ in that R (1) admits that causal explanations have some merit, even if "phenomenological," or superficial, and this merit can be accounted for in terms of Woodward's view on causation and causal explanation. By contrast, D (1) strips causal claims even of this virtue. Next, I examine the cases of ecological research that will help show the limitations of R (1), R (2), D (1) and D (2).

17.3 Causal Explanations and Mechanism Description in the Study of Competition and Pollination

Causal relationships and explanations are central in ecological research that employs structural equation modeling, or SEM for short. SEM is used to infer causes from observational, statistical data, to test causal hypotheses and to help formulate new hypotheses concerning causal structures (Shipley 2002; Grace 2006). As such, the SEM method leads to causal explanations of ecological phenomena. That this method is important in ecology is evidenced by its recent use in the investigation of various ecological problems, such as effects of natural selection (Scheiner et al. 2000), individual and environmental variability in observed populations (Cubaynes et al. 2012), the effect of competition on the life-history and fecundity of wild and hybrid or cultivated plant populations (Campbell and Snow 2007; Pantone

$$y_1 = \alpha_1 + \gamma_{11}x_1 + \zeta_1 \qquad (1.2)$$

$$y_2 = \alpha_2 + \beta_{21}y_1 + \gamma_{21}x_1 + \zeta_2 \qquad (1.3)$$

$$y_3 = \alpha_3 + \beta_{32}y_2 + \gamma_{31}x_1 + \zeta_3 \qquad (1.4)$$

Fig. 17.1 Structural equations that underlie a path diagram shown in Fig. 17.2. From Grace (2006, p. 11). *Structural equation modeling and natural systems.* Copyright © 2006 Cambridge University Press. Reprinted with permission

et al. 1992), the effects of seeding density on plant and panicle density and on final yield (Lamb et al. 2011), the strength of the interactions among species in natural communities, such as the strength of the direct and indirect effect of birds on other species of an intertidal community (Wootton 1994), the importance of male-male competition, female choice and male-female conflict in water striders (Sih et al. 2002), plant species richness in coastal wetlands (Grace and Pugesek 1997), the effect of plant biomass on seed number and germination success (Allison 2002), the factors that determine reproductive success and plantlet survival (Iriondo et al. 2003), the interspecific relationships between functional traits in succession (Vile et al. 2006), the ecological structures and the role of ecological processes (Arhonditsis et al. 2006), the direct and indirect effects of climate and habitat diversity on butterfly diversity (Menéndez et al. 2007), the environmental drivers of disease emergence (Plowright et al. 2008), and the effect of humans on terrestrial food webs (Muhly et al. 2013).

The first step in SEM is to conjecture causal relationships among variables in light of available knowledge about the phenomenon under scrutiny and to formulate a path model incorporating the conjectured causal relationships. Causal relationships are described by parameters that show the magnitude of the direct or indirect effect that independent variables (observed or latent) exert on dependent variables (observed or latent). In graphical representations, arrows of varying thickness indicate the magnitude of the effect exerted by independent variables. The proposed model of the conjectured causal relationships is then tested for fit with the observed data using a χ^2 test of model fit, and it is rejected if it does not agree with the data. Structural equations are used to calculate the effect of independent variables on the dependent variable. For example, structural equations (Fig. 17.1) calculate the values of the dependent variables linked by causal relations in a path model (Fig. 17.2).[1] The equations are interpreted causally in the sense that manipulations of x yield changes in the value of y, provided that there is no other path from y to x.

---

[1] In this model, boxes stand for observed variables, while arrows designate directional relationships. The latter are represented by equality signs in the structural equations. γ designates effects of x variables on y variables, β stand for effects of ys on other ys, and ζ indicate error terms for response variables (Grace 2006, p. 11).

Fig. 17.2 A path diagram
associated with structural
equations depicted in
Fig. 17.1. From Grace (2006,
p. 11). *Structural equation
modeling and natural
systems.* Copyright © 2006
Cambridge University Press.
Reprinted with permission

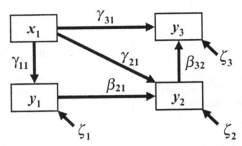

Figure 1.1. Graphical representation of a structural equation model (Eqs. (1.2)–(1.4) involving one independent (x_1) variable and three dependent (y) variables.

Randal Mitchell (1992, 1994) used SEM to examine casual relationships and to test hypotheses about the causal relationships between floral traits, pollination visitation, plant size and fruit production. He first formulated a path diagram, which is a causal model, of various causal relations possible between floral traits, pollinator behavior and fruit production, and which were conjectured based on prior knowledge of the case (Fig. 17.3). He then changed this scheme and produced a total of six causal diagrams by deleting or adding to it causal relations, as shown in Fig. 17.3. The six models that Mitchell examined express six different conjectures regarding the causal relations among the foregoing factors. Using structural equations and testing the six conjectured causal models for fit with the observed data, he eliminated five conjectures and their models, and identified the causal diagram that better fits the data. That diagram expresses the basic hypothesis according to which plant traits (floral nectar production rate, corolla size, number of open flowers, and inflorescent height) affect pollinator behavior (approaches and probes per flower), which may influence plant reproductive success through fruit production (proportion fruit set and total fruit set) (Fig. 17.4).

The solved path diagram (Fig. 17.4) shows that the causal relationship established in this case is not the simple "C causes E," but rather the complex of positive causal relations of various strength: "[((C1, & C2, & C3 & C4 & C5 & U) cause B1) & ((B1 & U) cause B2) & ((C4 & B1 & B2 & U) cause PS) & ((DM & U) cause C4) & ((DM & U) cause TF) & (TF & PS & U)] cause E," where C1–5 are corolla length, corolla width, nectar production, inflorescence height, number of open flowers, respectively; U symbolizes unknown factors; DM represents dry mass; TF indicates total flowers, and PS is proportion fruit set; B1 stands for pollinator approaches, and B2 for probes per flower, and E is the effect total fruit. To underscore that Mitchell uses the causal approach, it is worth mentioning that he explicitly takes the model to be one about causal relationships and causal mechanisms, which is an expression that he uses to designate networks of causal relationships (Mitchell 1992 pp. 123, 124).

A related example is due to Eric G. Lamb and James F. Cahill (2008) who likewise used SEM to examine the importance of the intensity of root competition in a rough fescue grassland community in structuring plant species diversity or

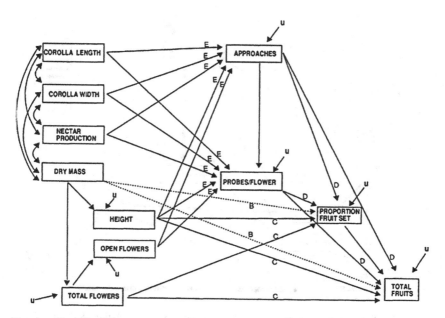

FIG. 1.—Hypothesized causal scheme for the relationships among plant traits, pollinator visitation, and reproduction through female function for *Ipomopsis aggregata* (model A). Each *single-headed arrow* indicates an effect of one variable on another; each *double-headed arrow* connects variables allowed to correlate with one another for reasons not considered in the diagram; *u* represents unexplained variation. The basic model includes all *solid arrows*. *Letters* near individual arrows indicate paths that can be deleted (or, for model B, added) to produce alternative nested models. For example, the paths labeled *C* are deleted from model A to produce model C. The *dotted arrows* are added for model B only and are not included in model A or other alternatives. Models F and G are not considered here.

Fig. 17.3 Initial model containing various possible causal paths between floral traits, pollinator behaviour and fruit production. From Mitchell (1994, p. 875). Effects of floral traits, pollinator visitation, and plant size on *Ipomopsis aggregata* fruit production. *The American Naturalist* 143(5):870–889; published by The University of Chicago Press for The American Society of Naturalists. Reprinted by permission of The University of Chicago Press

community composition.[2] They used structural equation modeling to examine how competition influences species richness, composition, and evenness, by situating these characteristics of communities within a wider set of environmental and

---

[2]Lamb and Cahill define *intensity of competition* as "the degree to which competition for a limited resource reduces plant performance below the physiological maximum achievable in a given environment." *Importance of competition* is "the effect of competition relative to other environmental conditions. ... competition can be considered important if variation in the intensity of competition is the cause of predictable variation in plant community structure" (Lamb and Cahill Jr, 2008, p. 778).

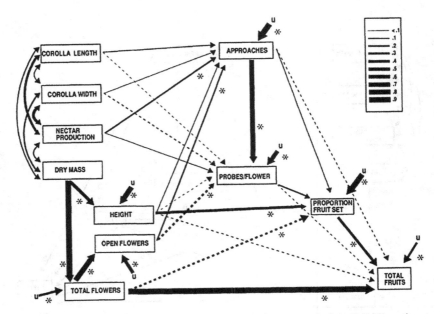

Fig. 2.—Solved path diagram for model A for the Almont population. *Solid lines* denote positive effects; *dashed lines* denote negative effects. Width of each line is proportional to the strength of the relationship (see legend), and paths differing significantly ($P < .05$) from zero are indicated with an *asterisk*. Actual values for path coefficients appear in table 2.

Fig. 17.4 Solved path diagram supporting the basic hypothesis. From Mitchell (1994, p. 879). Effects of floral traits, pollinator visitation, and plant size on *Ipomopsis aggregata* fruit production. *The American Naturalist* 143(5):870–889; published by The University of Chicago Press for The American Society of Naturalists. Reprinted by permission of The University of Chicago Press

plant conditions that are known to influence competition intensity and community structure, as shown in the path model (Fig. 17.5). This initial model contains only species richness as the dependent variable of primary interest because including evenness and composition in a single model would make it too complex. Instead, separate models for evenness and species composition were formulated.

A χ^2 test of model fit showed that the model does not fit the data adequately. To address this issue, Lamb and Cahill added new paths to models. The resulting models are represented in Fig. 17.6. In the species richness model, they introduced direct paths from site conditions to shoot biomass, soil moisture to species richness and from nitrogen treatment to soil moisture (Fig. 17.6A). The starting value of the path from total nitrogen to the site conditions variable was modified, which led to an adequate fit of the model for species evenness (Fig. 17.6B). To address the fit of the model for species composition, they added a path from community composition to light interception (Fig. 17.6C). Evenness, richness and composition had varying influence on root and shoot biomass and this in turn affected the coefficients of variables linked by paths to shoot and root biomass.

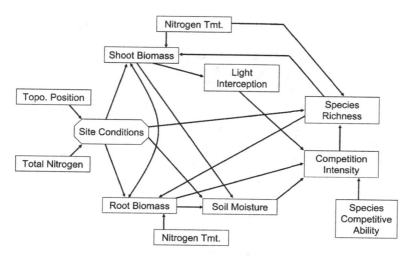

Fig. 17.5 Initial structural equation model. From Lamb and Cahill Jr (2008, p. 781). When Competition Does Not Matter: Grassland Diversity and Community Composition. *The American Naturalist* 171(6):777–787; published by The University of Chicago Press for The American Society of Naturalists. Reprinted by permission of The University of Chicago Press

Lamb and Cahill formulate these findings in terms that indicate a causal interpretation. They say that nitrogen treatment and soil moisture are the factors that "positively *influence*" species richness. Competitive intensity is the factor that in addition to nitrogen treatment and soil moisture "positively *influence*" species evenness. As for community composition, it is *affected* by environmental conditions, i.e., site conditions, and, in its turn, is linked to shoot and root biomass. In all three cases, species competitive ability, which is based on phytometer species identity, *influences* competition intensity. Furthermore, "[e]nvironmental conditions *strongly controlled* shoot and to a lesser extent root biomass, and a combination of environmental conditions and plant biomass *exerted strong control* on light interception and soil moisture" (Lamb and Cahill Jr 2008, pp. 782–784). Moreover, when reviewing the contribution of other authors to the problem they study, Lamb and Cahill say: "...competition can be considered important if variation in the intensity of competition is the *cause* of predictable variation in plant community structure."; "Plant community structure is generally *under the control* of complex networks of interaction among factors ranging from soil and environmental conditions to disturbance regimes, herbivory, litter and standing shoot biomass."; "...competition is an important *factor controlling* plant community diversity and competition in rough fescue grassland."[3]

In the next section, I examine the implications that the research by Mitchell, Lamb and Cahill has for the theses of Raerinne and Darden.

---

[3]Italics added throughout the paragraph for emphasis.

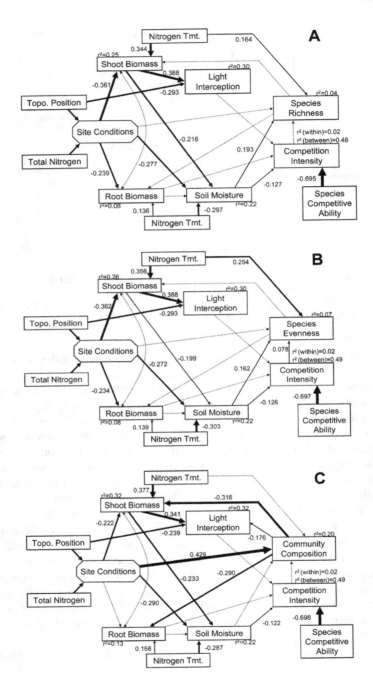

17.4 R (1) & R (2), D (1) & D (2) *versus* Ecological Research

Research of Mitchell, and of Lamb and Cahill shows that neither R (1) nor D (1) adequately characterize the causal claims that they make and their explanatory role. Being good examples of the use of SEM to articulate explanations of ecological phenomena, their work indicates that explanations in ecology do not consist only of simple causal claims, but complex causal claims that can be assimilated to descriptions of mechanisms. I look first at the example of Mitchell, and then turn to Lamb and Cahill.

The causal claim that Mitchell makes is not about a simple, binary "C causes E" relationship, as both R (1) and D (1) would make us expect, but rather about the more complex "[((C1, & C2, & C3 & C4 & C5 & U) cause B1) & ((B1 & U) cause B2) & ((C4 & B1 & B2 & U) cause PS) & ((DM & U) cause C4) & ((DM & U) cause TF) & (TF & PS & U)] cause E," and that is expressed by a causal model (Fig. 17.4). Although this claim captures a dependency relationship, it is not a superficial, or impoverished one as the relationship that would only relate two variables, say, the probes per flower to the proportion fruit set. Instead, Mitchell's causal model is a complex model that cites six properties of flowers, two types of pollinator behaviors, and causal relationships, and that links all of these causal factors in a certain way to account for the relationship between pollinator behavior and fruit production. The simple number of causal factors and their causal structure that Mitchell cites to account for the phenomenon under scrutiny — amount of total fruits — is on a par with descriptions of mechanisms that are not superficial or impoverished. Another reason for speaking against the superficiality of Mitchell's complex causal claim, and for the inadequacy of R (1) and D (1) in this case, is the fact that it is a result of testing six causal diagrams that expressed six different hypotheses about the causal relationships among factors responsible for fruit production. Those tests ruled out five of the conjectured causal links. The remaining sixth diagram is a complex causal claim that does not just express an observed correlation.

The work of Lamb and Cahill offers another ground for the inadequacy of R (1) and D (1) in the context of causal explanations in ecology. They examined three causal relations that both Raerinne and Darden would deem as simple, superficial

---

Fig. 17.6 Solved structural equation models for species richness (**A**), species evenness (**B**), and plant community composition (**C**). Dotted arrows represent paths that are not significant, while continuous arrows denote significant paths. The thickness of arrows indicates the degree of significance, which is also shown by coefficients. Thicker arrows represent more significant paths, while thinner ones stand for less significant ones. This graphical representation shows the causal relations and the factors that affect the dependent variables of community structure From Lamb and Cahill Jr (2008, p. 783). When Competition Does Not Matter: Grassland Diversity and Community Composition. The American Naturalist 171 (6): 777–787; published by The University of Chicago Press for The American Society of Naturalists. Reprinted by permission of The University of Chicago Press.

and impoverished: "competition intensity controls species richness," "competition intensity controls species evenness," and "competition intensity controls community composition." However, Lamb and Cahill examined each of these relationships in the context of a network of interactions among environmental and community factors (soil and topographical position, nitrogen treatment) to assess the importance of competition intensity. Their research established that competition intensity affects species evenness, but not richness or community composition. They also established how significant are the paths connecting other factors, such as site conditions, soil moisture, etc. to the dependent variables of interest. Last, but not the least, their work showed that for the initial model of causal paths to account for the observed data, they had to add several paths. In particular, in all three models (Fig. 17.6) they added a path from topographical position to light interception, and another path from soil moisture to the dependent variable, and a path from nitrogen treatment to soil moisture. The final model for community composition adds a link between light interception and community composition. The addition of these paths that represent in the model causal relations is an indication of the fact that competition intensity is not the only cause of species evenness, richness, or community composition and that other factors are instrumental as well. This example shows that ecologists examine even simple causal relations in a complex and structured network of causal factors. Simple causal relations are explanatory precisely because they are situated in such a causal network. Considering Lamb and Cahill's simple causal relationships in this context, they turn out to be anything but superficial or impoverished.

I claimed earlier that I accept R (2), but disagree with Raerinne on the empirical support for this thesis. R (2) is a normative statement, but the practice of ecologists, he argues, shows that "most explanation in ecology are undetermined by data or lacking in data" and "there are no known or confirmed mechanistic explanations" (Raerinne 2011, p. 267). Fortunately, causal diagrams that are part and parcel of SEM and exemplified by Mitchell, Lamb and Cahill vindicate R (2). Raerinne accepts Woodward's counterfactual account of representations of mechanisms as account of mechanistic explanation. Since R (2) simply expresses Woodward's conception of explanation, I argue that R (2) is correct by showing that the aforementioned causal models satisfy Woodward's counterfactual account of mechanisms and stress the empirical support of the models.

Woodward defines representations of mechanisms as follows:

(MECH) a necessary condition for a representation to be an acceptable model of a mechanism is that the representation (i) describe an organized or structured set of parts or components, where (ii) the behavior of each component is described by a generalization that is invariant under interventions, and where (iii) the generalizations governing each component are also independently changeable, and where (iv) the representation allows us to see how, in virtue of (i), (ii) and (iii), the overall output of the mechanism will vary under manipulation of the input to each component and changes in the components themselves. (2002, p. S375)

The causal path diagrams that Mitchell, Lamb and Cahill use satisfy MECH. The gist of MECH is that a mechanism should be decomposable into parts or modules that can be independently changed and the overall output of the mechanism varies as a result of changes to the modules. Each diagram used in the ecological examples I examined represents an organized set of components and their behaviors (i). In Mitchell's example, for instance, the solved path diagram is an organized set of features of plants and behaviors of pollinators. Each component and behavior is described by generalizations that link them to other components and/or behaviors (ii). The behavior 'approaches' is described as dependent on corolla length, corolla width, nectar production, plant height and open flowers. The generalization describes the link between approaches and the rest of components and behaviors as invariant under interventions. One can change, say, the corolla length, and that will affect approaches, yet the relationship between the two will stay invariant, as long as changes to corolla length are within a certain range. Furthermore, one can change the link between nectar production and approaches independently from the link between corolla width and approaches (iii). And the entire causal diagram allows us to see how the overall output of total fruits varies as a result of manipulating components and behaviors that make up the organized set of components and behaviors that the causal diagram represents (iv). Since Woodward takes **MECH** to specify the conditions for a model to be an acceptable representation of a mechanism, and causal graphs used by ecologists satisfy **MECH**, as explained above, it follows that the causal diagrams are models of mechanisms, and the explanations articulated by their means are mechanistic explanations.

The use of causal models in ecology also addresses Raerinne's concern about the lack of empirical support of mechanistic explanations in ecology. He does not elaborate on the standards of confirmation or of the relationship between data and mechanistic explanation, but the cases that I considered offer reasons to be optimistic about the empirical support of mechanistic explanations. Mitchell tested six models for fit with observational data, rejected five of them and settled on the one that better accounted for the data (Fig. 17.4). Lamb and Cahill tested their models for fit with observational data as well, and had to modify them, producing a version that better fits the data (Fig. 17.6). In addition, all models in both cases were formulated in light of prior empirical knowledge about the organisms under scrutiny.

Employment of causal models in ecological explanations offers several reasons to question D (2). I explain these reasons against the backdrop of assuming that "causal talk" is more than thinking in terms of "C causes E" and articulating such reasoning, but it comprises causal modeling as illustrated in Sect. 17.3. First, causal models satisfy important features of mechanistic explanation that Darden defends. Second, causal talk cannot be replaced with talk of activities, sub-mechanisms, stages, and set-up conditions. Here is the more detailed examination of these reasons.

According to Darden's characterization of mechanisms, mechanisms (a) produce a phenomenon, (b) consist of entities and activities, (c) that are organized spatially and temporally, (d) description of the mechanism goes through recharacterization and reevaluation, and (e) mechanisms are sought for explanation, prediction, and control. In line with (a), Mitchell's causal models show what factors produce reproductive success in plants, i.e., total fruit sets, while the models by Lamb and Cahill reveal what factors are responsible for variation in species diversity and community composition. That is, the phenomena for which causal models are sought are total fruit sets, in one case, and species diversity and community composition, in the other case. As required by (b), causal models represent both entities and activities. Mitchell's models list approaches and probes per flower, nectar production, which are activities, and the entities flowers, corolla, and fruit set. Moreover, to offer a more detailed account, the models contain properties of entities: length, width, and height. Similarly, Lamb and Cahill list light interception as an activity of main interest, and the entities coupled with their properties: shoot biomass, soil moisture, total nitrogen, etc. Causal models under scrutiny focus primarily on the causal organization of entities and activities. The models are careful to specify which entity, property or activity is at the receiving end and which one exerts the causal influence, for any change in this organization can result in a causal model that does not account for the phenomenon under scrutiny. It matters for the adequacy of the model whether dry mass affects total fruits directly, or via height and total flowers (Fig. 17.4). Likewise, it matters whether it is shoot biomass that affects soil moisture, rather than vice versa (Fig. 17.6). While the causal organization is a constitutive one and does not stress the spatial and temporal organization, the latter are implied, and should they play an important role in producing a phenomenon, they can be easily incorporated in causal models. For example, Mitchell's solved model (Fig. 17.5) indicates temporal organization when it implies that probes per flower have to occur *before* a plant can produce fruits. Lamb and Cahill's model (Fig. 17.6) shows that spatial organization can be explicitly incorporated in the model as suggested by the variable topological position. Causal models contain those organizational aspects that researchers find relevant in the cases they investigate. Figures 17.5 and 17.6 emphasize causal organization, while other causal models can incorporate spatial and/or temporal organization if deemed relevant. What is important is that entities and activities are organized, and this matches the spirit of (c). Description of the aforementioned causal models goes through recharacterization and reevaluation, as described by (d). Any formulation of a causal model begins with a tentative model that is modified following tests for fit with data, even if the terminology used to refer to the two types of models is different from the one applied to the case of mechanisms. Mitchell calls the tentative model a *hypothetical causal scheme*, while the final one is a *solved path diagram*. Darden uses *sketch* and *schemata*, correspondingly. While a sketch of a mechanism contains black boxes for components to be identified, a hypothetical causal schema contains more causal relations than there are, or misses some, yet both are similar in that they explore possible structures and are tentative. Furthermore, a solved path diagram is the final destination of an investigation that uses SEM, just as a scheme filled in with descriptions of the relevant parts and entities is the end result of mechanistic accounts.

Neither Mitchell nor Cahill and Lamb discuss the use of their models for the purpose of predicting outcomes of intervention in nature or for controlling nature. Their primary goal is to use causal models to explain reproductive success in plants and why root competition is not important in determining species richness and community composition. Yet since findings of ecology are used in practical applications, such as conservation and restoration which involve prediction and control, their causal models can be seen as suitable for such applications. In fact, other ecologists use causal models for prediction, explanation and management, as shown by the work of James B. Grace (Grace and Pugesek 1997; Grace 2008, 2006). Consequently, causal models are sought for explanation, prediction, and control, just as (e) requires of mechanisms. Causal talk using the language of causal models is far from being poor; it satisfies the desiderata of the mechanistic view.

I turn next to showing that causal talk understood in the broader sense as illustrated above cannot be replaced with talk of activities, sub-mechanisms, and set-up conditions. In fact, the latter require the former.

MDC characterize mechanisms using qualitative models of them, yet models of this kind have limitations: they do not contain quantitative information that enables prediction. (For a related objection see Gebharter and Kaiser [2014, pp. 82–83]). Darden (2013) admits the use of computational simulation models for quantitative predictions (p. 23), but these models are not causal. Causal models used in SEM, however, combine both qualitative and quantitative virtues. They are able to represent all the relevant characteristics of mechanisms along with path coefficients that are necessary for prediction and explanation. The MDC view cannot do this, since it does not accept causal models as necessary elements of final mechanistic explanation, but requires causes to be specified as activities, and is not working with path coefficients. For the MDC view to be more comprehensive, it has to integrate causal models and path coefficients.

Woodward questioned the ability of the mechanistic view such as the one proposed by MDC to account for the overall relationship between start and termination conditions using bottom out activities. The overall relationship is not an activity, and it is not plausible to claim that it is productive if the start condition is connected to the termination condition via a series of intermediate activities (Woodward 2002, pp. S372–S373). This objection is particularly important in connection with examples from ecology where the relationship between start and termination conditions is the focus of investigation rather than the intermediate activities, or is as important as the latter. Ecologists are interested in how changes in start conditions, such as availability of nutrients, prey, predators, or changes in environmental conditions, or in initial densities of populations affect termination conditions such as competitive exclusion, or lack thereof, increase or decrease in the abundance of a population, or co-occurrence of two species. To show this, I will consider an example of experimental research on competition by David Tilman and David Wedin (1991). They examined the mechanisms of nitrogen competition among four grass species by planting *Agrostis scabra* in pair with three other grass species: *Agropyron repens*, *Schizachyrium scoparium* and *Andropogon gerardi*. Grass pairs were subjected to several environmental conditions and treatments. In particular, they modified the soil composition and produced eight mixtures containing different proportions of topsoil; they used three seedling ratios of grasses of different species (80 % and 20 %, 20 % and 80 %, 50 % and 50 %); and three levels of nitrogen treatment, which was the only limiting resource. Two seedling densities (3,000 and 600 seedlings/m^2) were used to examine the competition between two grass species: *Agrostis scabra* and *Agropyron repens*, but only one seedling density (3,000 seedlings/m^2) was used to study pairwise competition between three species: *Agrostis scabra* and *Schizachyrium scoparium* and *Andropogon gerardi*. Except in a few cases, the common outcome of these experiments was the competitive displacement of *Agrostis*. When paired with *Schizachyrium* or *Andropogon*, *Agrostis* was displaced independent of initial seedling ratios and despite the fact that it inhibited the growth of the other two species in 1986 and 1987 (Fig. 17.7). *Agropyron* almost displaced *Agrostis* on nitrogen level (N-level) 3, but persisted on levels 1 and 2 (Fig. 17.8), which points to the two species having similar competitive abilities. Tilman and Wedin explain the dynamics of competition in mechanistic terms. *Schizachyrium* and *Andropogon* displaced *Agrostis* because they have higher root biomass and are better nitrogen competitors than *Agrostis*. The former species are poor colonists, for they produce few seeds. By contrast, *Agrostis* allocates resources to seed production and is as a result a successful colonist of abandoned fields and occupies them in the first two years. *Agropyron* is a good colonist as well due to high allocation to rhizomes through which it spreads. The determinant factor that allows *Agropyron* to displace *Agrostis* is that it produces rhizomes that can penetrate deep litter, while *Agrostis* cannot do that.

Description of mechanisms responsible for the dynamics of competition does not eliminate the need to specify an overall causal relationship, as a closer examination of the work of Tilman and Wedin shows. They investigate how changes in the start conditions – planting of seeds of two different species – affect the termination condition of competitive exclusion. This overall relationship is causal in the manipulationist sense of causation. Displacement is an effect of the initial planting of two species with different competitive abilities. Had only one species been present, or had one intervened to eliminate one of the two species, there would have been no competitive exclusion. MDC requires specifying causes as activities. However, there is no productive activity that links the start condition directly to the termination condition of competitive displacement, and MDC lacks an alternative concept of causation that would account for the overall causal relationship. Yet it is important to acknowledge this causal relationship, since it is the focus of Tilman and Wedin's examination, and it is required for understanding their research. They describe the productive activities that plants engage in, as well as the mechanisms that they constitute to account for the overall relationship that they determine experimentally. This relationship also guides the identification of productive activities and mechanisms. Had they investigated a different phenomenon, they would have either identified different activities and mechanisms, or used them differently in their account. Moreover, this overall causal relationship illustrates numerous other similar overall causal relationships that ecologists scrutinize, such as the quality of the environment and the type of interaction between plants; biodiversity and the risk of cascading extinctions; the distance between islands and mainland and rate of immigration or extinction; and the presence of mycorrhizal fungus and species composition and diversity.

Description only of individual activities that make up the productive continuity does not reveal another aspect of the overall causal relationship that Tilman and Wedin see as important. They observe that the long-term outcome of competition,

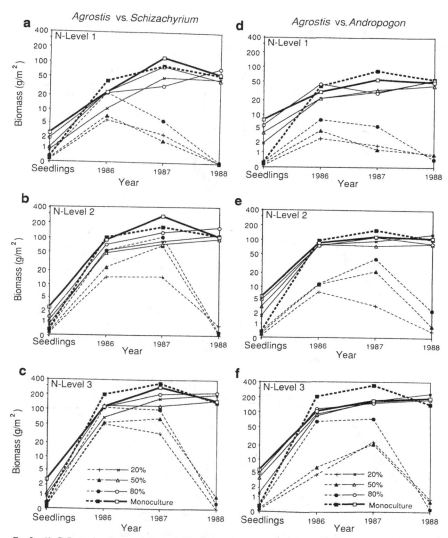

FIG. 2. (A–C) Dynamics of monocultures and of seed vs. seed competition between *Agrostis scabra* (broken lines and solid symbols or +) and *Schizachyrium scoparium* (solid lines and open symbols or ×) at three different soil nitrogen levels. Seedling biomass is the seed embryo mass per unit area for May 1986, based on the observed germination rate in each plot. All other values are aboveground biomass at the time of harvest. All values shown are averages over all soil mixtures within an N level. The four curves shown for each species are results for monocultures (squares and thicker lines), for plots with initial seed abundance (by number of seed) of 20% for a species (+, ×), for plots with initial seed densities of 50% for both species (triangles), and for plots with initial seed density of a species of 80% (circles). Note that a plot that had 20% of one species had 80% of the other, meaning that the 20% line for one species is associated with the 80% line for the other species. Independent of initial seedling ratio, *Agrostis* was displaced by *Schizachyrium*. (D–F) Competition between *Agrostis* (broken lines and solid symbols) and *Andropogon* (solid lines and open symbols), with results presented as described above.

Fig. 17.7 Dynamics of competition between *Agrostis* and *Schizachyrium*, and between *Agrostis* and *Andropogon* on three levels of nitrogen and in plots with different seed densities. From Tilman and Wedin (1991, p. 1042). Dynamics of nitrogen competition between successional grasses. *Ecology* 72(3):1038–1049. Copyright by the Ecological Society of America

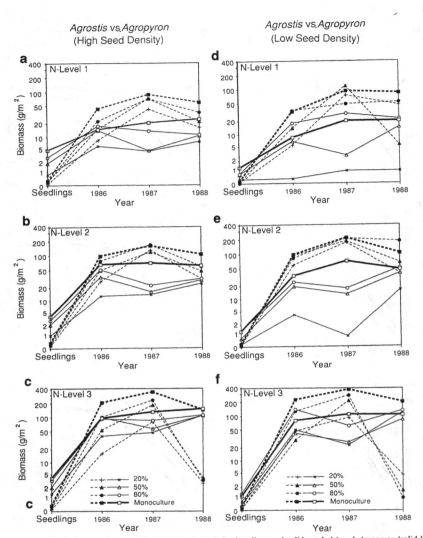

Fig. 4. (A–C). Seed vs. seed competition between *Agrostis* (broken lines and solid symbols) and *Agropyron* (solid lines and open symbols) when planted at high seed density (3000 seeds/m²). See legend to Fig. 2 for details of data presentation. (D–F). Seed vs. seed competition between *Agrostis* (broken lines and solid symbols) and *Agropyron* (solid lines and open symbols) when planted at low seed density (600 seeds/m²).

Fig. 17.8 Dynamics of competition between *Agrostis* and *Agropyron* at high and low seed density and on three nitrogen levels. From Tilman and Wedin (1991, p. 1045). Dynamics of nitrogen competition between successional grasses. *Ecology* 72 (3):1038–1049. Copyright by the Ecological Society of America

i.e., displacement of *Agrostis* was independent of changes in initial conditions, such as seed densities and abundances although they influenced the dynamics of pairwise interaction (Tilman and Wedin 1991, p. 1046). *Schizachyrium* displaced *Agrostis* regardless of whether the initial abundance of the latter in a plot was at 80 %, 50 %, or 20 % (Fig. 17.7a–c). When planted both at high and low seed density, *Agrostis* reached ultimately very low biomass, less than 5 g/m^2, which amounts to displacement. Moreover, Figs. 17.7 and 17.8 also show that the variation in nitrogen level had an effect on the dynamics of interaction between pairs of species, but did not cancel competitive displacement. For example, plots with initial seed density of 50 % of *Agrostis* had on N-level 2 a biomass of 20 g/m^2 in 1986, of about 60 g/m^2 in 1987, but 0 g/m^2 in 1988. On N-level 3, however, *Agrostis* had a biomass of 50 g/m^2 in 1986, about 70 g/m^2 in 1987, but only 1 g/m^2 in 1988 (Fig. 17.7b,c). Articulated in terms of Woodward's (2006) account of insensitivity of causation, this is an overall causal relationship between initial conditions and competitive displacement that is invariant and insensitive to changes in seed densities, seed abundances, and nitrogen level. Yet, as I already showed, the mechanistic conception focused on activities does not have the means to account for the overall causal relationship. Arguably, it could offer a schema of the overall causal relationship, schemas being truncated abstract descriptions that can be completed with additional descriptions of known parts and activities. This solution is unlikely to work, because the mechanistic view does not have the notions of insensitivity and invariance. Even if it assumed them, a schema of the overall causal relationship would be constructed by removing details about it, but the notion of insensitivity and invariance does not remove the detail. Instead, it specifies the changes to which the relationship is insensitive and invariant.

From the foregoing it follows that the causal talk cannot be replaced in with talk of activities, sub-mechanisms, stages and set-up conditions. Instead, causal talk has to complement the latter.

17.5 Conclusion

In the foregoing sections, I showed that four theses on the relationship between causal relations and mechanisms, and between causal and mechanistic explanations that can be found in the articles by Raerinne and Darden are not applicable to some important cases of ecological research. Ecologists do more than just cite simple causal dependencies, and even when they focus on simple causal relationships, they are investigated as part of complex causal networks. As a result, the explanations that the causal models articulate are not superficial or trivial. Rather, ecologists' explanations often consist of complex causal claims articulated by means of intricate causal models. Furthermore, causal talk cannot be replaced with a mechanistic discourse *sensu* MDC. Instead, it is necessary to produce a more complete account of the mechanisms underlying the phenomena under scrutiny. Causal models used in SEM represent the features of mechanisms as required by MDC and, in addition, incorporate quantitative information required for prediction and explanation. They also capture the overall causal relationship between start and termination conditions, as well as its invariance and insensitivity.

Woodward (2011) argues that a more adequate characterization of the notion of mechanisms, that is necessary to account for scientific explanation, could result from integrating aspects of both causal and mechanistic perspectives. The foregoing examination lends support to Woodward's proposal. Causal perspective is a necessary element in formulating mechanistic explanations of ecological and of other similar phenomena. If the term *causal* is reserved for counterfactual accounts that seek to establish dependency relationships between two events, use causal graphs and SEM, but without consideration of the intimate connection between the cause and its effect; and *mechanistic* is reserved for accounts that look at the productive activities that link the cause and the effect, then the ecological examples show that both are needed to furnish an explanation.

Acknowledgements I am grateful to Iulian D. Toader for advice on improving the final version of the article, and for patience while the improvements materialized. I also thank Diane Dunham for suggestions on how to correct deficiencies in my writing.

References

Allison VJ (2002) Nutrients, arbuscular mycorrhizas and competition interact to influence seed production and germination success in Achillea millefolium. Funct Ecol 16(6):742–749

Arhonditsis GB, Stow CA, Steinberg LJ, Kenney MA, Lathrop RC, McBride SJ, Reckhow KH (2006) Exploring ecological patterns with structural equation modeling and Bayesian analysis. Ecol Model 192(3):385–409

Bechtel W, Abrahamsen A (2005) Explanation: a mechanist alternative. Stud Hist Philos Biol Biomed Sci 36:421–441

Campbell LG, Snow AA (2007) Competition alters life history and increases the relative fecundity of crop–wild radish hybrids (Raphanus spp.). New Phytol 173(3):648–660

Craver CF (2007) Explaining the brain: mechanisms and the mosaic unity of neuroscience. Oxford University Press, New York

Cubaynes S, Doutrelant C, Grégoire A, Perret P, Faivre B, Gimenez O (2012) Testing hypotheses in evolutionary ecology with imperfect detection: capture-recapture structural equation modeling. Ecology 93(2):248–255

Darden L (2013) Mechanisms versus causes in biology and medicine. In: Chao H-K, Chen S-T, Millstein RL (eds) Mechanism and causality in biology and economics. Springer, Dordrecht, pp 19–34

Gebharter A, Kaiser MI (2014) Causal graphs and biological mechanisms. In: Kaiser MI, Scholz OR, Plenge D, Hüttemann A (eds) Explanation in the special sciences. Springer, Dordrecht, pp 55–85

Glennan S (1996) Mechanisms and the nature of causation. Erkenntnis 44:49–71

Glennan S (2002) Rethinking mechanistic explanation. Philos Sci 69:S342–S353

Grace JB (2006) Structural equation modeling and natural systems. Cambridge University Press, Cambridge

Grace JB (2008) Structural equation modeling for observational studies. J Wildl Manag 72(1):14–22

Grace JB, Pugesek BH (1997) A structural equation model of plant species richness and its application to a coastal wetland. Am Nat 149:436–460

Iriondo JM, Albert MJ, Escudero A (2003) Structural equation modelling: an alternative for assessing causal relationships in threatened plant populations. Biol Conserv 113(3):367–377

Lamb EG, Cahill JF Jr (2008) When competition does not matter: grassland diversity and community composition. Am Nat 171(6):777–787

Lamb E, Shirtliffe S, May W (2011) Structural equation modeling in the plant sciences: an example using yield components in oat. Can J Plant Sci 91(4):603–619

Machamer P, Darden L, Craver CF (2000) Thinking about mechanisms. Philos Sci 67(1):1–25

Menéndez R, González-Megías A, Collingham Y, Fox R, Roy DB, Ohlemüller R, Thomas CD (2007) Direct and indirect effects of climate and habitat factors on butterfly diversity. Ecology 88(3):605–611

Mitchell RJ (1992) Testing evolutionary and ecological hypotheses using path analysis and structural equation modelling. Funct Ecol 6(2):123–129

Mitchell RJ (1994) Effects of floral traits, pollinator visitation, and plant size on *Ipomopsis aggregata* fruit production. Am Nat 143(5):870–889

Muhly TB, Hebblewhite M, Paton D, Pitt JA, Boyce MS, Musiani M (2013) Humans strengthen bottom-up effects and weaken trophic cascades in a terrestrial food web. PLoS One 8(5):e64311

Pantone DJ, Baker JB, Jordan PW (1992) Path analysis of red rice (Oryza sativa L.) competition with cultivated rice. Weed Sci 40(2):313–319

Plowright RK, Sokolow SH, Gorman ME, Daszak P, Foley JE (2008) Causal inference in disease ecology: investigating ecological drivers of disease emergence. Front Ecol Environ 6(8):420–429

Raerinne J (2011) Causal and mechanistic explanations in ecology. Acta Biotheor 59(3–4):251–271

Scheiner SM, Mitchell RJ, Callahan HS (2000) Using path analysis to measure natural selection. J Evol Biol 13(3):423–433

Shipley B (2002) Cause and correlation in biology: a user's guide to path analysis, structural equations and causal inference. Cambridge University Press, Cambridge, UK

Sih A, Lauer M, Krupa JJ (2002) Path analysis and the relative importance of male–female conflict, female choice and male–male competition in water striders. Anim Behav 63(6):1079–1089

Tilman D, Wedin D (1991) Dynamics of nitrogen competition between successional grasses. Ecology 72(3):1038–1049

Vile D, Shipley B, Garnier E (2006) A structural equation model to integrate changes in functional strategies during old-field succession. Ecology 87(2):504–517

Waskan J (2011) Mechanistic explanation at the limit. Synthese 183(3):389–408

Woodward J (2000) Explanation and invariance in the special sciences. Br J Philos Sci 51:197–254

Woodward J (2002) What is a mechanism? A counterfactual account. Philos Sci 69:S366–S377

Woodward J (2003) Making things happen: a theory of causal explanation. Oxford University Press, Oxford/New York

Woodward J (2006) Sensitive and insensitive causation. Philos Rev 115(1):1–50

Woodward J (2011) Mechanisms revisited. Synthese 183:409–427

Wootton JT (1994) Predicting direct and indirect effects: an integrated approach using experiments and path analysis. Ecology 75(1):151–165

Chapter 18
Against Harmony: Infinite Idealizations and Causal Explanation

Iulian D. Toader

18.1 Introduction

The idea that some of the causal factors that are responsible for the causal production of a natural phenomenon are explanatorily irrelevant is an old one. It goes back to J. L. Mackie and Alan Garfinkel, who both cite Mill as a precursor.[1] But it has been recently revamped by Michael Strevens.[2] According to him, the omission or distortion of irrelevant causal factors, that is factors considered to make no difference to the occurrence of a phenomenon, brings about scientific understanding, by increasing the explanatory power of a causal model of that phenomenon. On Strevens' view, as we will see below in more detail, omission is a means for the optimization of the causal model, while distortion is a means for its idealization. These two procedures are alleged to bring about understanding in harmony with each other, in the sense that both the optimized and the idealized models of a natural phenomenon can represent the causal relationships of the system of interest.[3]

In this paper, I first spell out this claim of methodological harmony and then offer an argument against it, based on the standard explanation of phase transitions in statistical mechanics. Briefly put, I contend that this explanation makes clear the fact that idealization does not merely distort irrelevant causal factors. Rather, it eliminates relevant causal factors as well, that is factors that are considered to

---

[1] See Mackie (1980), ch 3, and Garfinkel (1981), ch. 5. See also Cartwright (1983).

[2] See Strevens (2008).

[3] An optimized model is called "canonical" in Strevens (2008), but I will not follow that terminology here.

I.D. Toader (✉)
Department of Theoretical Philosophy, University of Bucharest, Bucharest, Romania
e-mail: itoad71@gmail.com

© Springer International Publishing Switzerland 2015
I. Pârvu et al. (eds.), *Romanian Studies in Philosophy of Science*, Boston Studies
in the Philosophy and History of Science 313, DOI 10.1007/978-3-319-16655-1_18

make a difference to the occurrence of phase transitions. On the assumption that an optimized model can represent the causal relationships of a system, this contention implies that there is no harmony between optimization and idealization.

My argument goes against a suggestion recently made by Craig Callender: "It is possible that we could understand the standard explanation of phase transitions as a distortion that nonetheless successfully represents the causal relationships of the system. Perhaps the thermodynamic limit is legitimatized by the fact that surface effects aren't a difference-maker (in the sense of Strevens) in the systems of interest."[4] But, I submit, if surface effects are considered to make no difference to the occurrence of phase transitions, then the idealized model obtained by ignoring these effects fails to represent the causal relationships of the systems of interest. More generally, the argument not only rejects the harmony between optimization and idealization, but also emphasizes an essential misalignment between our current explanatory practices and the causal structure of the world. I end the paper by pointing to what I think is the deep philosophical significance of this misalignment – a fundamental tension between scientific objectivity and understanding.

18.2 Optimization and Idealization, à la Strevens

In this section, I spell out the claim that optimization and idealization, as Strevens conceives of them, are in harmony with each other. I start by presenting these two procedures, then I explain the relations between idealized models and the optimized ones from which they are derived: extensional equivalence and intensional inequivalence. Finally, I argue that harmony may be seen as grounded in extensional equivalence.

According to Strevens, an optimized model requires basically three things: first, a veridical causal model that represents a causal mechanism, i.e., a mechanism that is responsible for the causal production of the natural phenomenon one wants to understand; secondly, the elimination of explanatory irrelevancies, i.e., of everything that fails to make a difference to whether the phenomenon obtains or not; and lastly, a procedure of optimization, whereby actual parameter values are replaced with ranges of values. This is the procedure that one deploys for eliminating non-difference-makers from a veridical causal model and building an optimized one. Before we look at an example, let's see what this elimination might amount to.

Since a veridical causal model is a set of true statements about the physical system wherein the phenomenon of interest obtains, more exactly, the statements describing the causal relationships of the system, the elimination of non-difference-makers via optimization amounts to adjusting this set such that some statements that give actual values for some parameters are replaced by other statements that, instead, provide ranges of values for the same parameters. This replacement, by

---

[4] See Callender and Menon (2013), 210.

itself, does not decrease the veridicality of the causal model. Instead, as Strevens points out, it increases its generality, or its abstract character. Thus, on this view, an optimized model is able to represent the causal relationships of the system. This ability, together with its generality, makes the optimized model, under certain conditions, explanatorily more powerful than the initial veridical model.

Consider, as an example, the Boylean behavior of gases as expressed by Boyle's law. An optimized model that would explain this behavior starts from its veridical causal model, i.e., from the set of true statements describing the trajectory of every single molecule in the gas (or the statistical description provided by its Maxwell-Boltzmann distribution), and then eliminates everything that does not make a difference to whether Boylean behavior obtains or not. For example, according to Strevens, the true statements that specify exact values for the volume of gas molecules and for the long range intermolecular forces are replaced by other statements that, instead, specify certain ranges (between zero and some value that is not too high). Same for intermolecular collisions.

An idealized model requires, according to Strevens, two things: an optimized model, and a procedure of idealization, whereby ranges of values, which in the optimized model had replaced actual parameter values, are themselves replaced by extreme values (typically, "zero" or "infinite").[5] Since an optimized model is a set of true statements describing the causal relationships of the system, the elimination of non-difference-makers via idealization amounts to adjusting this set such that some statements that give ranges of values for some parameters are replaced by other statements that, instead, provide extreme values for the same parameters. This replacement decreases the veridicality of the causal model, because the latter statements, as Strevens points out, are not true about the physical system. However, he further argues, an idealized model simplifies the optimized model and also increases its concrete character.[6] Despite its decreased veridicality, an idealized model is nevertheless able to represent the causal relationships of the system, and it is explanatorily more powerful than the optimized model.

Consider again the example of Boylean behavior. An idealized model that would explain this behavior starts from its optimized model, i.e., from the set of true statements that describe exactly just everything that makes a difference to whether Boylean behavior obtains and specify certain ranges for some parameters, e.g., for the volume of gas molecules and the long range intermolecular forces. Idealization replaces some of these statements by others that provide extreme values for the same parameters. For example, the statements that specify certain ranges (between zero and some value that is not too high) for the volume of gas molecules and the long

---

[5] "The idealized model can therefore be understood as the conjunction of the optimized model and certain further, false claims about reality." (Strevens, 2008, 325)

[6] "[B]ecause it replaces ranges with definite values, [an idealized model] is also often simpler than the [optimized] model." (Strevens, 2008, 321) "Better to think of the [idealized] model as formed by way of a structural simplification of the [optimized] model, and more particularly, by a kind of concretization of certain of its abstract aspects, achieved by the substitution of particular values for specified ranges." (324)

range intermolecular forces are replaced by statements that, instead, assign zero to both parameters. Same for intermolecular collisions.

Now, what exactly is the relation between an optimized model and an idealized model of the same natural phenomenon? This comes out quite clearly in the formulation of one constraint on the idealization procedure: "the setup of an idealized explanatory model for a phenomenon must entail the setup of the [optimized] explanatory model for that explanatory target, so guaranteeing that the elements of reality falsely represented by the model are non-difference-makers." (Strevens, 2008, 325) What this means is that if a causal factor is a non-difference-maker in an idealized model, then it has to be a non-difference-maker in the optimized model, too. In other words, if a parameter in an idealized model is given an extreme value, then the very same parameter must be given a range of values in that optimized model from which the idealized one is obtained. But the implication also holds in the other direction, for "an idealized model in one sense does the same explanatory work as the corresponding [optimized] model: both models correctly specify all the difference-makers and all the non-difference-makers." (319) This suggests, I think, that the two models are to be considered *extensionally equivalent*.

This relation of extensional equivalence constrains, for example, the construction of an idealized model of Boylean behavior: "collisions make no difference to Boylean behavior and thus are irrelevant to the explanation of Boyle's law. The idealized model, then, adopts the optimal explanatory policy on collisions." (316) More generally, Strevens suggests, extensional equivalence constrains the construction of any idealized model: "All idealizations . . . work in the same way: an idealization does not assert, as it appears to, that some non-actual factor is relevant to the explanandum; rather, it asserts that some actual factor is irrelevant. The best idealized models will be equivalent in one explanatorily central sense to the corresponding optimized models: when understood correctly, both kinds of models cite the same relevant factors and no irrelevant factors." (316)

One should further note that idealized models and the optimized ones from which they are derived are, at the same time, *intensionally inequivalent*: "they represent these explanatory facts using two different conventions: the fact that certain pervasive causal influences play no role in bringing about the explanatory target is left implicit in an optimized model – what is irrelevant is passed over silently – while it is made explicit in an idealizing explanation's flagrant introduction of fictional physical factors." (316) This suggests that, for example in the case of Boylean behavior, an idealized model only partially adopts the optimal explanatory policy on collisions, for although collisions are correctly (i.e., in accordance with extensional equivalence) identified as non-difference-makers, their properties in the idealized model are different than those in the optimized one. Another way of expressing the relation of intensional inequivalence is by noting that "the [optimized] model identifies non-difference-makers by failing to mention them, whereas an idealized model identifies non-difference-makers by conspicuously distorting their properties." (320sq)

We are now in a position to spell out the claim that optimization and idealization are in harmony with each other, in the sense that both the optimized and the idealized models of a natural phenomenon can represent the causal relationships of the system of interest. Harmony, in this sense, is grounded in extensional equivalence: an idealized model and the optimized model from which it is derived can each represent the causal relationships of the system *because* the two models specify the same difference-makers and the same non-difference-makers. In other words, since the parameters that are given extreme values in the idealized model are exactly those parameters that are given ranges of values in the optimized model, and optimization allows representation of the causal relationships of the system, then it follows that idealization allows such a representation as well.

That Strevens is committed to this view can be seen, for example, in the following remark: "[An idealizing claim] fills out certain details left unspecified by the [optimized] explanatory model. Thus, the claim is explanatorily irrelevant and so cannot stand in the way of the causal entailment of the explanatory target." (318) In other words, the idea that an idealizing claim, which assigns an extreme value to a certain parameter, is explanatorily irrelevant is inferred from the idea that the optimized claim that provides a range of values for the very same parameter is explanatorily irrelevant. But this inference needs extensional equivalence in order to go through. Furthermore, the idealized model would not be able to represent the causal relationships in the system if extensional equivalence did not hold.

In the next section, I offer an argument against the claim of methodological harmony. My argument does not reject the grounding of harmony in extensional equivalence, but questions the idea that an idealized model is actually extensionally equivalent to the optimized model from which it is derived. The argument emphasizes that it is not the case that both models cite the same relevant causal factors, and concludes that there is no harmony between optimization and idealization, in the sense already mentioned.

18.3 Against Methodological Harmony

The argument against methodological harmony, offered in this section, is based on reflection on whether the standard explanation of phase transitions in statistical mechanics may be considered causal, in Strevens' sense, as suggested by Callender (see quotation above). What I want to show is that this idealized explanation cannot represent the causal relationships of the system because it fails to cite explanatorily relevant causal factors, i.e., factors that make a difference to whether phase transitions occur or not. But this failure invalidates the harmony claim.

Examples of phase transitions are evaporation, magnetization, superconductivity, and the like. As is well known, the classical thermodynamic description of a physical system says that, at certain temperatures and pressures, the system can be in multiple phases at equilibrium. For example, at a temperature of -38.83 C and a pressure of 0.2 mPa, mercury exists at equilibrium in three phases, solid, liquid, and gas. Also, water, ice and steam coexist at 0.01 C and 611 Pa. In statistical mechanics, in order to account for phase transitions, that is, to derive the equations that govern the behavior of a physical system undergoing such transitions, one typically considers the system in the thermodynamic limit. In other words, the standard approach is to stipulate that the physical system has an infinite number of degrees of freedom, i.e., an infinite number of particles and an infinite volume, but a finite density (or, equivalently, one stipulates that the system has zero surface effects). Unless the system is considered in the thermodynamic limit, the partition function that describes its behavior does not display any singularity, i.e., any non-analyticity in the free energy.

Now, is this explanation causal or not? If it were causal, in Strevens' sense, i.e., in the sense that it cites all and only difference-makers, then one should endorse the following Strevensian inference: *The idealizing claim that surface effects are zero fills out certain details left unspecified by the optimized explanatory model. Thus, the claim that surface effects are zero is explanatorily irrelevant and so cannot stand in the way of the causal entailment of the explanatory target.* However, this inference is invalid, on account of the fact that surface effects *do* make a difference to the occurrence of phase transitions: no natural phenomenon, including phase transitions, could ever occur in a system with zero surface effects! For if these effects were zero, then the system would be infinitely large. But there is no such system. An infinitely large system is a fiction, and no natural phenomenon could ever occur in a fictional system! Or, perhaps more accurately, no fictional system can possibly causally produce a natural phenomenon. To believe otherwise is to allow, e.g., that JFK was possibly murdered by the average American; or that Sherlock Holmes could catch the Lane Bryant gunman. Thus, I think that the lesson that the standard approach to phase transitions in statistical mechanics teaches us is that there are non-causal explanations of natural phenomena that occur in finitely large systems.[7]

I suspect that most people would agree that there is no infinitely large system. Nevertheless, some might point out that this assumes that we know that physical systems are not infinitely large, that they do not actually have an infinite volume. But then what justifies this assumption, if not a privileged insight into the very nature of things? The point here seems to be that the correct lesson to be learned from the standard approach to phase transitions in statistical mechanics is that physical systems, or at least physical systems undergoing such transitions, are actually infinitely large, rather than that such systems should be considered *as if* they were infinitely large. Some would consider this second lesson literally nonsense.[8] But suppose that this is not so, that this lesson is in fact correct, i.e., that (at least some) physical systems (those undergoing phase transitions) are actually infinitely large. It

---

[7]To be sure, in the standard explanation of phase transitions, surface effects *are* considered to be zero. But this is not because their actual value does not make a difference to the occurrence of phase transitions. Rather, they are considered to be zero *despite* their difference-making property, and because the actual value stands in the way of the (non-causal) standard explanation.

[8]As John Norton recently put it: "One might casually speak of 'an infinitely large sphere' as the limit system. But that talk is literally nonsense. A sphere is the set of points equally far away from some center. An infinitely large sphere would consist of points infinitely far away from the center. But there are no such points. All points in the space are some finite distance from the center." (Norton, 2012, 213)

follows that the claim that surface effects are zero is a true statement. Note, however, that surface effects cannot be a non-difference-maker, as required by the causal characterization of the standard explanation, because according to this puzzling lesson no phase transition could occur in a system with non-zero surface effects. So even if such phenomena occurred in infinitely large systems, their explanation would not be causal, in Strevens' sense.

There is a third, no less puzzling, lesson drawn from the standard approach, one that agrees that no phase transition could occur in a system without zero surface effects, but at the same time, denies that such systems exist: "The properties of systems containing infinitely many particles are qualitatively different from those of finite systems. In particular, phase transitions cannot occur in any finite system; they are solely a property of infinite systems. [...] Since a phase transition only happens in an infinite system, we cannot say that any phase transitions actually occur in the finite objects that appear in our world." (Kadanoff, 2009, 782–784) Note, however, that the standard explanation is insufficient to justify the view that phase transitions cannot occur in finite systems, or that we cannot say that they do so. For one should not neglect non-standard explanations of phase transitions that attempt to derive the equations that govern the behavior of a physical system undergoing such transitions without taking the thermodynamic limit.[9] Besides, for the same reason as above, even if such phenomena occurred only in infinitely large systems, their explanation would not be causal, in Strevens' sense.

One can hopefully see by now why Callender's suggestion that the thermodynamic limit may be legitimized by the fact that surface effects aren't a difference-maker is false. It is false precisely because surface effects *are* a difference maker. As I have argued, the idealizing claim that surface effects are zero stands in the way (to use Strevens' terms once again) of the causal entailment of phase transitions. This is why, as I pointed out, the standard explanation should be considered non-causal. But this is also why there is no methodological harmony between optimization and idealization. For, whereas in an optimized model surface effects may be assigned a range of finite, non-zero values, and this does not stand in the way of the causal entailment of phase transitions, in an idealized model surface effects are considered zero, but this does stand in the way of the causal entailment of phase transitions. To put it differently, and perhaps more precisely, no-harmony is due to the extensional inequivalence of the idealized and the optimized models. They are extensionally inequivalent because the optimized model can consider surface effects as non-difference-makers, whereas the idealized model cannot do so while preserving at the same time the ability to represent the causal relationships in the physical system.

I take this to be a sound argument and I want to discuss, in the next section, its philosophical significance, i.e., the significance of the no-harmony between idealization and optimization.

---

[9]For a discussion of such explanations, see Callender and Menon (2013).

18.4 The Philosophical Significance of No-Harmony

So far I have argued, against Strevens, that there is no methodological harmony between optimization and idealization. The no-harmony claim is justified by the fact that an idealized model is extensionally inequivalent to the optimized model from which it is derived, since it is not the case that these models cite the same relevant causal factors. As I pointed out, unlike an optimized model, the idealized model derived from it fails to cite causal factors that make a difference to whether the phenomenon obtains.

My criticism is similar to one recently made by Angela Potochnik: "According to [Putnam (1975), Garfinkel (1981), and Strevens (2008)], explanations should neglect causal information that is irrelevant to the occurrence of the event to be explained. Simply put, their idea is that if different underlying causal dynamics could have led to the same event, then the actual underlying dynamics are not relevant to explaining the event. I urge a further-reaching omission of causal information. Some of the causal factors that I suggest are to be omitted from an explanation in fact *are* relevant to the occurrence of the event to be explained. Consider the genetic causes of the redshank's foraging preference. Those genotypes are an essential part of the process of cumulative evolution that leads to the redshank's foraging habits. [...] Yet, as causally important as these genotypes are, their particular specification is not relevant to the causal pattern of optimal foraging. Despite their causal relevance, then, the genetic details do not belong in an explanation that focuses on the optimal foraging pattern." (Potochnik, 2010, 222)

The question that I want to touch upon in this section concerns the philosophical significance of the no-harmony claim. More exactly, I want to argue that the no-harmony between optimization and idealization is a symptom of a more general tension, which characterizes a view that I elsewhere called *Weylean skepticism* – the view that objectivity and understanding are opposite epistemic ideals of science, a view that I attributed to the mathematician, theoretical physicist, and philosopher Hermann Weyl.[10] More precisely, according to this view, to the extent that science strives to attain objectivity, it may do so only at the expense of understanding; and vice versa, to the extent that it aims at understanding, it may do so only by sacrificing objectivity. As Weyl metaphorically put it: "Both roads run, as it were, in opposite directions." (Weyl, 1949, 283)

Why do I think that no-harmony is a symptom of this particular type of skepticism? Roughly, this is because I believe that whereas optimization is indeed motivated by the ideal of understanding, idealization should be regarded as driven primarily by the ideal of objectivity. Thus, if Weylean skepticism is true, that is if objectivity and understanding do run in opposite directions, it is only to be expected that there is no methodological harmony between idealization and optimization. But let me briefly explain what I mean here by objectivity and understanding.

---

[10]For an extensive discussion, see Toader (2011).

Objectivity is taken to be an epistemic achievement whereby one realizes that a statement correlates with experience-independent facts. This is to be distinguished from objectivity as intersubjectivity, which is often based on consensus among the members of particular research communities. Objectivity, in the sense adopted here, requires a "view from nowhere" (Nagel, 1989), which Weyl believed could only be provided by Hilbert's formal axiomatics. As I explained elsewhere at length, objectivity requires that scientific concepts be freely created by the mind, that is ultimately introduced via Hilbert-style axiomatic systems (Toader, 2013). This allows the use of partly non-contentual discourse and, as we will see below, answers a general question with regard to a necessary condition for objectivity: "How is the aspiration to objectivity *ever* to be satisfied if our response to an issue, arising in whatever discourse, can never be freed from a dependence upon propensities of spontaneous reaction – those involved in the appreciation of content – whose own status in point of objectivity is called into question?" (Wright, 1994, 229)

Understanding is taken to be an epistemic achievement whereby one realizes *why* a statement is true, rather than merely *that* it is true. This is what is usually called factive, i.e., truth-entailing, understanding, and is to be distinguished from a purely subjective intuitive state of cognizance. Weyl thought that the best account of what is required for understanding, in this sense, was offered by Husserl's phenomenological conception of evidence. This implies that scientific concepts be produced by abstraction from our experience (where both the method and the basis for abstraction are conceived of in phenomenological terms), rather than freely created by the mind.[11] This conception allows only the use of contentual discourse and, as we will see presently, answers a general question with regard to a necessary condition for understanding: "[U]nderstanding is a cluster of epistemic virtues. How could there be intellectual virtue at work without transparent appreciation of the content of one's theoretical beliefs, content that characterizes the causes of the phenomenon to be explained?" (Trout, 2002, 229)

Now, the view that optimization is motivated by the ideal of scientific understanding is relatively straightforward. I think that Strevens got this right. Optimization, as we have seen already, does not decrease the veridicality of a causal model, but rather increases its generality, while preserving its ability to represent the causal relationships of the physical system of interest. To put it differently, optimization does not introduce any non-contentual elements of discourse in a model, and in particular, it does not introduce any statements that fail to represent some causal relationships of the physical system. Thus, optimization may indeed lead to factive understanding, since an optimized model can allow a transparent appreciation of causal content. I think this justifies Strevens' view about understanding Boylean behavior: "[O]nce all the factors [that do not make a difference] have been abstracted away, what remains in the [optimized] model are just those high level properties of the mechanics of kinetic theory that make a difference to gases' Boylean behavior. It is by appreciating that these and only these properties are the difference-makers, that you understand Boyle's law." (Strevens, 2008, 377)

---

[11]See Toader (2013), for more details.

But it is not so easy to realize that idealization is driven by the ideal of scientific objectivity. For, as we have seen, idealization decreases the veridicality of a causal model. Despite this fact, an idealized model is, according to Strevens, able to represent the causal relationships of the system. However, as I argued above, this is false: unlike an optimized model, the idealized model derived from it fails to represent these causal relationships, since it fails to cite causal factors that make a difference to whether the phenomenon obtains. For example, the standard explanation of phase transitions in statistical mechanics requires that surface effects are neglected and, thus, fails to cite such effects as difference makers. Just to repeat the point made in the previous section, these effects are difference makers because no natural phenomenon can actually obtain in a system with zero surface effects. Such a system would be infinitely large – a fiction! – but no fictional system can possibly causally produce a natural phenomenon. If this is correct, then the statement that surface effects are zero lacks any content that characterizes the causes of phase transitions. More generally, the implication is that (infinite) idealization introduces non-contentual elements of discourse in a model, i.e., statements that fail to represent any causal relationships of the system. Thus, idealization may indeed be seen as driven by our craving for objectivity (to use Putnam's expression), insofar as an idealized model provides independence from the subjective propensities involved in the appreciation of causal content (or, at least, more independence than an optimized model ever could).

This view raises, of course, important questions that remain to be addressed. In particular, one needs to justify the apparent assumption that introducing non-contentual elements of discourse in a model is an effective way of providing independence from subjective propensities involved in the appreciation of causal content. Furthermore, one needs to address the worry that mere independence from subjective propensities is obviously insufficient for objectivity, in the sense of this term adopted earlier in this section. Similarly, one would have to say something about cases in which transparent appreciation of causal content may not be enough for scientific understanding. It is no less important to emphasize that there are various ways in which one could try to argue that Weylean skepticism is false. One such attempt can be seen, I think, already in Carnap's 1928 classic, *Der logische Aufbau der Welt*, which embraces the epistemic ideal of scientific objectivity as intersubjectivity, but ultimately fails to show that the roads of objectivity and understanding run in the same direction.[12] Another attempt can be seen in Nagel's very influential 1989 book, *The View from Nowhere*, which makes a case for objectivity as a method of understanding.[13] A careful discussion of such attempts, as well as others in more recent literature, must be postponed for another day.

---

[12]For discussion, see Toader (2015).

[13]Objectivity, Nagel wrote, "is not the test of reality. It is just one way of understanding reality." (Nagel, 1989, 26) He believed that, for most x, the more objectively x is viewed, the better x is understood. By contrast, the Weylean skeptic believes that, for most x, the more objectively x is viewed, the less well understood it is; and vice-versa, the better it is understood, the less objectively it is viewed.

Acknowledgements This paper was presented at the Hungarian Academy of Sciences in Budapest and the University of Bucharest. Thanks to both audiences and especially to Balázs Gyenis, Gábor Hofer-Szabó, Mircea Flonta, and Ilie Pârvu, for helpful questions and suggestions. Thanks also to Sorin Bangu for comments on a previous draft. For numerous discussions, during my Notre Dame years, on philosophy of science and mathematics, and more particularly on Weyl's views, I am greatly indebted to Paddy Blanchette, Mic Detlefsen, Don Howard, Chris Porter, and Sean Walsh. Financial support for writing the paper was provided through the 7th European Community Framework Programme Marie Curie Actions (CIG 293899 and EU-IAS-FP 246561), the University of Bucharest, and the Institute for Advanced Study at Central European University in Budapest.

References

Callender C, Menon T (2013) Turn and face the strange. . . Ch-ch-changes. Philosophical questions raised by phase transitions. In: Batterman R (ed) The Oxford handbook of philosophy of physics. Oxford University Press, New York, pp 189–223

Cartwright N (1983) How the laws of physics lie. Oxford University Press, New York

Garfinkel A (1981) Forms of explanation: rethinking the questions in social theory. Yale University Press, New Haven

Kadanoff L (2009) More is the same; phase transitions and mean field theories. J Stat Phys 137:777–797

Mackie JL (1980) The cement of the universe: a study of causation. Oxford University Press, Oxford

Nagel T (1989) The view from nowhere. Oxford University Press, New York/Oxford

Norton J (2012) Approximation and idealization: why the difference matters. Philos Sci 79:207–232

Potochnik A (2010) Explanatory independence and epistemic interdependence: a case study of the optimality approach. Br J Philos Sci 61:213–233

Putnam H (1975) Philosophy and our mental life. In: Philosophical papers, vol 2, ch 14. Cambridge University Press, London, pp 291–303

Strevens M (2008) Depth: an account of scientific explanation. Harvard University Press, Cambridge

Toader ID (2011) Objectivity sans intelligibility. Hermann Weyl's symbolic constructivism. PhD dissertation, University of Notre Dame

Toader ID (2013) Concept formation and scientific objectivity: Weyl's turn against Husserl. HOPOS: J Int Soc Hist Philos Sci 3:281–305

Toader ID (2015) Objectivity and understanding: a new reading of Carnap's Aufbau. Synthese. doi:10.1007/s11229-014-0648-2

Trout JD (2002) Scientific explanation and the sense of understanding. Philos Sci 69:212–233

Weyl H (1949) Philosophy of mathematics and natural science. Princeton University Press, Princeton

Wright C (1994) Truth and objectivity. Harvard University Press, Cambridge/London

Printed in the United States
By Bookmasters